In the 50 years since the inception of the *Society for General Microbiology*, the study of pathogenic microbes and the development of methods for their control have been a focus of attention for many microbiologists. This volume reviews the immense progress which has been made during the past half-century, opening with the text of Sir Alexander Fleming's 1946 Linacre Lecture 'Chemotherapy: yesterday, today and tomorrow', and then drawing together contributions which consider the development of key antimicrobial compounds, both naturally occurring and synthetic, active against bacteria, viruses, fungi and protozoa. The broader issues of antimicrobial production, screening, improvement and resistance are also considered. Topics such as why epidemics still occur, and the need for new antibiotics, highlight the fact that, despite the advances, the fight against infection continues unabated.

FIFTY YEARS OF ANTIMICROBIALS:
PAST PERSPECTIVES AND FUTURE TRENDS

SYMPOSIA OF THE
SOCIETY FOR GENERAL MICROBIOLOGY

Series editor (1991–1996): Dr Martin Collins, Department of Food Microbiology, The Queen's University of Belfast
Volumes currently available:

FIFTY YEARS OF ANTIMICROBIALS: PAST PERSPECTIVES AND FUTURE TRENDS

EDITED BY

P. A. HUNTER, G. K. DARBY AND N. J. RUSSELL

FIFTY-THIRD SYMPOSIUM OF THE
SOCIETY FOR GENERAL MICROBIOLOGY
HELD AT THE UNIVERSITY OF BATH
APRIL 1995

Published for the Society for General Microbiology

CAMBRIDGE
UNIVERSITY PRESS

Published by the Press Syndicate of the University of Cambridge
The Pitt Building, Trumpington Street, Cambridge CB2 1RP
40 West 20th Street, New York, NY 10011-4211, USA
10 Stamford Road, Oakleigh, Melbourne 3166, Australia

First published 1995

Printed in Great Britain at the University Press, Cambridge

A catalogue record for this book is available from the British Library

Library of Congress cataloguing in publication data

Society for General Microbiology. Symposium (53rd : 1995)
    Fifty years of antimicrobials: past perspectives and future trends:
fifty-third symposium of the Society for General Microbiology held at
. . . April 1995 / edited by P. A. Hunter, G. K. Darby and N. J. Russell.
    p.   cm.
"Published for the Society of General Microbiology."
ISBN 0-521-48108-2 (hbk.)
1. Anti-infective agents—Congresses.
2. Microbial metabolites—Congresses.   I. Hunter, P. A.
II. Darby, G. K.   III. Russell, N. J., 1936–
RM409.S64   1995
615′.3—dc20   94-25258   CIP

ISBN 0 521 48108 2 hardback

PN

CONTENTS

CONTRIBUTORS

AHARONOWITZ, Y. Telaviv University, George S. Wise Faculty of Life Sciences, Department of Molecular Microbiology and Biotechnology, Ramat Aviv, 69378 Tel Aviv, Israel

BISHOP, D. H. L. NERC Institute of Virology and Environmental Microbiology, Mansfield Road, Oxford OX1 3SR, UK

CHU, D. T. W. Anti-Infective Research Division, Department 47N, Building AP9A, One Abbott Park Road, Abbott Park, Illinois 60064-3500, USA

COHEN, G. Telaviv University, George S. Wise Faculty of Life Sciences, Department of Molecular Microbiology and Biotechnology, Ramat Aviv, 69378 Tel Aviv, Israel

CORY, J. S. NERC Institute of Virology and Environmental Microbiology, Mansfield Road, Oxford OX1 3SR, UK

CROFT, S. L. Department of Medical Parasitology, London School of Hygiene and Tropical Medicine, University of London, Keppel Street, London WC1E 7HT, UK

DARBY, G. K. Department of Molecular Sciences, Wellcome Research Laboratories, Langley Court, Beckenham, Kent BR3 3BS, UK

DEMAIN, A. L. Professor of Industrial Microbiology, Department of Biology, 77 Massachusetts Avenue, Cambridge, Massachusetts 02139, USA

FIEDLER, H. P. Institute of Biology II, Auf der Morgenstelle 28, D-72076 Tübingen, Germany

HIRST, M. L. NERC Institute of Virology and Environmental Microbiology, Mansefield Road, Oxford OX1 3SR, UK

HUNTER, P. A. Burnthouse, Burnthouse Lane, Cowfold, Horsham, West Sussex RH13 8DH, UK

KERR, K. G. Department of Microbiology, The Old Medical School, University of Leeds, George Street, Leeds LS2 9JT, UK

LACEY, R. W. Department of Microbiology, The Old Medical School, University of Leeds, George Street, Leeds LS2 9JT, UK

MILLING, R. J. Department of Biochemistry, AgrEvo UK Ltd, Chesterford Park, Saffron Walden, Essex CB10 1XL, UK

Possee, R. D. NERC Institute of Virology and Environmental Microbiology, Mansefield Road, Oxford OX1 3SR, UK

Pudney, M. Wellcome Research Laboratories, Langley Court, Beckenham, Kent BR3 3BS, UK

Rolinson, G. N. Parkgate House, Newdigate, Dorking, Surrey RH5 5DZ, UK

Russell, A. D. Welsh School of Pharmacy, Biochemistry Unit, University of Wales College of Cardiff, Redwood Building, King Edward VII Avenue, Cardiff CF1 3XF, UK

Russell, N. J. School of Molecular and Medical Biosciences, Biochemistry Unit, University of Wales College of Cardiff, Redwood Building, King Edward VII Avenue, Cardiff CF1 3XF, UK

Russell, P. E. Department of Biochemistry, AgrEvo UK Ltd, Chesterford Park, Saffron Walden, Essex CB10 1XL, UK

Ryley, J. F. 2 Wych Lane, Adlington, Macclesfield, Cheshire SK10 4NM, UK

Shen, L. L. Anti-Infective Research Division, Department 47N, Building AP9A, One Abbott Park Road, Abbott Park, Illinois 60054-3500, USA

Wright, K. Department of Biochemistry, AgrEvo UK Ltd, Chesterford Park, Saffron Walden, Essex CB10 1XL, UK

Zähner, H. Institute of Biology II, Auf der Morgenstelle 28, D-72076 Tübingen, Germany

EDITORS' PREFACE

It seems particularly apt that, on the occasion of the 50th anniversary of the institution of the Society for General Microbiology, the topic of this Symposium should concern antimicrobials, since the first President of the Society was none other than Sir Alexander Fleming. In the Linacre Lecture delivered at Cambridge on May 6th 1946, entitled 'Chemotherapy, yesterday, today, and tomorrow', and reproduced in full in this volume, Sir Alexander said:

> '. . . in the last ten years, more advances have been made in the chemotherapy of bacterial infections than in the whole history of medicine.'

What of more recent progress? In 1985, the topic of the 38th SGM Symposium was *The Scientific Basis of Antimicrobial Chemotherapy*, and the editors (Greenwood and O'Grady) refer in their preface to the 'extraordinary advances' made in the field since a previous Symposium in 1967. They also, however, commented that, in spite of these advances in scientific knowledge,

> '. . . the major burden of untreated infection is carried by the Third World; and . . . that the spectacular successes of antibacterial agents have not been matched by equal improvements in the armamentarium of drugs active against protozoa, helminths, fungi and viruses, agents of disease that, on a global scale, far outweigh bacteria in importance.'

Sadly, this is a theme which recurs in this volume, as we see again that the economically deprived developing countries still bear the burden of numerous infections. With some microbial pathogens, even where good targets have been identified and active compounds found, the cost of developing such agents is frequently considered uneconomic. There are numerous references in this volume to the development of resistance by diverse microbes to many antimicrobials; this may be connected with the over-use or inappropriate use of antimicrobial agents.

Currently, concern is also being expressed over the re-emergence of infections previously believed to have been controlled adequately, and although some infections such as smallpox have been eradicated, there are still epidemics of other microbial infections both in developed and less developed countries. Frequently, these epidemics and recrudescences of older pathogens are seen to be related to the economic status of the population at risk. The spread of AIDS and the increasing numbers of immunocompromised patients worldwide have changed, and continue to change the spectrum of infectious disease organisms. In addition, the problems in treating these patients with inadequate immune systems have emphasized deficiencies in many of the available drugs. It would seem that, whenever the scientists believe that they have conquered the microbe, the microbes have unerring capacities to fight back.

Against such a historical backcloth, it is the intention of this celebratory Symposium (and book) to consider the progress made during the past half-century across a wide field of antimicrobials, both naturally occurring and synthetic, including biocides, that are active against bacteria, viruses, fungi and protozoa. It is impossible to be comprehensive, but, instead, important representative examples are the subject of individual contributions. Others focus on broader issues of antimicrobial production, screening and improvement, on resistance, on why epidemics still occur, and on whether new antibiotics are needed.

It is hoped that this information will provoke and stimulate researchers in academia, industry and medicine, who are involved directly or indirectly in the battle against infection, to discover novel, more efficient antimicrobials. The ever-fluctuating balance between pathogenic microbes and man demands periodic re-appraisal: as we approach a new century, it is timely to consider what can be learned from the explosive growth in antimicrobials since the first clinically useful antibiotic, penicillin, was introduced at about the same time as the Society for General Microbiology was born.

CHEMOTHERAPY:
YESTERDAY, TODAY AND TOMORROW

SIR ALEXANDER FLEMING

Formerly Professor of Bacteriology, University of London,
St Mary's Hospital, W2

THE LINACRE LECTURE DELIVERED AT CAMBRIDGE ON MAY 6,
1946

The title which I gave for this lecture was too ambitious. It would require a whole series of lectures to do it justice. I must limit it to chemotherapy of bacterial infections, but even that subject is too large for one lecture, so I propose to restrict myself for the most part to chemotherapeutic happenings of which I have first-hand knowledge.

Bacterial infections have existed since time immemorial, and physicians, in all ages, have tried to deal effectively with them. It has been our good fortune to have lived in an era when many of these infections have, for the first time, been brought under control, and there is promise of more advances in the near future.

Until the middle of last century there was practically no knowledge of the bacterial nature of the infections, so before that time everyone was working in the dark, and many and curious were the prescriptions used in the combat against bacterial infections; but we shall get no profit by discussing these.

In the consideration of almost every branch of bacteriology we go back to Pasteur and the latter half of the nineteenth century. Pasteur proved that certain fermentations were due to the action of microbes and that microbes were living objects which did not arise *de novo* from the putrescible material but were descendants of previously existing organisms. Pasteur himself did not do any serious work on chemotherapy, but his earlier bacteriological work stimulated Lister, who had putrefaction of wounds very much at heart, to engage himself on the subject.

There are some who use the word chemotherapy in a very limited sense to cover those methods in which the chemical is administered in such a way that it gets into the blood and attacks the infecting microbes through the circulation in concentrations sufficient to destroy them or modify their growth. This is too narrow a definition, and I shall use chemotherapy to cover any treatment in which a chemical is administered in a manner directly

Originally published by Cambridge University Press, 1946.

injurious to the microbes infecting the body. In this latter sense antiseptic treatment comes under chemotherapy – call it local chemotherapy if you like. The same general laws govern the treatment whether it is local or systemic but there are certain particulars in which one has to draw a distinction. There are many chemicals used locally which are so poisonous to the human organism as a whole that they cannot be used for systemic treatment, but large numbers of these, although they had considerable vogue in the past when perhaps there was nothing better, are practically useless as chemotherapeutic agents except in the prophylactic sense. On the other hand, there are some chemicals, e.g. the sulphonamides, which are powerful agents for systemic treatment but which are frequently of little use when locally applied to a suppurating area, as their action is neutralized by substances occurring in the pus.

If a chemical is to be effective in the treatment of established infection it is necessary that, in addition to killing or inhibiting the growth of the microbes on the surface, it should be able to diffuse into the tissues to reach the microbes there. In a septic wound there are, of course, microbes in the cavity of the wound, but far more important are those which have invaded the walls.

Lister, like all other surgeons of 90 years ago, was struggling with the problem of septic wounds. To him, Pasteur's work came as a ray of light in the darkness. Putrefaction was due to living microbes which were introduced from outside. He set to work to prevent them being introduced. He cleansed his hands and his instruments, and treated them with chemicals, and he used a carbolic spray to kill bacteria in the air and prevent them reaching his operation wounds. In this way he revolutionized surgery.

That was prophylactic chemotherapy. Lister himself recognized that carbolic acid, which was his standby as an antiseptic, was very poisonous to the tissues. There are times, however, when if it is possible to kill all the bacteria in an infected area with a chemical it may be worth while to sacrifice certain tissues locally, but the usefulness of such toxic chemicals is very strictly limited.

As a result of the success of Lister's antiseptic treatment, a large variety of chemicals were introduced as local chemotherapeutic agents for the treatment of localized infections. In time, bacteriology was put on a sound basis, and the antibacterial effect of these chemicals could be tested in the laboratory. In the early days of the laboratory investigation of these antiseptics (or local chemotherapeutic agents), little attention was paid to anything but their action on bacteria in a watery medium. This resulted in very high values being given to substances like mercuric chloride, which could be diluted about half a million times before it lost its power of inhibiting the growth of bacteria when tested in watery medium, but which was largely 'quenched' in the presence of serum or blood.

But none of these chemicals had much effect in destroying bacteria once they had invaded the tissues. I commenced medicine in the early years of this century. Then Lister's methods were rather discredited – asepsis had taken the place of antiseptics for prophylactic chemotherapy, but for the treatment of infections which were already established, a great variety of chemicals were used. Carbolic acid, boric acid, mercuric chloride, silver salts, iodine, etc., were used extensively on septic conditions, but there did not in most cases appear to be any striking benefit except perhaps in some superficial infections. These chemicals were used quite empirically and were, I suppose, a relic of the antiseptic days when they had proved valuable in prophylaxis. They failed in treatment. They were either non-diffusible, or if they were diffusible they poisoned the tissues more than the bacteria. One of the first things I learnt in the casualty room was not to put a carbolic compress on a septic finger or carbolic gangrene was likely to result.

Then came the war of 1914–18. The aseptic surgeons were suddenly presented with masses of wounds, all of which became infected. The primary infection was from the soil and the soldiers' clothing and was largely anaerobic, but after a week or more in hospital this was replaced by the usual septic infection of civil life – mainly staphylococci, streptococci and coliform and diphtheria bacilli.

Into these wounds all manner of chemicals were poured in an attempt to destroy the infecting microbes. It was not so difficult sometimes to get rid of the majority of the microbes in the cavity of the wound – they could for the most part be washed out by simple irrigation with normal saline – but none of the chemicals had much action on the bacteria in the infected wound walls.

I might here show you a simple experiment which illustrated the inability of chemicals to sterilize even the cavity of a wound. From a test-tube some small processes were drawn to imitate the irregular processes in the cavity of a war wound. The tube was now filled with serum and infected with the usual bacteria which were found in wounds. Here we had an irregular infected cavity, but there was no possibility of the microbes invading the walls (Fig. 1). The 'artificial wound' was 'dressed' by inverting the tube and allowing the fluid to escape. This was replaced with an antiseptic which was allowed to remain in the tube for various times up to 24 hours, after which it was poured out and replaced by serum. The tube was then incubated, and next day there was a copious growth of bacteria with all the chemical antiseptics tested. The antiseptic had been unable to diffuse into the processes and kill the bacteria there, so as soon as the antiseptic was removed they grew out again and contaminated the whole tube.

Another observation made in the 1914–18 war is of some importance in local chemotherapy. In 1917 probably the most favoured method of treatment of a septic wound was the Carrell Dakin treatment. Dakin's fluid (sodium hypochlorite) was instilled into a wound every 2 hours. I had an opportunity of studying the length of time that Dakin's fluid remained active in a wound; I found a cup-shaped wound into which I could put a fluid and

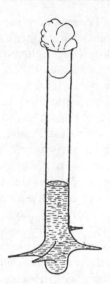

Fig. 1. Artificial wound.

withdraw the whole of it after any interval. When Dakin's fluid was left in such a wound for 10 minutes its potency had diminished below the limit at which it was antiseptic in serum. It followed from this that, for 1 hour and 50 minutes out of every 2 hours, there was no effective chemical antiseptic in the wound. But Dakin's fluid had another quite unexpected action. After it had been applied, it caused a marked increase in the transudation of fluid from the walls of the wound which persisted for some time after fluid was removed.

Fig. 2 illustrates this increased transudation. Incidentally the number of living bacteria was not reduced in the exudate after the application of Dakin's fluid for $4\frac{1}{2}$ hours.

I suggest that the chief virtue of Dakin's fluid was not direct antiseptic action, but that it lay in this power of stimulating the exudation of fluid from the infected walls of the wound, thus draining the oedematous tissues just as did hypertonic saline solution, which was another favourite dressing for a septic wound. But Dakin's fluid *in vitro* in the absence of serum or pus was a powerful antibacterial agent, so all its benefits in treatment were ascribed to its direct antiseptic action.

It is desirable in the investigation of the action of these chemicals to see how long they remain active in the body, and before we class them as chemotherapeutic agents we should see, if possible, whether the apparently beneficial effect is due to a direct antibacterial action.

I have said that Lister recognized the local toxic action of his favourite antiseptic, carbolic acid, but the toxic action of many of its successors was not so obvious and was sometimes forgotten.

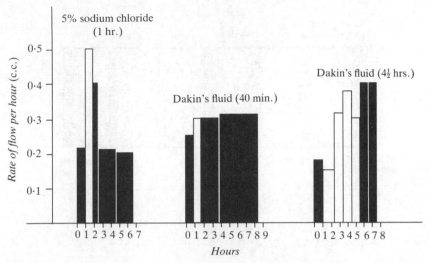

Fig. 2. Increased transudation. Dakin's fluid. Black columns = before and after. White columns = during application.

Later, and especially in the war of 1914–18, some notice was taken of the action of these chemicals on cells and especially on leucocytes, as it was not difficult to test the effect of a chemical on leucocytic function. Most usually it was the phagocytic power of the leucocytes which was tested, but the methods adopted did not always give a true picture. The effect of the chemical on bacteria was tested by its power to inhibit growth and its effect on leucocytes by its power to inhibit phagocytosis. This at first sight seems a perfectly good method, but actually the chemical acts on the bacteria during a period of hours, whereas in the phagocytic experiments the maximum time of action was 15 minutes. When blood is mixed with bacteria phagocytosis takes place very rapidly, and even in 5 minutes the cells will take up large numbers of microbes. That being so, if a chemical is added to the mixture which does not have a *rapid* lethal action on the leucocyte the latter continues to phagocyte the bacteria, and if the observation is ended after 15 minutes quite a false idea is obtained as to the destructive action of the chemical on the leucocyte. Acriflavine is a good example of this. It is a slow-acting bactericidal and leucocidal agent. If its antileucocytic power is tested only for 15 minutes it has been found that it requires a dilution of 1 in 500 to reduce the amount of phagocytosis by 50%. If, however, the chemical is allowed to remain in contact with the blood for 5 hours before the phagocytic test is made, it is found that a 1 in 500 000 dilution will cause a 50% reduction in the amount of phagocytosis. As it takes a 1 in 200 000 dilution to inhibit the growth of bacteria the 'Therapeutic index' calculated after 15 minutes' exposure of leucocytes is 400, whereas if the time of exposure had been 5

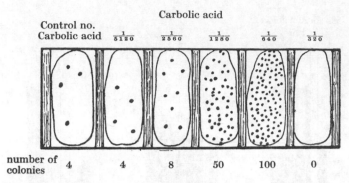

Fig. 3. Effect of carbolic acid on bacteria and leucocytes in human blood.

hours (a more reasonable time) it would have been 0.4 – a considerable difference.

In 1924 I adapted Wright's slide-cell method to show in one experiment the action of chemicals on bacteria and on leucocytes. Dilutions of the chemical were made in normal saline, and to these was added an equal volume of human blood suitably infected with the test bacteria (staphylococci or streptococci). These mixtures were run into slide cells, sealed and incubated.[1] Normal human blood kills off about 95% of the bacteria, but if the leucocytes are removed the bactericidal power disappears. In the case of all the chemical antiseptics in use there was a range of concentration where the chemical destroyed the leucocytes without interfering with the growth of the bacteria. This destruction of the leucocytes removed the natural antibacterial power of the blood, and resulted in an increased growth of the bacteria. This I regard as the most important series of experiments I have ever done, and it had a certain bearing on more recent advances of which I shall speak later.

Fig. 3 shows the result obtained with carbolic acid. There is an antileucocytic zone of concentration resulting in increased growth of the bacteria, and at a concentration of 1 in 640 the bactericidal power of the blood is completely destroyed and every microbe grows.

Let us now come to chemotherapy in the narrower sense, i.e. the attack on the infective microbe through the circulation.

There was the mercury treatment of syphilis. Mercury was swallowed, injected or rubbed into the skin. There were no tests proving that it ever reached the circulation in concentrations inimical to the spirochaete, but from the clinical results we may presume that something happened after a strenuous course of mercury which influenced the disease. With no more

[1]With slow-acting antiseptics the chemical was mixed with the blood and allowed to stand for a suitable interval, after which the bacteria were added and the mixture incubated.

evidence we might assume that potassium iodide had some direct action on the infective agent of syphilis, actinomycosis and other diseases.

In the early days of antiseptics it was shown that formalin inhibited the growth of the tubercle bacillus in quite high dilution. An eminent physician in the days of my youth recommended that formalin should be injected intravenously for the treatment of tubercle. I should like to show what happens when formalin is added to blood infected with *Staphylococcus*, an organism at least as susceptible to its action as is the tubercle bacillus.

The experiment is similar to the one I have already described with carbolic acid and the result is much the same; there is a range of concentration which encourages growth by destroying the leucocytes. Actually the amount administered was very much less than that necessary to influence the growth of the bacteria in any way. It was fortunate, perhaps, that sufficient of the chemical could not be injected to destroy the leucocytes (Fig. 4A).

Exactly the same may be said of quinine, which was recommended as an injection for the treatment of streptococcal septicaemia (Fig. 4B).

Eusol, too, was recommended in 1915 as an intravenous injection for *Streptococcus* septicaemia. I show you the result of mixing Eusol with infected human blood (Fig. 4C). If several litres could have been injected, the leucocyctes would have suffered, but fortunately that was not possible.

Mercuric chloride has been recommended as an intravenous injection for streptococcal septicaemia. This is more interesting. When it is tested with *Staphylococcus* it gives a result similar to those quoted above, but when *Streptococcus pyogenes* is used as the test organism an extraordinary result is obtained (Fig. 5). Here 1/20 000 destroys leucocytic action and allows the streptococci to grow, but weaker dilutions completely stop growth of the streptococci – but only if the leucocytes are present. The leucocytes cannot do this by themselves (vide control) nor can the mercuric chloride (vide 1/20 000 cell), but in some way the combined effect of the two can completely inhibit growth. If conditions are carefully adjusted, this inhibitory effect can be seen with a dilution of almost 1 in half a million – a quantity which can almost be reached by a therapeutic dose of the drug.

Scientific chemotherapy dates from Ehrlich and scientific chemotherapy of a bacterial disease from Ehrlich's Salvarsan, which in 1910 revolutionized the treatment of syphilis. The story of Salvarsan has often been told, and I need not go further into it except to say that it was the first real success in the chemotherapeutic treatment of a bacterial disease. Ehrlich originally aimed at 'Therapia magna sterilisans', which can be explained as a blitz sufficient to destroy at once all the infecting microbes. This idea was not quite realized, and now the treatment of syphilis with arsenical preparations is a long-drawn-out affair. But it was extraordinarily successful treatment, and stimulated work on further chemotherapeutic drugs. While they had success in some parasitic diseases the ordinary bacteria which infect us were still unaffected.

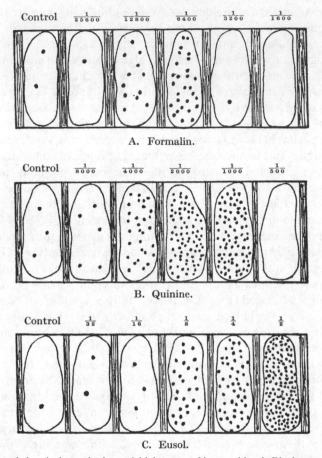

A. Formalin.

B. Quinine.

C. Eusol.

Fig. 4. Effect of chemicals on the bactericidal power of human blood. Black spots represent bacterial colonies.

Some aniline dyes were shown by Churchman many years ago to have remarkable selective properties as antibacterial agents, and they became prominent as antiseptics in septic wounds after Browning in 1917 described the action of acriflavine. This substance has been recommended as a chemotherapeutic agent for intravenous injection, but it proved to have some toxicity for the liver. When it is injected, it very rapidly disappears from the blood and all the tissues are stained except the nervous system. The following figures give the antibacterial power of the blood before and at intervals after an intravenous injection of acriflavine, and these are con-

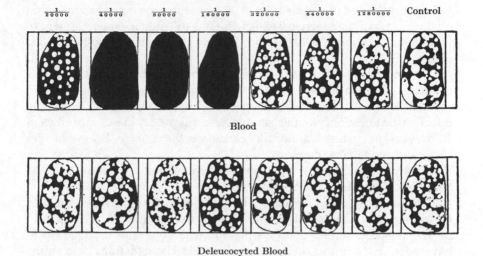

$\frac{1}{20000}$ $\frac{1}{40000}$ $\frac{1}{80000}$ $\frac{1}{160000}$ $\frac{1}{320000}$ $\frac{1}{640000}$ $\frac{1}{1280000}$ Control

Blood

Deleucocyted Blood

Fig. 5. Action of mercuric chloride on haemolytic streptococci in blood.

trasted with the result obtained after an intravenous injection of hypertonic sodium chloride:

Bactericidal power of blood after intravenous injection (in rabbit)

A. Acriflavine		B. NaCl	
20 cc of 1/1000		3 cc of 10%	
	Number of colonies		Number of colonies
Before injection	87	Before	25
1 min after injection	62	2 min after	28
3 min after injection	87	30 min after	8
45 min after injection	92	2 h after	0
		6 h after	0

Whereas the increased antibacterial power after acriflavine was slight and evanescent, there was a very considerable increase after sodium chloride which lasted for hours. This could not be due directly to the sodium chloride, which in itself was not antibacterial, and the only change that could be discovered was some rise in the opsonic power of the serum. This could not be true chemotherapy, but it illustrates another of the factors which have to be borne in mind in the investigation of the action of these drugs.

Then Sanocrysin was introduced as a chemotherapeutic agent against the tubercle bacillus, and it was said that after administration so many tubercle bacilli were destroyed that an antitubercular serum had to be given to prevent poisoning with the toxins of the dead bacilli. This was another failure; Fry showed that Sanocrysin in the concentrations used had no action on the growth of the tubercle bacillus, but it is still used for the treatment of tuberculosis, although it is not with the idea of direct chemotherapeutic action.

Long before this it had been noticed that some microbes were antagonistic to others – Pasteur himself was the first to show this – and some microbic substances or antibiotics had been used for local treatment for their direct effect on the infection. Notable among these was pycocyanase – a product of *B. pyocyaneus* – which was introduced early in the century. It was not very successful and fell into disuse.

I have said something of what happened in the past – let us say up to 10 years ago. That was yesterday.

Now we have to go on to a consideration of the chemotherapeutic happenings of today, and by today I mean the last decade. Things have moved indeed, and it is safe to say that in the last ten years more advances have been made in the chemotherapy of bacterial infections than in the whole history of medicine.

TODAY

It was in 1932 that a sulphonamide of the dye chrysoidine was prepared, and in 1935 Domagk showed that this compound (Prontosil) had a curative action on mice infected with streptococci. It was only in 1936, however, that its extraordinary clinical action in streptococcal septicaemia in man was brought out. Thus just 10 years ago and 26 years after Ehrlich had made history by producing Salvarsan, the medical world woke up to find another drug which controlled a bacterial disease. Not a venereal disease this time, but a common septic infection which unfortunately not infrequently supervened in one of the necessary events of life – childbirth.

Before the announcement of the merits of the drug, Prontosil, the industrialists concerned had perfected their preparations and patents. Fortunately for the world, however, Tréfouel and his colleagues in Paris soon showed that Prontosil acted by being broken up in the body with the liberation of sulphanilamide, and this simple drug, on which there were no patents, would do all that Prontosil could do. Sulphanilamide affected streptococcal, gonococcal and meningococcal infections as well as *B. coli* infections in the urinary tract, but it was too weak to deal with infections due to organisms like pneumococci and staphylococci.

Two years later Ewins produced sulphapyridine – another drug of the same series – and Whitby showed that this was powerful enough to deal with pneumococcal infections. This again created a great stir, for pneumonia is a condition which may come to every home.

The hunt was now on and chemists everywhere were preparing new sulphonamides – sulphathiazole appeared, which was still more powerful on streptococci and pneumococci than its predecessors, and which could clinically affect generalized staphylococcal infections.

Since then we have had sulphadiazine, sulphamerazine, sulphamethazine and others. But of these we need not go into detail, so much has already been written about them. Meantime there had appeared other sulphonamide compounds, such as sulphaguanidine, which were not absorbed from the alimentary tract, and these were used for the treatment of intestinal infections like dysentery.

The sulphonamides were very convenient for practice, in that they could be taken by the mouth. The drug was absorbed into the blood, where it appeared in concentration more than was necessary to inhibit the growth of sensitive bacteria. From the blood it could pass with ease into the spinal fluid, so it was eminently suited for the treatment of cerebrospinal infections. The sulphonamides were excreted in high concentration in the urine, so that although they were unable to control generalized infections with coliform bacilli they rapidly eliminated similar infections of the urinary tract. In contrast to the older antiseptics they had practically no toxic action on the leucocytes. There were disadvantages in that they were not without toxicity to the patient. Many suffered from nausea and vomiting, in some the bone marrow was affected with resultant agranulocytosis, and in others the drug was excreted in such a concentration that it crystallized out in the kidney tubules with serious results.

However, for the first time we had something which did control many common bacterial infections.

It was found, however, that the action was inhibited by certain substances. Early in the work on sulphanilamide it was noticed that, *in vitro*, complete bacteriostasis was obtained with a small inoculum, while if the inoculum was large the microbes grew freely. Fig. 6 illustrates this clearly.

It was shown that haemolytic streptococci, one of the most sensitive organisms, contained a substance which inhibited the action of sulphanilamide. Then it was found that pus, peptone, and products of tissue breakdown, would inhibit the action. For these reasons the local application of the sulphonamides to septic areas has not been quite so successful.

The following experiment illustrates the effect of streptococci in inhibiting the action of sulphanilamide. A large number of haemolytic streptococci, sensitive to sulphanilamide, were suspended in a 1% solution of the drug and allowed to extract for an hour or two. They were then centrifuged out and the supernatant fluid was boiled to kill any remaining organisms. Serial dilutions of this fluid were made and incubated with blood infected with a small number of the same haemolytic streptococci. (If the streptococci grew, the blood was haemolysed and served as an indicator.) In the strongest concentrations of the fluid the streptococci grew freely, but after it was diluted growth was completely inhibited, and this inhibition was manifest until the dilution reached 1 in 200 000 (Fig. 7). This was the dilution at which

Fig. 6. Comparison of sensitivity of a haemolytic streptococcus to penicillin and sulphanilamide by the gutter method.

Dilutions of streptococci streaked between gutters containing penicillin and sulphanilamide. With penicillin there is little difference whatever the size of the inoculum. With sulphanilamide there is no inhibition of the undiluted culture.

the original solution of sulphanilamide would have failed to inhibit if it had not been treated with streptococci.

The streptococci, therefore, produced something which did not destroy the sulphanilamide but merely inhibited its action.

This experiment furnishes an instance of a watery fluid which is not in itself antibacterial, but which becomes strongly bacteriostatic by the simple process of dilution with water.

Soon after the sulphonamides came into practice, also, it was discovered that some strains of what were generally sensitive microbes were resistant to their action. The result of widespread treatment was that the sensitive strains were largely displaced by insensitive strains. This was especially noticeable in gonococcal infections, and after a few years something like half of the gonococcal infections were sulphonamide insensitive.

This could be due to one of two things; the sensitive organisms might have been eliminated by treatment with the drug, while the insensitive ones

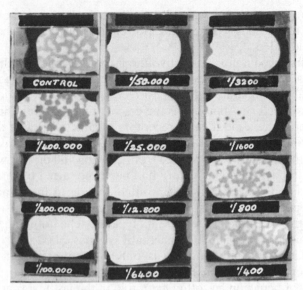

Fig. 7. Inhibition of sulphanilamide by streptococci.

Sulphanilamide 1 per cent, was treated with haemolytic streptococci (which were subsequently removed). It was then diluted with normal saline and mixed with an equal volume of human blood infected with haemolytic streptococci and grown in slide cells. The dark areas represent haemolysis due to growth of streptococci.

persisted and were passed on from one individual to another; or that by insufficient treatment with the drug a sensitive microbe might have acquired a resistance or 'fastness' to the drug.

It is not difficult in the laboratory to make sensitive bacteria resistant to the sulphonamides, but this is not peculiar to the sulphonamides. There is probably no chemotherapeutic drug to which in suitable circumstances the bacteria cannot react by in some way acquiring 'fastness'.

In the first year of the war the sulphonamides had the field of chemotherapy of septic infections to themselves, but there were always the drawbacks I have mentioned. Later another type of sulphonamide, 'Marfanil', was introduced in Germany which for systemic administration had relatively little potency, but which was not inhibited by pus or the usual sulphonamide inhibitors. This was largely used in Germany throughout the war, but there is no doubt from what was seen in German hospitals when they were overrun that their methods of dealing with sepsis were far behind ours.

The sulphonamides did not directly kill the organisms – they stopped their growth, and the natural protective mechanisms of the body had to complete their destruction. This explained why in some cases of rather long-continued streptococcal septicaemia sulphanilamide failed to save the patient, although the *Streptococcus* was fully sensitive to the drug; the protective mechanism of the body – the opsonic power and phagocytes – had become worn out and failed.

Fildes introduced a most attractive theory of the action of chemotherapeutic drugs. It was that these drugs had a chemical structure so similar to an 'essential metabolite' of the sensitive organism that it deluded the organism into the belief that it was the essential metabolite. The organism therefore took it up, and then its receptors became filled with the drug so that it was unable to take up the essential metabolite which was necessary for its growth. Thus it was prevented from growing and died or was an easy prey for the body cells. This theory had been supported by many experimental facts and may give a most profitable guide to future advances in chemotherapy.

But another completely different type of chemotherapeutic drug appeared, namely, penicillin. This actually was described years before the sulphonamides appeared, but it was only concentrated sufficiently for practical chemotherapeutic use in 1940.

The story of penicillin has often been told in the last few years. How, in 1928, a mould spore contaminating one of my culture plates at St Mary's Hospital produced an effect which called for investigation: how I found that this mould – a *Penicillium* – made in its growth a diffusible and very selective antibacterial agent which I christened Penicillin; how this substance, unlike the older antiseptics, killed the bacteria but was non-toxic to animals or to human leucocytes; how I failed to concentrate this substance from lack of sufficient chemical assistance, so that it was only 10 years afterwards, when chemotherapy of septic infections was a predominant thought in the physician's mind, that Florey and his colleagues at Oxford embarked on a study of antibiotic substances, and succeeded in concentrating penicillin and showing its wonderful therapeutic properties; how this happened at a critical stage of the war, and how they took their information to America and induced the authorities there to produce penicillin on a large scale; how the Americans improved methods of production so that on D day there was enough penicillin for every wounded man who needed it, and how this result was obtained by the closest co-operation between Governments, industrialists, scientists and workmen on both sides of the Atlantic without thought of patents or other restrictive measures. Everyone had a near relative in the fighting line and there was the urge to help him, so progress and production went on at an unprecedented pace.

Penicillin is the most powerful chemotherapeutic drug yet introduced. Even when it is diluted 80 000 000 times it will still inhibit the growth of *Staphylococcus*. This is a formidable dilution, but the figure conveys little except a series of many naughts. Suppose we translate it into something concrete. If a drop of water is diluted 80 000 000 times it would fill over 6000 whisky bottles.

We have already seen that all the older antiseptics were more toxic to leucocytes than to bacteria. The sulphonamides were much more toxic to bacteria than to leucocytes, but they had some poisonous action on the whole human organism. Here in penicillin we had a substance extremely toxic to some bacteria but almost completely nontoxic to man. And it not only stopped the growth of the bacteria, it killed them, so it was effective even if the natural protective mechanism of the body was deficient. It was effective, too, in pus and in the presence of other substances which inhibited sulphonamide activity.

Penicillin has proved itself in war casualties and in a great variety of the ordinary civil illnesses, but it is specific, and there are many common infections on which it has no effect. Perhaps the most striking results have been in venereal disease. Gonococcal infections are eradicated with a single injection and syphilis in most cases by a treatment of under 10 days. Subacute bacterial endocarditis, too, was a disease which until recently was almost invariably fatal. Now with penicillin treatment there are something like 70% recoveries.

So far in this country, penicillin has been under strict control, but soon it will be on sale in the chemists' shops. It is to be hoped that it will not be abused as were the sulphonamides. It is the only chemotherapeutic drug which has no toxic properties – in the ordinary sense of the word it is almost impossible to give an overdose – so there is no medical reason for under-dosage. It is the administration of too small doses which leads to the production of resistant strains of bacteria, so the rule in penicillin treatment should be to give enough. If more than enough is given there is no harm to the patient but merely a little waste – but that is not serious when there is a plentiful supply.

But I am not giving you a discourse on penicillin. Suffice it to say that it has made medicine and surgery easier in many directions, and in the near future its merits will be proved in veterinary medicine and possibly in horticulture.

The spectacular success of penicillin has stimulated the most intensive research into other antibiotics in the hope of finding something as good or even better.

Gramicidin and tyrothricin

But even before penicillin was publicized another antibiotic had been introduced by Dubos in 1939. This was a substance made by the *Bacillus brevis*, which had a very powerful inhibitory action on the Gram-positive bacteria. This substance was originally named gramicidin, but later the name was changed to tyrothricin, when it was found to be a mixture of two antibiotic substances – true gramicidin and tyrocidine. Gramicidin has proved to be a very useful local application to infected areas. It has an inhibitory power on bacteria far in excess of its antileucocytic power, but unfortunately it is toxic when injected, so that it cannot be used for systemic treatment. If penicillin had not appeared it is likely that gramicidin or tyrothricin would have been much more extensively used, but penicillin, which is quite non-toxic, can be used either locally or systemically for almost every condition which would be benefited by gramicidin.

Streptomycin

Waksman in 1943 described this antibiotic, which is produced by *Streptomyces griseus*. This substance has very little toxicity and has a powerful action on many of the Gram-negative organisms. It has been used in tularaemia, undulant fever, typhoid fever, and *B. coli* infections, but the greatest interest has been in its action on the tubercle bacillus. *In vitro* it has a very powerful inhibitory action on this bacillus, and in guinea-pigs it has been shown to have a definitive curative action. In man, however, the clinical results have not been entirely successful, but in streptomycin we have a chemical which does have *in vivo* a definite action on the tubercle bacillus and which is relatively non-toxic. This is a great advance and may lead to startling results. One possible drawback may be that bacilli appear to acquire rapidly a fastness to streptomycin, much more rapidly than they do to penicillin or even the sulphonamides.

Many other antibiotics have been described in the last five years. Most of them are too toxic for use, but there are some which so far have promise in preliminary experiments. Whether they are going to be valuable chemotherapeutic agents belongs to the future.

TOMORROW

Let us now consider the future. There are now certain definite lines on which research is proceeding in antibacterial chemotherapy.

Fildes's theory of the action of chemotherapeutic drugs has already led to certain results – not sufficiently powerful to have made wonderful advances in practical therapeutics – but the work goes on, and from it at any time some new antibacterial chemical combination may emerge. All this is dependent on further fundamental research on the essential metabolites necessary for the growth of different bacteria.

Bacteriologists and mycologists are, by more or less established methods, investigating all sorts of moulds and bacteria to see if they produce antibiotic substances. The chemist concentrates or purifies the active substance, and then the experimental pathologist tests the concentrate for activity and toxicity. There are teams of workers who are thus investigating every bacillus and every mould in the collections which exist in various countries. This is useful team work and may lead to something of practical importance, but it is reminiscent of the momentous German researches lacking in inspiration but which by sheer mass of labour bear some fruit. This plodding along appeals to many, but no one can expect the results to be revolutionary.

I have already mentioned some antibiotics which have given promise, and there are others which in preliminary experiments seem to have possibilities. It seems likely that in the next few years a combination of antibiotics with different antibacterial spectra will furnish a 'cribrum therapeuticum' from which fewer and fewer infecting bacteria will escape.

Then the work on antibiotics has led to the discovery of many new chemical combinations possessing antibacterial powers. Most of the antibiotics have certain disadvantages – many of them are too toxic – but it may not be beyond the powers of the organic chemists to alter the formula in such a way that the antibiotic power is retained, but the toxic power reduced to such an extent that these substances can be used therapeutically.

The most important chemotherapeutic want in Britain at the moment is a central institute for fundamental research in microbiology. In the past we have no reason to be ashamed of the results we have achieved in this direction, but the researches have been done in diverse institutions, often in the worker's spare time, and there have been times when the facilities for extending new discoveries did not exist. In Britain we are far behind the United States and, indeed, some of the smaller countries of the Continent of Europe in this respect, and it seems to me very necessary that here we should have a complete institute comprising bacteriologists, mycologists, protozoologists, biologists, biochemists, organic chemists, experimental pathologists and pharmacologists. It would probably be a good investment as well as an advertisement for British biological sciences.

Take penicillin as an example. Discovered here in 1928 in a hospital bacteriological laboratory where the chemical facilities were lacking, then worked on in a biochemical laboratory where the bacteriological co-operation failed, it had to wait 10 years before it was concentrated sufficiently to show its remarkable chemotherapeutic properties. Even then there were difficulties in developing it in this country, with the result that the Americans have reaped a large part of the reward.

Had we had a microbiological institute such as I have envisaged, penicillin could have been developed here years ago, and this one substance could well have paid the cost of the institute over a long period of years, to say nothing of a certain amount of alleviation of suffering in the interim period. It is not too late to be found such an institute. Enormous sums are being spent on national medical service – a large amount could well be spent on a national agricultural and veterinary research service, and we have unique association with the tropics in all matters. A small fraction of this money would found a microbiological institute which would, I am quite sure, pay handsome dividends in discoveries of importance in medicine, agriculture and industry. I am not the first to cry out for such an institute. The Royal Society's report on the needs of research in fundamental science after the war (issued in January 1945) makes the following recommendation:

'There is an urgent need for the establishment of an institute of general microbiology which should be the focal point for microbiological research in the Empire and should also have advisory and consultative functions.'

One misfortune is that it is nobody's baby. The Medical Research Council, the Agricultural Research Council and the Department of Scientific and Industrial Research, and other bodies may recognize the importance of such an institute, but as its activities spread over medicine, veterinary medicine, agriculture and industry, none of them feels justified in shouldering the burden.

With supertax so high there were possibilities of obtaining considerable sums of money for such an institute from the very rich under a seven-year covenant which allowed the institution to recover income tax and supertax. But I noticed the other day in the papers that a suggestion had been made in the House of Commons that steps would be taken to stop that 'racket'.

If far-sighted wealthy men are prevented from providing funds and if the Government will not provide funds there is no hope for any such an innovation as an institute for fundamental research in microbiology. Then we shall have to watch other more enlightened countries getting ahead and resign ourselves to be mere followers instead of leaders.

As to chemotherapeutic research in general I should like to conclude with a quotation from Mervyn Gordon: 'No research is ever quite complete. It is the glory of a good bit of work that it opens the way for something still better, and thus rapidly leads to its own eclipse. The object of research is the advancement, not of the investigator, but of knowledge.'

NEW DEVELOPMENTS IN NON-AZOLE ANTIFUNGALS FOR HUMAN DISEASE

PAMELA A. HUNTER

Burnthouse, Burnthouse Lane, Cowfold, Horsham, West Sussex RH13 8DH, UK

> The ability of fungi to cause disease in humans appears to be an accidental phenomenon.
> J.W. Rippon, 1988

INTRODUCTION

Historically, although dermatophytoses are among the commonest diseases in man and other animals, fungal infections in humans generally have had less impact on mankind than fungal infections of plants, or bacterial and viral infections. There are no great pandemics; no equivalents to the plagues, potato famines, yellow fevers, cholera, and Black Deaths that have caused such devastation. Fifty years ago the dermatophytes, thrush caused by *Candida albicans*, occasional cases of cryptococcosis, and a few rare, serious deep-seated mycoses were the only fungal infections of which most clinicians were aware. The last decade has seen an increase both in the occurrence of fungal infections and in awareness of them. There has also been a change of emphasis in the type of infection, many of the dermatophytoses are now well contained, but a whole range of other fungal diseases has become more common, particularly in hospital patients. This review will consider the reasons for these changes; discuss the problems of antifungal chemotherapy; describe the drugs available; and touch briefly on their modes of action. Finally, some thought will be presented on the future prospects.

FUNGAL DISEASES IN MAN AND THEIR EARLY TREATMENT

Fifty years ago deep-seated mycoses were rare and were confined to infections produced by a few species of dimorphic pathogenic fungi, *Coccidioides immitis*, *Blastomyces dermatidis*, *Paracoccidioides brasiliensis*, *Histoplasma* species and *Sporothrix schencki*. These organisms are true pathogens with the ability to invade healthy tissue and produce disease. They are, however, of relatively low virulence. Rippon (1988) has estimated that more than 90% of infections are entirely asymptomatic and are quickly resolved. In addition they have a restricted geographical distribution, mostly occurring in the tropics. Human to human transmission is not believed to occur.

Dermatophytoses on the other hand are common world-wide, the majority being caused by the dermatophyte species *Trichophyton*, *Microsporon*, and *Epidermophyton*. Some of these species have become adapted to a parasitic mode of life, depending on animal to animal contact for survival. They cause infections of the keratinised tissues (skin, hair, nails) and these range from mild ringworms and athlete's foot, through the more severe cases such as favus, to the rare debilitating infections that may even be life-threatening (Rippon, 1988).

The earliest treatments of fungal infections were aimed primarily at the superficial mycoses; most agents being non-specific and many keratolytic or drying agents. One of the oldest preparations known as Whitfield's ointment, a mixture of benzoic and salicylic acids, has only fallen out of use relatively recently. Sulfur was a common ingredient in many of the early remedies. Other agents included gentian violet and potassium permanganate (also used for superficial *Candida* infections), aluminium chloride, and undecanoic acid. Some of these agents are still available today as over-the-counter preparations for athlete's foot.

Agents to treat systemic mycoses were few and far between until the advent of the modern chemotherapeutic era. Oral potassium iodide has been used for treating subcutaneous sporotrichosis for several decades, and has remained the drug of choice until very recently. Hydroxystilbadine was used in the 1950s to treat blastomycosis, and sulfonamides had some value in the treatment of paracoccidioidomycosis and to a lesser extent, histoplasmosis (Speller, 1980).

The 1950s and 1960s saw the discovery of a number of specific and highly active antifungal agents, both natural products and synthetic compounds, many of which remain in use today. The major agents that are currently used are:

- *amphotericin B*: a polyene discovered in 1956, produced by *Streptomyces nodosus*. The first effective compound available for systemic use against serious mycotic infections.
- *griseofulvin*: an antibiotic produced by *Penicillium* species (including *P. griseo-fulvum*). An oral agent with high activity against dermatophytes introduced in 1958.
- *flucytosine* (5-fluorocytosine, 5FC): a synthetic compound whose antifungal properties were recognised in 1964. It is active orally against *C. albicans* and *Cryptococcus neoformans* but the occurrence of resistant strains limits its value.
- imidazoles: synthetic broad-spectrum agents, many of them only available for topical use. Miconazole and clotrimazole were first described in 1969. Ketoconazole, first described in 1979, was the first azole to have significant oral activity.

THE CHANGING PATTERN OF FUNGAL INFECTIONS

As noted above, serious fungal infections have grown in clinical importance in recent years, and they are likely to continue to increase (Bodey, 1988;

Walsh, Jarosinski & Fromtling, 1990). They are associated particularly with immunosuppressed conditions such as cancer chemotherapy, leukaemia, AIDS, and transplantation. It has been estimated by Bodey (1988) that 30–40% of liver transplant patients and 20–25% of patients with severe leukaemia get fungal infections. There are however, other factors that predispose patients to iatrogenic and nosocomial fungal infections and these are listed in Table 1 (Stevens, 1987; Bodey, 1988; Perfect *et al.*, 1991; Pfaller, 1992; Warnock, 1991). In addition, increasing travel and tourism to endemic areas increases the pool of people exposed to the dimorphic pathogens.

Table 1. *Factors, in addition to immunosuppression, affecting the increase in fungal infections*

- Aggressive and invasive surgery
- Indwelling catheters
- Increased use of broad spectrum antibacterial agents
- Hyperglycaemia and acidosis
- Adreno-corticosteroid therapy
- Chronic malnutrition and debilitation
- Increased use of intensive care units (neonates and adults)
- Parenteral alimentation
- Alcoholism and drug abuse

The commonest species causing infection are generally *Candida albicans*, although non-albicans species are increasing (Komshian *et al.*, 1989), *Aspergillus* species, and *Cryptococcus neoformans*. Increasingly however, infections are caused by a wide number of other fungi of environmental origin not normally regarded as pathogens (Perfect *et al.*, 1991). Such organisms as *Acremomium, Alternaria, Curvularia, Drechslera, Fusarium, Phialophora, Rhizopus,* and *Trichosporon* are no longer a rarity (Anaissie *et al.*, 1989; Bodey, 1988; Perfect *et al.*, 1991). This increase has been described by Deepe and Bullock (1990) as an 'insidious epidemic of mycotic infections that has been emerging for the past 15 years ...' Similarly, Rippon (1988) comments on the increasing number of species isolated from human infections for the first time, and notes 'Organisms not formerly associated with human infection have quite literally come out of the walls ... to infect such patients.'

DRUGS AVAILABLE AND THEIR TARGETS

The drugs available to treat this growing number of infections are still surprisingly few, and it is significant that a number of the antifungal agents used in clinical medicine are agents discovered several decades ago. Significant, because it is an indication not of the problems of finding compounds with good antifungal activity, but of finding agents with sufficient activity and selectivity to have clinical utility. The problem lies in finding an agent which will have a greater effect on the eukaryotic fungal cell than on the eukaryotic mammalian cell. Since the fundamental organisation of all eukaryotes is essentially similar, the targets available that differ sufficiently between fungi and mammalian cells are fewer than between prokaryotes and mammalian systems, and thus many interesting agents prove too toxic for systemic use.

Membrane active compounds

Polyenes

The polyene amphotericin B is regarded by many as the drug of choice for the seriously ill patient, in spite of a number of major drawbacks (Clements & Peacock, 1990; Sarosi, 1990). It is nephrotoxic, poorly tolerated, and has such low solubility in biological fluids that it has to be formulated as a colloidal suspension in bile salts and administered as a slow infusion. Nevertheless, it is still claimed by many to be the 'Gold Standard' against which other agents should be compared (Sabra & Branch, 1990; Sarosi, 1990). It has a broad spectrum of activity and is fungicidal against most species, at least *in vitro*. The newer azoles have still not replaced it for many indications, particularly in the immunocompromised host.

The polyenes exert their effect by associating with the eukaryotic membrane sterols and disrupting membrane integrity, as can be shown by a rapid efflux of potassium ions from treated cells. The precise manner in which amphotericin inserts into the membrane is still a matter of some discussion. It is believed to have a higher affinity for ergosterol in the fungal membrane than cholesterol in the mammalian membrane, and thus has a greater effect on fungi at low concentrations than on the host cells. Some aspects of the early theory of its mode of action propounded by Gale in 1973 have been questioned subsequently, and some of the recent ideas have been reviewed by Bratjburg *et al.* (1990). There have been many studies investigating Gale's original proposition that amphotericin produces aqueous, structured pores in the membrane. These resulted in contradictory findings, depending on whether unilamellar or multilamellar, small or large vesicles were used as the model membrane. A current view is that amphotericin molecules self-associate and then insert into cholesterol-containing membranes, but associate directly with ergosterol in ergosterol-containing membranes (Bolard *et al.*, 1991).

As amphotericin B is such an active compound, there have been attempts to find ways of reducing its toxicity whilst retaining its antifungal activity. Various semisynthetic derivatives, particularly the methyl ester and the N-D-ornithyl ester, showed promise in animal studies, but unfortunately, although kidney toxicity was reduced dramatically, neurological effects

Amphotericin B ; $R = CO_2H$

Amphotericin B methyl ester ; $R = CO_2CH_3$

Amphotericin B ornithine salt ; $R = CO_2^-\ ^+NH_3(CH_2)_3CH(NH_2)CO_2H$

SB 49594 ; $R = CH_2OH$

were seen (Summarized by Hoeprich *et al.*, 1988; Bennett, 1991). It is not clear whether the effects seen with these esters at high doses would also occur if amphotericin could be administered at equivalent doses. More recently a novel semi-synthetic derivative, the ascorbate salt of 16-hydroxymethyl amphotericin B has been described (Hunter *et al.*, 1992*a*, 1992*b*). This compound is claimed to be water-soluble and of similar activity to amphotericin against fungi, but to have a lower propensity to lyse erythrocytes. This reduced effect against mammalian cells was reflected in a marked reduction in kidney damage in sub-acute toxicity tests.

An alternative approach to modifying the molecule is to modify the formulation. There have been numerous studies over the years demonstrating that liposomal or other lipid-rich vehicles can reduce the toxicity of the compound. Three commercial preparations are currently under clinical evaluation, but early reports on the first, a true liposomal preparation, have been mixed. It is too early to tell whether these formulations will prove of real value in the clinic (Bennett, 1991).

Other membrane active compounds

The fungal membrane, and in particular ergosterol biosynthesis, is the target for a number of other groups of synthetic antifungal agents. The azoles inhibit C-14 lanosterol demethylase, whereas the allylamines affect an earlier point inhibiting squalene epoxidase, and the morpholines a later point inhibiting delta 8-7 isomerase and delta 14 reductase.

Inhibition of lanosterol demethylase

The azoles are a highly successful group of compounds both in the field of plant protection and human medicine, and although this review is not dealing specifically with these compounds in detail, no discussion of antifungal agents would be complete without some mention. (The early development of imidazoles is reviewed in some detail by Davies in 1980 and the later compounds by Borgers in 1985).

Ketoconazole was the first imidazole to have useful oral activity, but it had limited value for the treatment of serious systemic fungal infections. The advent of triazoles marked a turning point for these compounds and two have been marketed in recent years, fluconazole, which is also available in an injectable form, and itraconazole. Both have undergone extensive clinical studies and have proved a valuable addition to the armamentarium, fluconazole being particularly active against cryptococcal infections, and itraconazole against *Aspergillus* infections (Bailey, Krakovsky & Rybak, 1990). None of these agents is fungicidal in action, which might be regarded as a drawback, particularly if treating seriously debilitated patients with impaired defence mechanisms. Several triazoles have been described in recent years that are claimed to have advantages over fluconazole and itraconazole. All have extended serum half-lives but clinical work on most has been suspended either on toxicological grounds, or because of formulation problems.

A natural product Ro 09-1470, that inhibits lanosterol demethylation has been described by Aoki *et al.* (1992). The compound was isolated from a species of *Penicillium*, and several strains of *Aspergillus sclerotiorum* were also found to produce a related compound (Matsukuma *et al.*, 1992). The compound was shown to produce an accumulation of C-14 methylated ergosterol intermediates in *C. albicans* and to inhibit lanosterol 14-demethylase in microsomal preparations from *Saccharomyces cerevisiae*. Activity against whole cells, however, was variable and a number of medically important species were resistant, including *C. albicans*, although

Clotrimazole   Ketoconazole

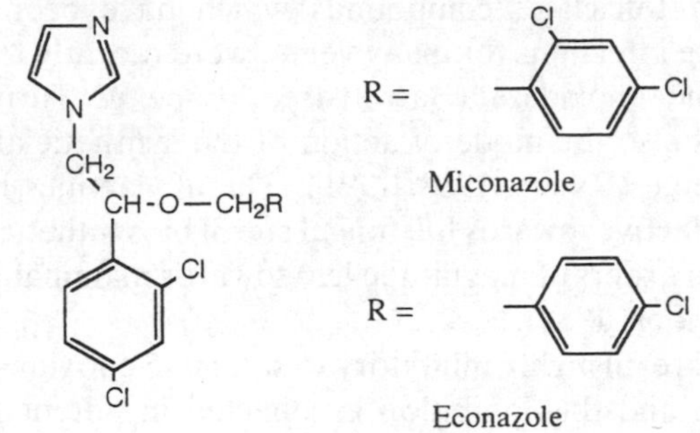

Miconazole

Econazole

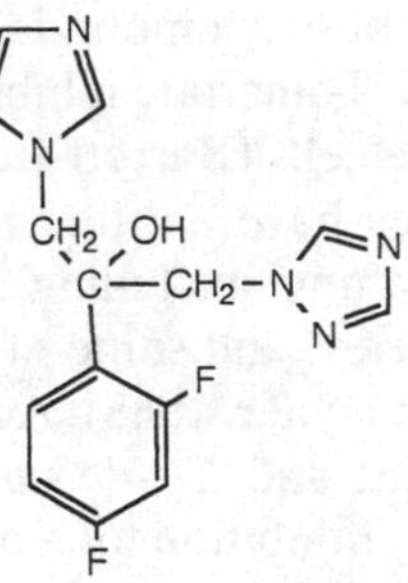

Fluconazole

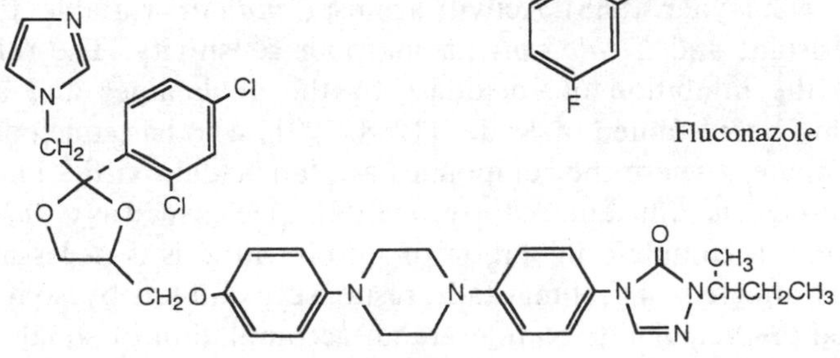

Itraconazole

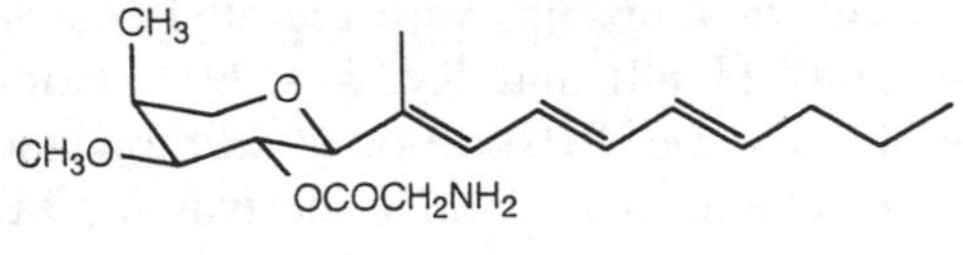

Ro 09-1470

partial inhibition against this species was seen using turbidimetric measurements.

Although this compound does not appear to have sufficient activity to be of clinical interest, its significance lies in that it is the first natural product to be reported as having such a mode of action.

Squalene epoxidation

Squalene is the first lipophilic intermediate in the sterol pathway and is converted to the 2,3-oxide by squalene epoxidase. The synthetic thiocarbamates tolnaftate and tolciclate, compounds which have been used for treating dermatophyte infections for many years, were recently found to be inhibitors of squalene epoxidase (Barrett-Bee, Lane & Turner, 1986; Ryder, 1991). This is also the mode of action of the synthetic allylamines naftifine and terbinafine (Ryder, 1988, 1991). The allylamines have been shown to be highly selective towards the fungal sterol biosynthetic pathway, having activity several orders of magnitude less towards mammalian cholesterol biosynthesis (Ryder, 1988).

These compounds are all highly inhibitory to squalene epoxidase isolated from dermatophytes, and this inhibition is reflected in potent antifungal activity. Although squalene epoxidases from other fungal species are inhibited by these compounds, this is not always reflected in activity against these species. Tolnaftate inhibits epoxidase from C. albicans, but is inactive against whole cells (Barrett-Bee et al., 1986).

Allylamines have antifungal effects (often fungicidal), against a broad spectrum of fungi including Aspergillus species, Cr. neoformans, some Candida species, and some of the dimorphic pathogenic fungi (Balfour & Foulds, 1992; Ryder, 1988). Activity against Candida is variable, C. glabrata being resistant and C. albicans having poor sensitivity. The relationship between the inhibition of epoxidase by the allylamines and antifungal activity has been studied by Ryder (1988, 1991) who has concluded that in dermatophytes, where the compounds are fungicidal, partial inhibition of sterol synthesis is sufficient to stop growth. In species such as C. albicans and C. glabrata, a complete inhibition of sterol synthesis is necessary before growth is affected. The fungistatic response exhibited by some fungi is attributed to their ability to tolerate an accumulation of squalene that is lethal to other species. Interestingly, filamentous fungi, including the mycelial stage of C. albicans, are generally more sensitive than yeasts (Ryder, 1991).

Structure–activity relationships with the allylamines have been studied extensively by Stütz (1988) and Ryder (1991). Considerable structural variation is possible and the allyl- portion of the molecule is not necessary for good activity, a related benzylamine derivative, butenafine, retaining activity similar to that of allylamines. Naftifine and terbinafine have both shown good clinical activity against dermatophyte infections, and terbinafine has the added advantage of having oral activity. It remains to be seen whether this valuable mode of action can be exploited to produce a compound with useful broad spectrum activity that translates to clinical efficacy.

Morpholines

The morpholines are characterized by a 2,6-dimethylmorpholine ring with a bulky N-substituent. A number have been developed as agricultural fungicides, and recently amorolfine, a derivative of fenpropimorph, has been

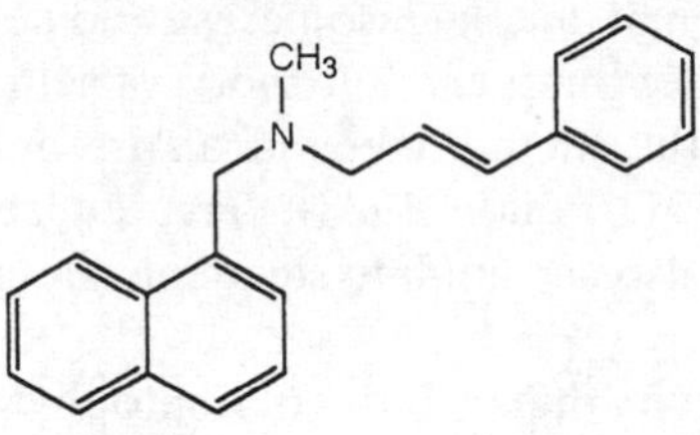

Naftifine

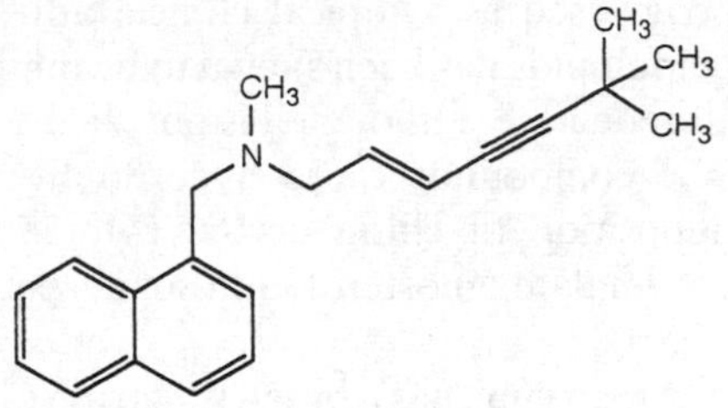

Terbinafine

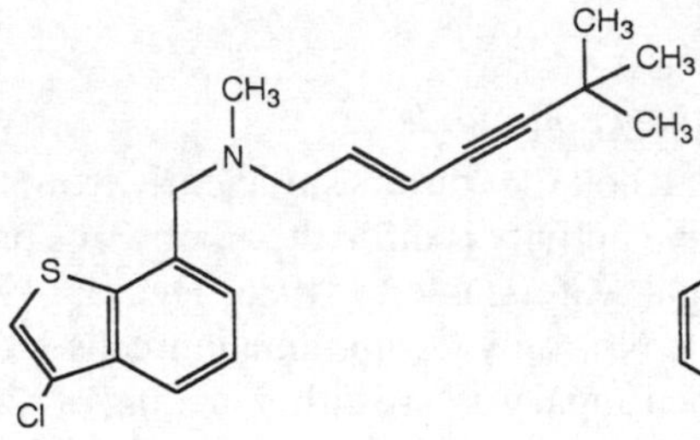

SDZ 87-469

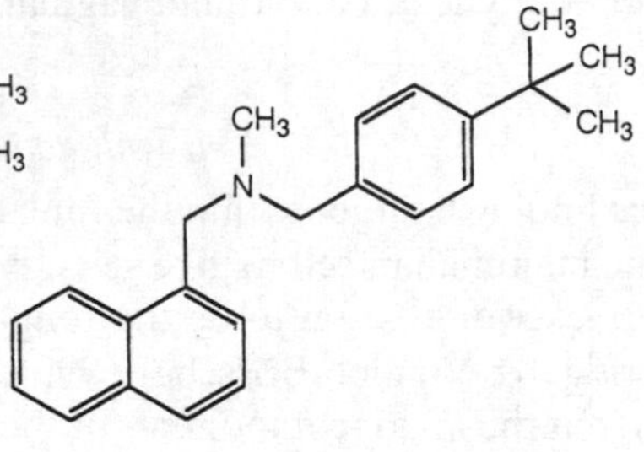

Butenafine

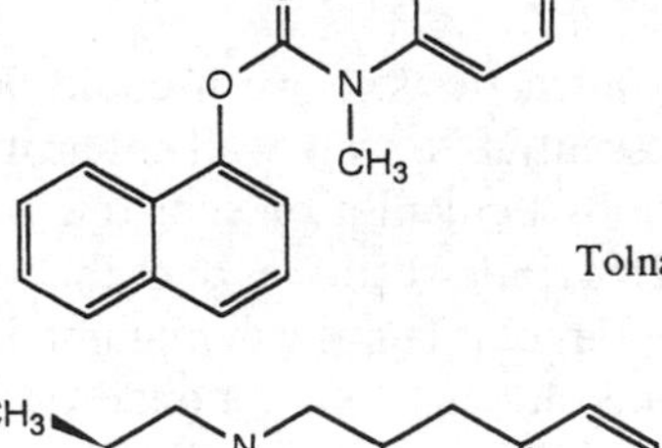

Tolnaftate

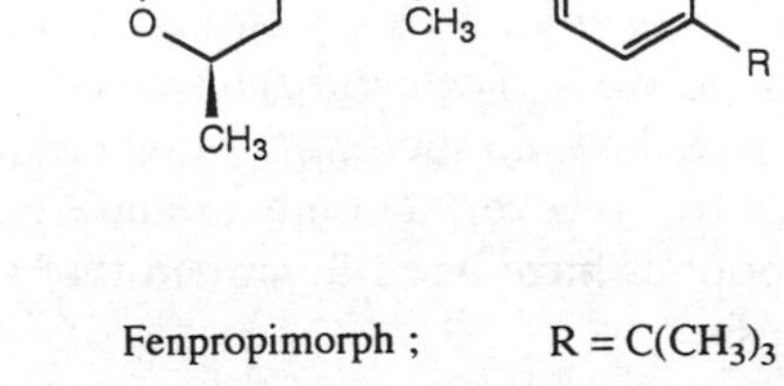

Fenpropimorph ; R = C(CH₃)₃

Amorolfine ; R = C(CH₃)₂CH₂CH₃

progressed as a topical clinical antifungal. The mode of action of these compounds has been shown to be inhibition of delta 8-7 isomerase and delta 14 reductase, although it is not clear whether this is the sole mode of action. The compounds cause hyperfluidity of the membrane and an irregular deposition of chitin occurs (Mercer, 1991; Polak, 1988). Treated cells accumulate ignosterol, and some species also accumulate squalene (Polak, 1990).

Amorolfine has a broad spectrum of activity that includes dermatophytes, *Candida* species, *C. neoformans*, dimorphic fungi, and dermatiaceous fungi. *Aspergillus* species and *Mucorales* are resistant. The compound is not absorbed orally and is undergoing clinical trials as a topical agent for dermatophytic infections and vaginal candidosis (Polak, 1990).

Cell-wall active compounds

Since finding a target within the fungal cell that differs sufficiently from that in the mammalian cell to give selective activity is difficult, an obvious point of attack would seem to be the fungal wall (Cassone, 1986; Hector, 1993; Kerridge & Vanden Bossche, 1990). Not only do mammalian cells lack a wall, but the major components of the fungal wall are either lacking or differ structurally from components of mammalian cells. The fungal wall is a multilayered, complex structure with a high polysaccharide content, the composition of which varies between species. Detailed knowledge of this structure is lacking for many fungi of medical importance, the bulk of the information being available on *C. albicans* and *S. cerevisiae*. It is clear, however, that fungi of clinical interest all contain variously linked glucans, chitin, and mannoproteins.

Such a target has great attractions, since it could be hypothesized that attacking a component essential to cell wall integrity would produce a similar effect to that given by penicillin on sensitive bacteria (Kerridge & Vanden Bossche, 1990). This could result, as does the action of penicillin, in lysis and cell death. As the target is lacking in mammalian cells, it should be possible to find a compound that lacks mammalian toxicity.

As is so often the case in antifungal research, the search for compounds active on the cell-wall has been more successful in plant pathology than in medical mycology. The polyoxins (*vide infra*) have been used as agrochemicals for many years, but to date, no such compound has reached the market for medical use. There are, however, a number of interesting developments in this area, and compounds have been described that are aimed at either chitin or glucan synthesis.

Chitin as a target

Chitin is an insoluble polymer composed of repeating *N*-acetyl glucosamine units with $1,4-\beta$-D glycosidic linkages. The proportion of chitin in the cell wall of fungi is only known accurately for a few species, but it is evident that it can vary widely both between species, and in different growth phases of the same species. In the dimorphic pathogenic fungi (*Coccidioides, Blasto-myces, Histoplasma* and *Paracoccidiodes*) chitin forms a substantial part of the wall of the pathogenic stage of the fungus (for review see Hector, 1993), whereas in *C. albicans*, chitin levels are far lower (approximately 1% of the wall), particularly in the yeast phase (Chiew, Shepherd & Sullivan, 1980).

Even in species such as *Candida* where the total chitin content is low and occurs mainly at the bud scar site, it is believed to play an important role in regulating growth and morphogenesis (Cassone, 1986). Disruption of chitin metabolism is possible by affecting either chitin synthases, the enzymes responsible for building up the polymer chains, or chitinases, the enzymes involved in reorganizing the chitin (Cabib, 1987; Gooday, 1990). Compounds have been described that affect these enzymes.

Chitin synthase inhibitors

The polyoxins and nikkomycins or neopolyoxins are a series of closely related natural products; they are peptido-nucleosides, and are structural analogues of uridine diphosphate *N*-acetyl glucosamine. They act as mimics of the building block of the chitin chains, UDP-*N*-acetyl glucosamine, and are highly specific competitive inhibitors of chitin synthase. They produce marked morphological changes in sensitive fungi, reminiscent of the effects of β-lactam antibacterials, eventually causing lysis of the cell. Some have been used successfully as agricultural fungicides.

The polyoxins were isolated from *Streptomyces cacaoi* and first described by Suzuki *et al.* (1965). They have been used commercially against the plant infections, rice sheath blast, a disease caused by *Pellicularia filimentosa*, and pear black spot, caused by *Alternaria kikuchiana*. For some years the polyoxins were believed to be inactive against medically important fungi, until it was realized that the compounds were dependent on a peptide transport system for uptake into the cell. Mitani and Inoue showed as early as 1968 that various nitrogenous substances, including dipeptides, could antagonize the activity of polyoxins against fungal plant pathogens, and subsequently various polyoxins were shown to produce both morphological abnormalities and cell death in medically important fungi (Becker *et al.*, 1983; Hector & Pappagianis, 1983). However, higher concentrations are required to inhibit these zoopathogenic fungi than are required to inhibit plant pathogenic species.

The closely related nikkomycins were isolated from *S. tendae* and first reported by Dähn *et al.* (1976). These metabolites were shown to have a similar mode of action and produce similar morphological effects to the polyoxins (Fiedler *et al.*, 1982; Müller *et al.*, 1981). Although the nikkomy-cins do not differ substantially from the polyoxins in their enzyme inhibitory properties, they are more active against whole cells (Gooday, 1990). This is

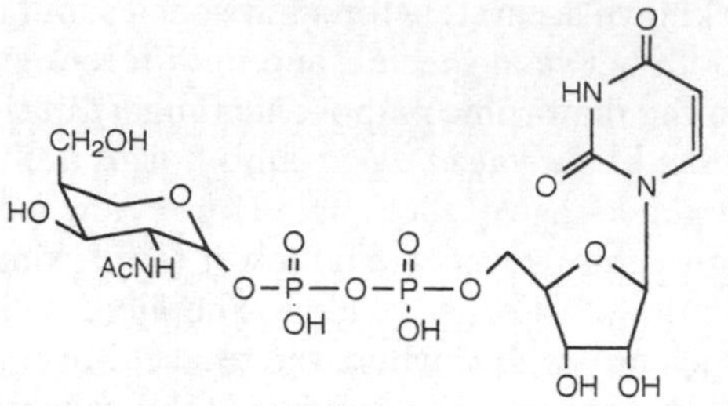

Uridine diphosphate *N*-acetylglucosamine

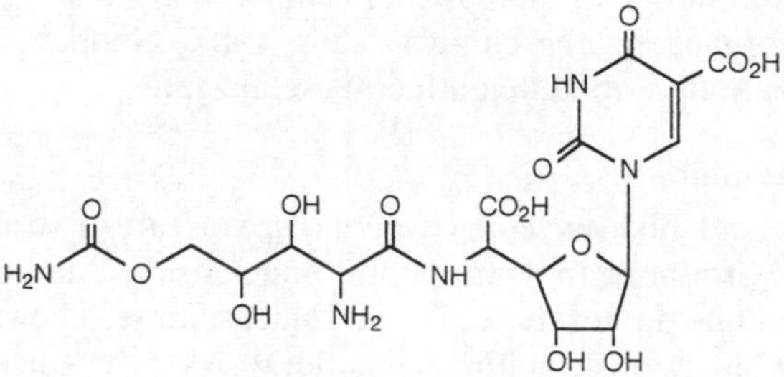

Polyoxin D

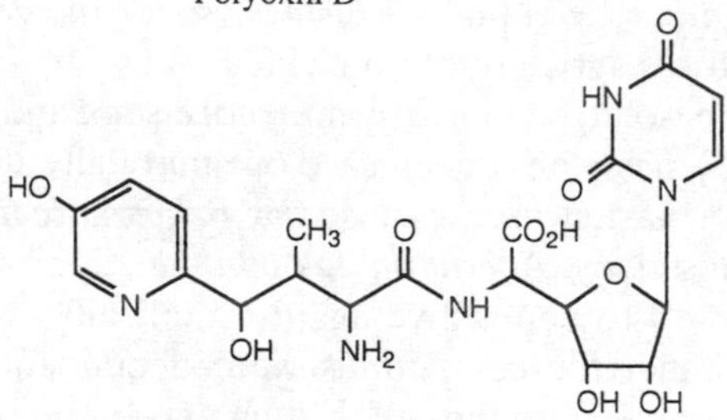

Nikkomycin Z

believed to be a consequence of better transport into the cell as the nikkomycins have been shown to be less sensitive to antagonism by peptides and are transported on a different peptide permease from that utilised by the polyoxins (Yadan *et al.*, 1984).

There have been numerous nikkomycins described by Zähner and co-workers, most isolated from mutants of *S. tendae*, and detailed structure–activity relationships have been determined. These provide a valuable insight into the structural properties necessary for enzyme inhibition, peptide transport and stability within the cell (Decker *et al.*, 1991).

In spite of the promise these compounds seem to have, there are very few reports of their being tested *in vivo*. Becker *et al*. (1988) found a mixture of nikkomycins X and Z (75:25) to have a moderate protective effect against a systemic *C. albicans* mouse infection. This activity *in vivo* was corroborated by Hector, Zimmer & Pappagianis (1990) who showed that nikkomycin Z had significant effects against *B. dermatitidis, H. capsulatum,* and *C. immitis* infections in mice. This activity correlated with good activity against the infecting organisms *in vitro*, and the authors point out that these fungi contain high levels of chitin in their pathogenic phase.

An interesting and unexplained point is that *Aspergillus* species are not susceptible to nikkomycin *in vitro* (*A. fumigatus* – Hector *et al*., 1990; *A. niger* – Nicholas, Williams & Hunter, 1994), although they have been reported to contain significant amounts of chitin (San-Blas, 1982). Nicholas *et al*. (1994), using the chitin-specific fluorochrome calcofluor white M2R, demonstrated that *A. niger* fluoresced strongly, indicating the presence of chitin. Since *Aspergilli* grow poorly in media free from antagonists, it is difficult to judge the significance of the apparent lack of activity of nikkomycin against these organisms *in vitro*, and there are no reports of nikkomycin or analogues being tested against *Aspergillus in vivo*.

Recent studies have revealed that a number of fungi produce multiple chitin synthases; most of the genes encoding these enzymes in both *C. albicans* and *S. cerevisiae* have been cloned and characterised (for review see Gow *et al.*, 1993; Robbins *et al.*, 1993). The enzymes in *Saccharomyces* have been shown to have different roles, CH1 is important in repair, while CH2 is the enzyme involved in septum formation. They also differ in their susceptibility to polyoxin and nikkomycin, CH1 being more sensitive to inhibition than CH2 (Cabib, 1991). Cabib (1991) points out that these different roles and sensitivities to inhibitors highlight the problems in screening for inhibition of chitin synthase, particularly for pathogenic fungi where less is known about the synthases.

A recent patent (Masubuchi, Okuda & Shimada, 1993) disclosed a novel microbial metabolite, xanthofulvin, produced by a species of *Eupenicillium*. This compound is unusual in two respects. It is claimed to have activity against CH2 obtained from an over-producing strain of *S. cerevisiae*, but shows no activity against CH1 at the concentrations tested (Table 2). In addition, the structure is quite unlike the polyoxins and nikkomycins, containing no nucleoside fragments and bearing no obvious resemblance to the substrate. Unfortunately the patent contains no data indicating whether the compound has activity against chitin synthases from pathogenic fungi, or against whole cells. However, this marked differential activity against the two synthases could make this compound a valuable tool for investigations on the role of the synthases. Extrapolating from what is known regarding the role of the *Saccharomyces* enzymes, it could be argued that inhibition of CH2 is more likely to produce an effect on whole cells than activity against CH1 (Cabib, 1991).

Table 2. *Inhibitory activity of xanthofulvin and polyoxin D against chitin synthases from* S. cerevisiae

	CHS I	CHS 2
Xanthofulvin	>200	2.2
Polyoxin D	0.26	10.3

IC_{50} values μM.

Two other microbial metabolites have been described that from their structure might be assumed to be inhibitors of chitin synthase, since they contain nucleoside residues. One compound, FR 900403, contains peptide–adenosine residues and is produced by a species of *Kernia* (Iwamoto *et al.*, 1990). It is claimed to be active against *C. albicans* with an MIC of 0.4 μg/ml, and to have activity in a mouse infection. It is not active against filamentous fungi. The other compound, FR 900848, contains a uridine residue with a fatty acid side chain, and produces swollen filaments and increased branching (Yoshida *et al.*, 1990). Activity against filamentous fungi is good with MIC values of 0.05 to 0.5 μg/ml against a number of species, but activity against yeasts is poor.

Chitinase inhibition

The role of chitinases in the cell wall of fungi has been the subject of much speculation, as little is known about their role or regulation (Cabib, 1987; Gooday, 1990). It is hypothesized that these enzymes are necessary for regulating the formation of crystalline chitin, modifying the microfibrils, and regulating the cross-linking of chitin in the wall (Gooday, 1990), all of which are necessary for growth, cell separation and the initiation of hyphal branches (Cabib, 1987). Kuranda and Robbins (1991) have cloned and sequenced the gene responsible for producing chitinase in *S. cerevisiae*, and have shown by disrupting this gene, that it is necessary for cell separation. Cabib, Silverman & Shaw (1992) suggest that opposing activities between chitinase and chitin synthase 1 are necessary for cell separation in *S. cerevisiae*. Since *S. cerevisiae* contains only minimal levels of chitin, it is interesting to speculate on the possible role of chitinase in fungi where chitin is believed to play a more major role.

Allosamidin, a natural product first reported by Koga *et al.* (1987), was isolated from an unidentified *Streptomyces* species using a screen for detecting inhibitors of insect chitinase. Allosamidin has inhibitory activity against insect, nematode and fungal chitinases; its inhibitory properties are reviewed by Gooday (1990). The compound is a pseudo-trisaccharide consisting of two novel sugar units (*N*-acetyl-D-allosamine; these are C-3

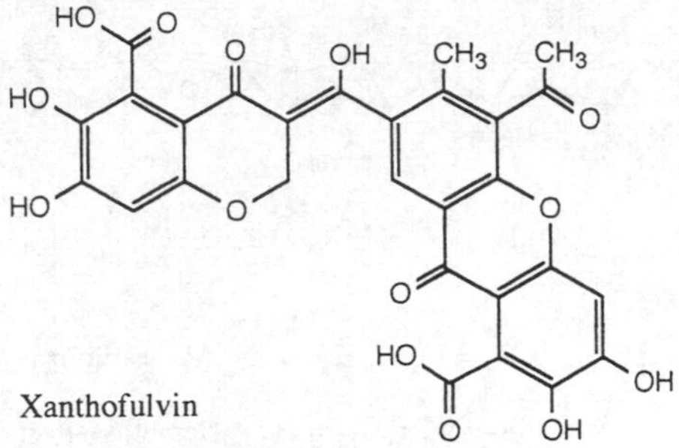

Xanthofulvin

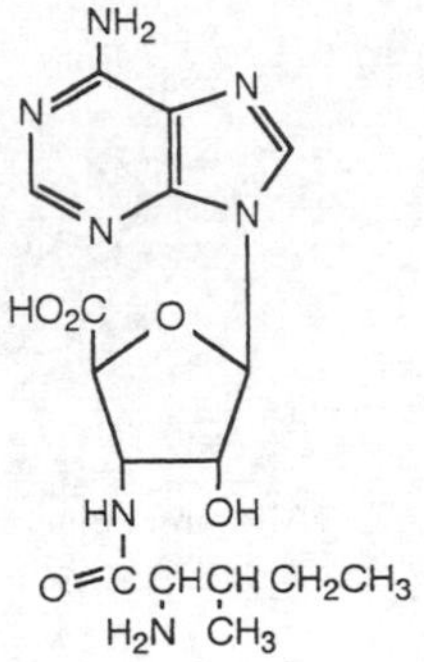

FR 900403

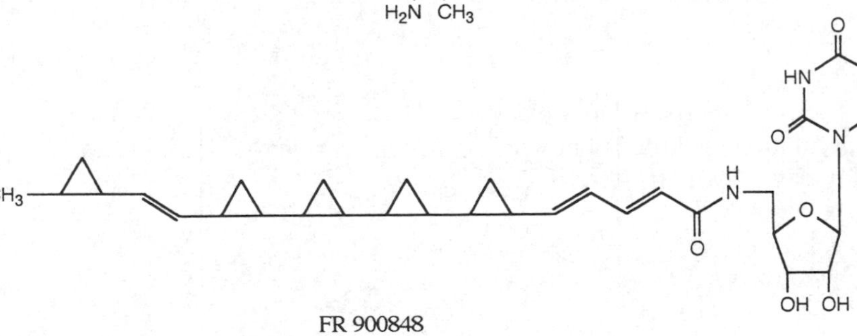

FR 900848

epimers of *N*-acetyl glucosamine) linked to a novel aminocyclitol derivative, allosamizoline. Although having good inhibitory activity against chitinase from *C. albicans*, this has not been reflected in any response against whole cells (Dickinson *et al.*, 1989; Hunter & Nicholas, unpublished observations).

Other unnamed *Streptomyces* cultures have been found to produce various allosamidins. Sakuda *et al.* (1990) and Zhou *et al.* (1993) describe the production of demethyl-allosamidin and didemethyl-allosamidin from culture No. AJ9463. Nishimoto *et al.* (1991) describe several derivatives including gluco-allosamidins and methyl allosamidin. They also prepared pseudodisaccharides by mild hydrolysis of the natural products. There are interesting differences in the spectrum of inhibition displayed by these compounds, removal of one of the methyl groups on the allosamizoline

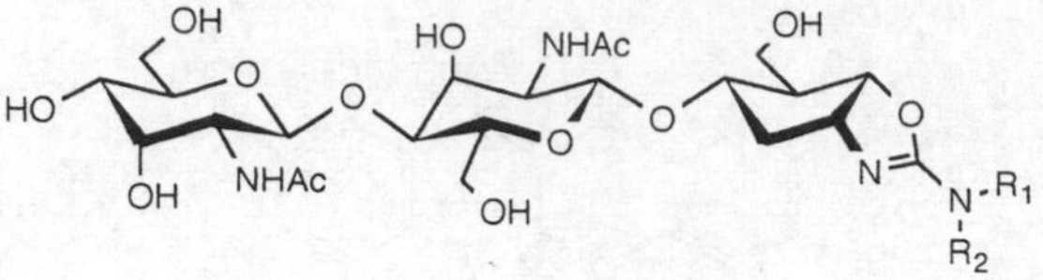

$R_1 = R_2 = CH_3;$ Allosamidin

$R_1 = CH_3, R_2 = H;$ Demethylallosamidin

$R_1 = R_2 = H;$ Didemethylallosamidin

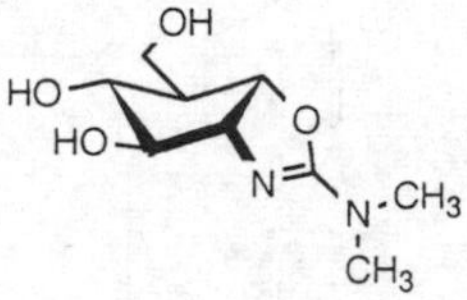

Allosamizoline

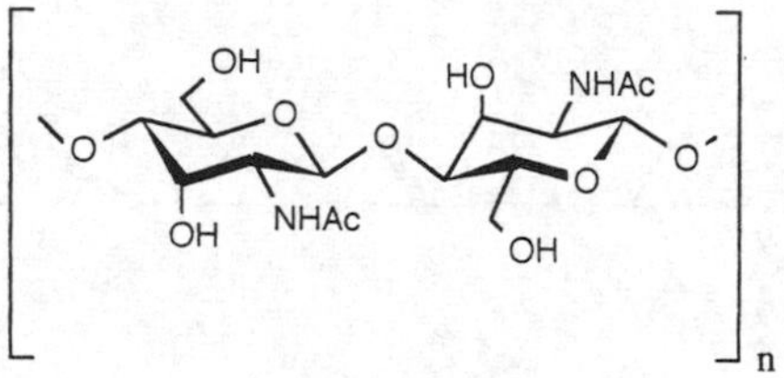

Repeating unit of chitin

enhanced activity against chitinases from both *S. cerevisiae* (100-fold) and *C. albicans* (10-fold), but activity against chitinase from a *Trichoderma* sp. was unaltered (Nishimoto *et al.*, 1991). Methylation at the 6″-position or the possession of gluco-stereochemistry had little effect on inhibitory activity. The pseudodisaccharides only retained activity against *C. albicans* chitinase.

Two groups have reported preparing synthetic analogues of the allosamizoline portion. Corbett, Dean & Robinson (1993) found that 6-membered carboxylic ring analogues had, at best, only weak inhibitory activity against

C. albicans chitinase. Terayama *et al.* (1993) prepared *N,N*-diacetyl-β-chitobiosyl allosamizoline and found a change in the spectrum of inhibitory activity, with the greatest effect being seen against silkworm chitinase; no mention being made of activity against whole cells. Various pseudodisaccharide compounds have also been synthesized with gluco-residues, but these are reported as having only very weak inhibitory activity against *C. albicans* chitinase. Other chitinases were not tested (Corbett, Dean & Robinson, 1994).

Progress in this area has been disappointing. It is clear that it is possible to change the spectrum of inhibitory activity, but detailed structure–activity relationships are hampered by the difficulty of synthesizing the component sugar residues. Lack of availability of the natural products has limited studies on whole cells with most of the compounds. The lack of activity against whole cells of *C. albicans* and *S. cerevisiae* (other than the impairment of cell division by demethyl allosamidin reported by Sakuda *et al.* (1990), might be explained by chitin playing only a minor role in these species, or could be a reflection of poor penetration or uptake through the wall (Dickinson *et al.*, 1989). Hunter and Nicholas (unpublished observations) examined several filamentous fungal species, including *Rhizopus, Aspergillus,* and *Trichophyton* and could detect no morphological changes in cultures exposed to allosamidin at concentrations up to 20 μg/ml. In these species, chitin and chitinase would be expected to play a more significant role in the organisation of the wall. There are no reports of such studies with analogues showing better activity than allosamidin against fungal chitinase.

Glucan as a target

The walls of many fungi contain varying amounts of glucose polymers known as glucans joined through α- and β- (1,3) or (1,6) linkages. The type of glucan linkage varies between species, thus in the yeasts *C. albicans* and *S. cerevisiae* β-linked glucans predominate (Cassone, 1986), whereas the yeast *Cr. neoformans* contains α-linked glucans (James *et al.*, 1990). Fungi belonging to the Order Mucorales do not contain glucans in their wall. Just as the chitin content of certain medically important fungal species varies in different morphological states (*vide supra*), so can the glucan content. *P. brasiliensis*, one of the dimorphic pathogenic fungi, has been shown to contain predominantly β-glucans in the mycelial stage, but these change to α-glucans, in the pathogenic yeast form (Dávila, San-Blas & San Blas, 1986). There are also indications that *C. immitis* might be similar (*vide infra*).

There are several naturally occurring, amphophilic compounds that have been found to have antifungal activity and to inhibit β-1,3-glucan. The major compounds of interest, together with their producing organisms are listed in Table 3. The continuing interest in this area is evident from the number of compounds that have been described recently.

Table 3. *Examples of the main classes of β-glucan synthase inhibitors described in the literature, and their producing organisms*

Compound	Producing organism	Reference
Echinocandin A	*Aspergillus nidulans* var. echinulatus	Benz *et al.*, 1974
Echinocandin C & D	*Aspergillus rugulosus*	Traber *et al.*, 1979
Papalacandins	*Papularia sphaerosperma*	Gruner & Traxler, 1977
	Dictyochaeta simplex	VanMiddlesworth *et al.*, 1991
	Phialophora cyclamis	Kaneto *et al.*, 1993
	Gilmaniella sp. FA4459	Aoki *et al.*, 1993
Aculeacin	*Aspergillus aculeatus*	Mizuno *et al.*, 1977
		Satoi *et al.* 1977
Chaetiacandin	*Monochaetia dimorphospora*	Komori *et al.*, 1985
Pneumocandins	*Zalerion arboricola*	Fromtling & Abruzzo, 1989.

Most are isolated as complexes of closely related compounds, and in some cases, separation and characterization of the individual components is not easy. The echinocandins and aculeacins are cyclic hexapeptides in which all of the amino acid residues contain one or more hydroxyl groups. All compounds contain a lipophilic side chain linked to the amino-group of an ornithine residue. the side chain is an extremely non-polar fatty acid: in the aculeacins derived from palmitic acid and in the echinocandins from linoleic acid. The papulacandins and chaetiacandins have a different structure being spirocylic diglycosides with two long-chain fatty acids of differing lengths.

The background of the discovery of the β-glucan inhibitory properties of these compounds and their effect on medically important fungi has been reviewed in some detail by Hector (1993), and will not be repeated here. Numerous synthetic and semi-synthetic analogues and derivatives of many of these compounds have been described. Cilofungin, a semi-synthetic derivative of echinocandin B, was progressed to studies in man following extensive laboratory studies, including detailed experimental animal tests (summarized by Gordee & Debono, 1989). Unfortunately, adverse effects were seen (Doebbeling *et al.*, 1990), and although these were attributed to the vehicle used (26% polyethylene glycol), clinical work on this compound was halted.

As potential clinical agents these inhibitors suffer from two major draw-backs: a narrow spectrum of activity against fungi, and poor solubility in pharmacologically acceptable solvents. Work on the mode of action of these compounds has paralleled an increase in knowledge of the nature of the fungal wall, although this is still woefully inadequate for many fungal species. It is clear however that the components inhibit the assembly of β-1,3 and β-1,6 linked glucans and not α-linked glucans. They exert a powerful effect on species such as *Candida* that contain large quantities of β-linked

	R_1	R_2	R_3		R_4
Echinocandin B_0	H	CH_2CONH_2	H	CH_3	
L-693,989	PO_3Na	CH_2CONH_2	H	CH_3	
Cilofungin	H	CH_3	CH_3		
Tetrahydroechinocandin B	H	CH_3	CH_3	CH_3	

glucans, but are ineffective on *Cryptococcus*, a species that has been shown to contain mainly α-1,3-glucan (James *et al.*, 1990).

The importance of the nature of the glucan is emphasized in the elegant studies with *P. brasiliensis* reported by Dávila *et al.* (1986) who showed that papulacandin B could only affect the mycelial stage and the transition of mycelium to yeast, stages in which β-glucans predominate. The pathogenic or yeast phase, in which α-glucans predominate, was unaffected. This finding has implications in the testing of these compounds since the glucan content of fungi in their pathogenic form or under *in vivo* conditions is rarely known. There are indications in which a similar situation may occur with

Papulacandin

another dimorphic pathogenic fungus, *C. immitis*. Galgiani *et al*. (1990) found that cilofungin had activity against the mycelial stage, producing characteristic damage to the hyphal wall, and also prevented arthrospores from germinating. The mycelium and arthrospores are the saprophytic stage of the fungus and when inhaled the arthrospores germinate to produce the pathogenic spherule. Cilofungin had no effect on spherules *in vitro* and was ineffective against a mouse infection.

The claims that the walls of *Pneumocystis carinii* may contain β-glucan (Matsumoto *et al*., 1989) led to the discovery that both echinocandins and papulacandins were active in a rat *P. carinii* pneumonia model (Schmatz *et al*., 1990; VanMiddlesworth *et al*., 1991). Subsequently, several new microbial metabolites related to echinocandin B have been isolated, and more recently, these have been named the Pneumocandins (Schmatz *et al*., 1992) in recognition of their good activity against this organism. The pneumocandins are distinguished from echinocandin B by having one of the two threonine residues replaced by hydroxyglutamine, and a 10,12-dimethylmyristoyl side chain.

One of the more active pneumocandins is B_o (L-688,786). This compound is at least as active as cilofungin against *C. albicans* infections, and is considerably more active in the rat *Pneumocystis* model (Schmatz *et al*., 1992). Pneumocandins are claimed to be less lytic to red blood cells than aculeacin and the echinocandins (Fromtling & Abruzzo, 1989).

These compounds still suffer from poor solubility, and as they are not absorbed orally, need to be administered parenterally. In view of the adverse effects seen with cilofungin administered in a polyethylene glycol carrier, the use of co-solvents is regarded with some suspicion. A new approach is the development of water-soluble prodrugs of the pneumocandins (Balkovec *et al*., 1992). From a series of potential prodrugs, the most active, with good activity *in vivo*, proved to be a phosphate ester of pneumocandin B_o (L 693-989).

In a series of posters at the 33rd Interscience Congress of Antimicrobial

Agents and Chemotherapy (New Orleans USA, October 1993), more water-soluble derivatives of pneumocandin B_o were described and claimed to have activity 10- to 40-fold greater than the phosphate ester against *C. albicans* and *P. carinii in vivo*. In addition, activity against *Aspergillus* is claimed *in vivo*, together with 'modest activity' against *Cr. neoformans*. At the same meeting cilofungin analogues were claimed to show considerable advantages, with activity both orally and parenterally against *C. albicans, A. fumigatus, P. carinii*, and *H. capsulatum* infections. If these results are confirmed, and the toxicology is favourable, these compounds hold out the promise of real progress in this area.

Other modes of action

Benanomicins and pradimicins

In 1988 novel microbial metabolites isolated from *Actinomadura* species were disclosed by two companies; the compounds were shown to be dihydrobenzo[a]naphthacenquinones, containing an alanine residue with a disaccharide attached. They differed only in a substituent on one of the sugar residues, and were named pradimicin A (Oki *et al.*, 1988) and benanomicin A and B (Takeuchi *et al.*, 1988).

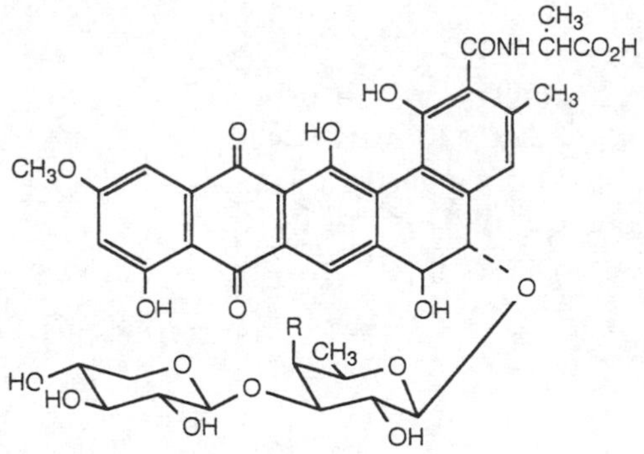

Benanomicin A; R = OH

Benanomicin B; R = NH$_2$

Pradimicin ;   R = NHCH$_3$

These compounds have a broad spectrum of antifungal activity that includes *Aspergillus, Candida, Cryptococcus* and dermatophyte species as well as a number of opportunistic fungi. Activity less than that displayed by amphotericin but superior to that of ketoconazole has been demonstrated in a range of mouse infections. There was no evidence of cytotoxicity against several cell lines, and acute toxicity was low (Oki *et al.*, 1990). The mode of action of these compounds is still not clear. In early studies it was observed that potassium leakage in *C. albicans* was associated with antifungal activity and this activity was dependent on calcium (Sawada *et al.*, 1990). Recent studies have established that they function rather like lectins and bind to yeast mannan in a highly stereo-specific fashion, forming a ternary complex with calcium and mannan (Ueki *et al.*, 1993).

Numerous papers have been published since 1993 describing various naturally occurring and semi-synthetic compounds, including those describing water-soluble derivatives. These have been obtained both by chemical modification and by directed biosynthesis (Furumai *et al.*, 1993). No comparative studies have been published as yet on these newer analogues, but a number appear to retain good activity in animal models. It is difficult to judge the current status of these compounds, but given their spectrum and novel mode of action, which appears to be specific, a therapeutic agent could emerge.

5-flurocytosine

The synthetic agent 5FC was made as part of a programme looking for cytostatic drugs, it was inactive in anticancer screens but proved to have

5-Flurocytosine

antifungal activity. Its chemotherapeutic antifungal activity was first described by Grunberg, Titsworth & Bennett (1964), and the compound was marketed as an oral antifungal agent in 1971. It is fungicidal against *Candida* and *Cryptococcus* species, and has some fungistatic activity against *Aspergillus* species and a number of dermatiaceous fungi. The compound acts as a prodrug of 5-flurouracil as it is taken into the fungal cell on a cytosine permease, and deaminated to 5FU. Subsequent metabolism affects both fungal RNA and DNA synthesis; 5-fluorouridine triphosphate replaces uracil in RNA, and 5-fluorodeoxyuridylate inhibits thymidylate synthase, thus affecting the biosynthesis of DNA (for details see Kerridge, 1986). Mammalian cells are believed to lack the ability to deaminate 5FC and thus are protected from damage by 5FU.

In spite of excellent activity against *Cryptococcus*, the development of resistance to 5FC during therapy has limited the value of this drug. It is generally used in combination with amphotericin B, but its use is gradually being eroded by the newer azoles.

Natural products as a source of antifungal agents

Natural products, in particular microbial metabolites, have been, and continue to be a fruitful source of novel structures. They may reveal unexpected modes of action, such as the pradimicins, xanthofulvin, and allosamidin described above. They often act as a lead for structural modification, or the producing organisms may be manipulated to produce derivatives, as exemplified by the work with the nikkomicins, echinocandins, and pneumocandins. Although many novel compounds are described each year, very few have sufficient activity to be of real interest. Many prove to be toxic or are too difficult to pursue.

Most of the natural products of note have been covered in detail above, with the exception of griseofulvin, one of the earliest antifungal antibiotics to be discovered. Originally found to be active against plant pathogens, it was observed to inhibit and produce distortions in germ tubes of *Botrytis allii* (Brian, Hemming & McGowan, 1945). The compound proved too expensive to develop as an agrochemical and it was not until 1958 that Gentles reported its efficacy as an oral agent against ringworm in experimental animals (Gentles, 1958). Griseofulvin rapidly became established as the drug of choice for dermatophyte infections.

Griseofulvin

The background to the development of the compound and investigation into its possible toxicity make interesting reading, especially in the present era of extensive toxicological and clinical evaluation programmes (Davies, 1980). In spite of its known effects on mitosis and nucleic acid synthesis, the drug has been used since 1959 with very few adverse reports. It is only more recently that its mode of action on the cytoskeleton has become clearer (Kerridge & Vanden Bossche, 1990). It interacts with tubulins acting as a mitotic poison. Although it inhibits tubulins from a number of fungi, it is only active against dermatophytes, and is presumed to be unable to penetrate other species (Kerridge, 1986). No analogues or derivatives with improved properties have been developed, but the allylamines and amorolfine are now eroding its position in the clinic.

Newer compounds of interest include the spartanomycins; these are anthraquinones isolated from *Micromonospora* species (Nair *et al.*, 1992)

Spartanamicin B R =   Spartanamicin A R =

and are reported to have a broad spectrum of activity. MIC values for spartanomycin B against a range of medically important fungi, including *C. albicans*, *Cr. neoformans*, and *Aspergillus* species were all between 1.0 and 0.2 μg/ml. There are no data yet on activity *in vitro*, toxicity, or mode of action.

The sampangines are copyrine alkaloids isolated from a West African tree, *Cliestopholis patens*, and were first reported by Clark *et al.* (1991). The parent compound and various derivatives have been synthesized and several are claimed to have good activity *in vitro* against *C. albicans* and *Cr. neoformans* and to have low acute toxicity. The compounds are believed to bind to DNA, but a recent report describing more derivatives also with good activity *in vitro*, contains no more information on the mode of action (Clark *et al.*, 1993). Information on activity *in vivo* and toxicity of these compounds is awaited.

A more unusual source of natural products is the shark; squalamine, an aminosterol, was isolated by Moore *et al.* (1993) from the stomach of the

Squalamine

common dogfish shark. It is the first antimicrobial steroid to be isolated from an animal, although aminosterols have been isolated from plants. The main interest in this compound is the possibility of chemical manipulation since both the amino side chain and the sulfate group should be amenable to derivatization.

RESISTANCE TO ANTIFUNGAL DRUGS

The development of resistance to antifungal agents in clinical medicine has been less of a problem than with antibacterial agents. Drugs such as amphotericin B, griseofulvin, and tolnaftate have been used for decades with no appreciable development of resistant strains being evident (Odds, 1992). Resistance, however, both primary and secondary, to 5FC is common and was evident from the early days of its use. The resistant strains generally have altered permeases resulting in a lack of penetration of the drug. A worrying recent development is the appearance of strains, particularly of *Candida* species, that are resistant to the azoles, especially to fluconazole (Odds, 1992; Johnson *et al.*, 1993). There are indications that the long-term use of fluconazole favours the development of resistance (Johnson *et al.*, 1993). In addition, there is increasing evidence that sustained use of fluconazole can produce a change in the microflora, with organisms that have greatly reduced susceptibility to fluconazole, such as *C. glabrata* and *C. krusei* emerging as the dominant species (Odds, 1992). It is not yet clear whether fluconazole-resistant strains will be cross-resistant with all other azoles.

WHAT OF THE FUTURE?

It is evident that with the increase in serious fungal infections, the toxicity and adverse effects seen with the most active and broad-spectrum compound amphotericin B, and the possibility of resistance developing to the azoles, new antifungal agents are needed. For the less serious, non-invasive diseases the need is less acute, but improvements are required for specific areas. Amorolfine and terbinafine both seem to offer major advantages in the treatment of onychomycosis, a condition traditionally responding very poorly to griseofulvin, and several of the new topical azoles have excellent activity against vaginal candidosis.

It is too early yet to judge the status of the pradimicins, echinocandins and pneumocandins. Work seems further ahead on the latter group, and if the newer water-soluble analogues prove to have low toxicity, then such a compound may enter the clinic. However, even if the activity against *Aspergillus* is confirmed, the spectrum of these compounds is limited to those species in which β-glucans predominate. The lack of activity against *Cryptococcus* would be regarded by many as a major disadvantage. The pradimicins would seem to offer a broader spectrum of activity, and water-soluble compounds have been developed, but little detailed evaluation *in vivo* has been published as yet, and information on toxicity is scanty. It is disappointing that, in spite of scientific advances, so little real progress has been made in developing a nikkomycin for clinical use. Theoretically these compounds offer the promise of breadth of spectrum, low toxicity, and fungicidal activity.

There is increasing interest in identifying new targets in fungi, preferably those that are unique to the fungus. Often, however, these targets occur in other eukaryotic cells, and selectivity relies on a high affinity for the fungal target. The advances in molecular biology that have made genetic manipulation feasible with *C. albicans* have opened up greater possibilities for research (Gow *et al*. 1993; Koltin, 1989), and as targets are identified, new screens are being developed to enable either natural products or novel synthetic compounds to be tested. Although the 'rational' approach of designing a molecule for a specific purpose has great scientific attraction, one should not forget that screening has resulted in a number of valuable drugs; naftifine was found by cross-screening, as was 5FC.

One area neglected as a target in medical mycology is the cytoskeleton. Griseofulvin, as noted above, is now known to interact with tubulin, but this mode of action has not been exploited as it has for plant pathogens. Benomyl has been used as a plant fungicide for many years and binds to fungal β-tubulins, but it is inactive against a range of medically important fungi.

NEW DEVELOPMENTS IN NON-AZOLE ANTIFUNGALS 45

Recent work (for review see Koltin, 1989) has shown that benomyl and related compounds do bind to *C. albicans* β-tubulin; the reason for the lack of activity against whole cells may be poor penetration or some other resistance mechanism. Although benomyl does have a higher affinity for fungal tubulin, selectivity could be a problem in this area since the proteins are highly conserved with close similarities between the eukaryotic tubulins. It is also possible that transport into the cell may be a deciding factor (Kerridge & Vanden Bossche, 1990).

A factor involved in protein synthesis, and unique to fungi, elongation factor 3, was found initially in *S. cerevisiae* and has been purified. Subsequently, this factor has been identified in a number of fungal species; its function is to stimulate the binding of aminoacyl-tRNA to the ribosome (Koltin, 1989). Since this target is believed to be lacking in mammalian cells, it has attracted considerable interest among those involved in designing screens for antifungal agents.

Topoisomerases have long been a target for antibacterial agents, but more recently it has been discovered that some compounds with activity against topoisomerase II have activity against fungi (Figgitt *et al.*, 1989). It had been thought previously that insufficient selectivity occurred between mammalian and fungal topoisomerases, but compounds are now being identified with differential inhibitory properties.

Many of these areas are still highly speculative and have yet to produce results; some are designed to use isolated enzymes, and a frequent stumbling block is lack of activity against whole cells. Nevertheless, it is encouraging that such interest is still shown, and gives hope that a novel, effective and non-toxic agent may emerge.

Anaissie, E., Bodey, G. P., Kantarjian, H., Ro, J., Vartivarian, S. E., Hopfer, F., Hoy, J. & Rolston, K. (1989). New spectrum of fungal infections in patients with cancer. *Reviews of Infectious Diseases*, **11**, 369–78.

Aoki, M., Andoh, T., Ueki, T., Masuyoshi, S., Suguwara, K. & Oki, T. (1993). BU-4794F, a new β-1,3-glucan synthase inhibitor. *Journal of Antibiotics*, **46**, 952–60.

Aoki, Y., Yamazaki, T., Kondoh, M., Sudoh, Y., Nakayama, N., Sekine, Y., Shimadu, H. & Arisawa, M. (1992). A new series of natural antifungals that inhibit P450 lanosterol demethylase. II. Mode of action. *Journal of Antibiotics*, **45**, 160–70.

Bailey, E. M., Krakovsky, D. J. & Rybak, M. J. (1990). The triazol antifungal agents: A review of itraconazole and fluconazole. *Pharmacotherapy*, **10**, 146–53.

Balfour, J. A. & Foulds D. (1992). Terbinafine. A review of its pharmacodynamic and pharmacokinetic properties, and therapeutic potential in superficial mycoses. *Drugs*, **43**, 259–84.

Balkovec, J. M., Black, R. M., Hammond, M. L., Beck, J. V., Zambias, R. A., Abruzzo, G., Bartizal, K., Kropp, H., Trainor, C., Scwartz, R. E., McFadden, D. C., Nollstadt, K. H., Pittarelli, L. A., Powles, M. A. & Schmatz, D. M. (1992).

Synthesis, stability, and biological evaluation of water-soluble prodrugs of a new echinocandin lipopeptide. discovery of a potential clinical agent for the treatment of systemic candidiasis and *Pneumocystis carinii* pneumonia (PCP). *Journal of Medicinal Chemistry*, **35**, 194–8.

Barrett-Bee, K. J., Lane, A. C. & Turner, R. W. (1986). The mode of antifungal action to tolnaftate. *Journal of Medical and Veterinary Mycology*, **24**, 155–60.

Becker, J. M., Covert, N. L., Shenbagamurthi, P., Steinfeld, A. S. & Naider, F. (1983). Polyoxin D inhibits growth of zoopathogenic fungi. *Antimicrobial Agents and Chemotherapy*, **23**, 926–9.

Becker, J. M., Marcus, S., Tullock, J., Miller, D., Drainer, E., Khare, R. K. & Naider, F. (1988). Use of the chitin-synthesis inhibitor nikkomycin to treat disseminated candidiasis in mice. *Journal of Infectious Diseases*, **157**, 212–14.

Bennett, J. E. (1991). Developing drugs for the deep mycoses: a short history. In: Bennett, J. E., Hay, R. J. & Peterson, P. K., eds. *New Strategies in Fungal Disease*. Churchill Livingstone.

Benz, F., Kneusel, F., Nuesch, J., Treichler, H., Voser, W., Nyfeler, R. & Keller-Schierlein, W. (1974). Echinocandin B, ein neuartiges polypeptid antibioticum aus *Aspergillus nidulans var echinulatus*: isolierung und baudsteine. *Helvetica Chimica Acta*, **57**, 2459–77.

Body, G. P. (1988). Fungal infections in cancer patients. *Annals of the New York Academy of Sciences*, **544**, 431–42.

Bolard, J., Legrand, P., Heitz, F. & Cybulska, B. (1991). One-sided action of Amphotericin B on cholesterol-containing membranes is determined by its self-association in the medium. *Biochemistry*, **30**, 5707–15.

Borgers, M. (1985). Antifungal azole derivatives. In: Greenwood, D. & O'Grady, F., eds. *The Scientific Basis of Chemotherapy*, Symposium 38, Society for General Microbiology. Cambridge University Press.

Bratjburg, J., Powderly, W. G., Kobayashi, G. S. & Medoff, G. (1990). Amphotericin B: current understanding of mechanisms of action. *Antimicrobial Agents and Chemotherapy*, **34**, 183–8.

Brian, P. W., Hemming, H. G. & McGowan, J. C. (1945). Origin of a toxicity to mycorrhiza in Wareham Heath soil. *Nature*, **155**, 637–8.

Cabib, E. (1987). The synthesis and degradation of chitin. *Advances in Enzymology*, **59**, 59–101.

Cabib, E. (1991). Differential inhibition of chitin synthetases 1 and 2 from *Saccharomyces cerevisiae* by polyoxin D and nikkomycins. *Antimicrobial Agents and Chemotherapy*, **35**, 170–73.

Cabib, E., Silverman, S. J. & Shaw, J. A. (1992). Chitinase and chitin synthase 1: counterbalancing activities in cell separation of *Saccharomyces cerevisiae*. *Journal of General Microbiology*, **138**, 97–102.

Cassone, A. (1986). Cell wall of pathogenic yeasts and implications for antimycotic therapy. *Drugs in Experimental Clinical Research*, **12**, 635–43.

Chiew, Y. Y., Shepherd, M. G. & Sullivan, P. A. (1980). Regulation of chitin synthesis during germ-tube formation in *Candida albicans*. *Archives of Microbiology*, **125**, 97–104.

Clark, A. M., Hufford, C. D., Peterson, J. R., Zjawiony, J., Liu, S. & Walker, L. A. (1991). Antifungal copyrine alkaloids: *in vitro* and *in vivo* antifungal activities of sampangines. In *Proceedings of the 31st Interscience Conference on Antimicrobial Agents and Chemotherapy, Chicago, USA*. Abstract No. 213.

Clark, A. M., Zhu, X., Walker, L. A., Zjawiony, J., Mukerjee, A. & Hufford, C. D. (1993). Evaluation of the *in vitro* antifungal activity of new sampangine analogs and studies on the mode of action of sampangine. *Proceedings of the 33rd*

Interscience Conference on Antimicrobial Agents and Chemotherapy, New Orleans, Abstract No. 373.

Clements, J. S. & Peacock, J. E. (1990). Amphotericin revisited: Reassessment of toxicity. *American Journal of Medicine*, **88** (5N), 22N-27N.

Corbett, D. F., Dean, D. K., & Robinson, S. R. (1993). The synthesis of pseudo-sugars related to allosamizoline. *Tetrahedron Letters*, **34**, 1525–8.

Corbett, D. F., Dean, D. K. & Robinson, S. R. (1994). Synthesis of pseudodisac-charides related to allosamidin. *Tetrahedron Letters*, **35**, 459–62.

Dähn, U., Hagenmaier, H., Höhne, H., König, W. A., Wolf, G. & Zähner, H. (1976). Stoffwechselprodukte von mikroorganismen. 154. Mitteilung. Nikkomy-cin, ein neuer hemmstoff der chitin synthese bei pilzen. *Archives of Microbiology*, **107**, 143–60.

Davies, R. R. (1980). Griseofulvin. In *Antifungal Chemotherapy*, (Speller, D. C. E. ed.). pp. 149–82, John Wiley & Sons.

Dávila, T., San-Blas, G. & San-Blas, F. (1986). Effect of papulacandin B on glucan synthesis in *Paracoccidioides brasiliensis*. *Journal of Medical and Veterinary Mycology*, **24**, 193–202.

Decker, H., Zähner, H., Heitsch, H., König, W. A. & Fiedler, H.-P. (1991). Structure–activity relationships of the nikkomycins. *Journal of General Micro-biology*, **137**, 1805–13.

Deepe, G. S. & Bullock, W. E. (1990). Immunological aspects of fungal patho-genesis. *European Journal of Clinical Microbiology and Infectious Diseases*, **9**, 567–79.

Dickinson, K., Keer, V., Hitchcock, C. A. & Adams, D. J. (1989). Chitinase activity from *Candida albicans* and its inhibition by allosamidin. *Journal of General Microbiology*, **135**, 1417–21.

Doebbeling, B. N., Fine, B. D., Pfaller, M. A., Sheetz, C. T., Stokes, J. B. & Wenzel, R. P. (1990). Acute tubular necrosis and anionic gap acidosis during therapy with cilofungin (LY-121019) in polyethylene glycol. *Proceedings of the 30th Interscience Conference on Antimicrobial Agents and Chemotherapy, Atlanta, USA*, Abstract No. 583.

Fiedler, H.-P., Kurth, R., Langharig, J., Delzer, J. & Zähner, H. (1982). Nikkomy-cins: microbial inhibitors of chitin synthase. *Journal of Chemical Technology and Biotechnology*, **32**, 271–80.

Figgitt, D. P., Denyer, S. P., Dewick, P. M., Jackson, D. E. & Williams, P. (1989). Topoisomerase II: a potential target for novel antifungal agents. *Biochemical and Biophysical Research Communications*, **160**, 257–62.

Fromtling, R. A. & Abruzzo, G. K. (1989). L-671,329, a new antifungal agent. III. *In vitro* activity, toxicity and efficacy in comparison to aculeacin. *Journal of Antibiotics*, **42**, 174–78.

Furumai, T., Saitoh, K., Kakushima, M., Yammoto, S., Suzuki, K., Ikeda, C., Kobaru, S., Hatori, M. & Oki, T. (1993). BMS-1821184, a new pradimicin derivative. Screening, taxonomy, directed biosynthesis, isolation and characteriz-ation. *Journal of Antibiotics*, **46**, 265–74.

Gale, E. F. (1973). Perspectives in chemotherapy. *British Medical Journal*, **4**, 33–8.

Galgiani, J. N., Sun, S. H., Clemons, K. V. & Stevens, D. A. (1990). Activity of cilofungin against *Coccidioides immitis:* differential *in vitro* effects on mycelia and spherules correlated with *in vivo* studies. *The Journal of Infectious Diseases*, 162–944–8.

Gentles, J. C. (1958). Experimental ringworm in guinea pigs: oral treatment with griseofulvin. *Nature*, **182**, 476–7.

Gooday, G. W. (1990). Inhibition of chitin metabolism. In: Kuhn, P. J., Trinci, A. P.

J., Jung, M. J., Goosey, M. W. & Copping, L. J., eds. *Biochemistry of Cell Walls and Membranes in Fungi*. Springer-Verlag.

Gordee, R. & Debono, M. (1989). Cilofungin. *Drugs of the Future*, **14**, 939–41.

Gow, N. A. R., Swoboda, R., Bertram, G., Gooday, G. W. & Brown, A. (1993). Key genes in the regulation of dimorphism of *Candida albicans*, In *Dimorphic Fungi in Biology and Medicine*, pp. 61–71. (Vanden Bossche, H., Odds, F. & Kerridge, D., eds). Plenum Press.

Grunberg, E., Titsworth, E. & Bennett, M. (1964). Chemotherapeutic activity of 5-fluorocytosine. *Antimicrobial Agents and Chemotherapy-1963*, American Society of Microbiology, Ann Arbor, pp. 566–68.

Gruner, J. & Traxler, P. (1977). Papulacandin, a new antibiotic, active especially against yeasts. *Experientia*, **33**, 137.

Hector, R. F. (1993). Compounds active against cell walls of medically important fungi. *Clinical Microbiology Reviews*, **6**, 1–21.

Hector, R. F. & Pappagianis, D. (1983). Inhibition of chitin synthesis in the cell wall of *Coccidioides immitis* by polyoxin D. *Journal of Bacteriology*, **154**, 488–98.

Hector, R. F., Zimmer, B. L. & Pappagianis, D. (1990). Evaluation of nikkomycins X and Z in murine models of Coccidioidomycosis, Histoplasmosis, and Blastomycosis. *Antimicrobial Agents and Chemotherapy*, **34**, 587–93.

Hoeprich, P. D., Flynn, N. M., Kawachi, M. M., Lee, K. K., Lawrence, R., Heath, L. K. & Schaffner, C. P. (1988). Treatment of fungal infections with semisynthetic derivatives of Amphotericin B. *Annals of the New York Academy of Sciences*, **544**, 517–46.

Hunter, P., Carter, E., Murdock, P., Williams, L. & Scott-Wood, G. (1992*a*). Comparative activity of a new polyene, BRL 49594A, and amphotericin B against fungal and mammalian cells. *Proceedings of the 32nd Interscience Conference on Antimicrobial Agents and Chemotherapy, Anaheim, USA*, Abstract No. 1042.

Hunter, P., Murdock, P., Randall, G., Anthony, S., Jones, S., Everett, A., Winch, C. & Burbidge, I. (1992*b*). Comparative activity *in vivo* and toxicity of a new polyene, BRL 49594A, and amphotericin B. *Proceedings of the 32nd Interscience Conference on Antimicrobial Agents and Chemotherapy, Anaheim, USA*, Abstract No. 1043.

Iwamoto, T., Fujie, A., Tsurumi, Y., Nitta, K., Hashimoto, S. & Okuhara, M. (1990). FR900403, a new antifungal antibiotic produced by a *Kernia sp*. *Journal of Antibiotics*, **43**, 1183–5.

James, P. G., Cherniak, R., Jones, R. G., Stortz, C. A. & Reiss, E. (1990). Cell wall glucans of *Cryptococcus neoformans* CAP 67. *Carbohydrate Research*, **198**, 23–38.

Johnson, E. M., Luker, J., Scully, C. & Warnock, D. W. (1993). Azole-resistant *Candida* species among HIV positive patients given long-term fluconazole for oral candidosis. *Proceedings of the 33rd Interscience Conference on Antimicrobial Agents and Chemotherapy, New Orleans, USA*, Abstract No. 741.

Kaneto, R., Chiba, H., Agematu, H., Shibamoto, N., Yoshioko, T., Nishida, H. & Okamoto, R. (1993). Mer-WF3010, a new member of the papulacandin family. I. Fermentation, isolation and characterization. *Journal of Antibiotics*, **46**, 247–50.

Kerridge, D. (1986). Mode of action of clinically important antifungal drugs. *Advances in Microbial Physiology*, **27**, 1–72.

Kerridge, D. & Vanden Bossche, H. (1990). Drug discovery: a biochemist's approach. *Handbook of Experimental Pharmacology*, **96**, 31–76.

Koga, D., Isogai, A., Sakuda, S., Matsumoto, S., Suzuki, A., Kimura, S. & Ide, A. (1987). Specific inhibition of *Bombyx mori* chitinase by allosamidin. *Agricultural and Biological Chemistry*, **51**, 471–6.

Koltin, Y. (1989). Targets for antifungal drug discovery. *Annual Reports in Medicinal Chemistry*, **25**, 141–8.

Komori, T., Yamashita, M., Tsurumi, Y. & Kohsaka, M. (1985). Cheatiacandin, a novel papulacandin. I. Fermentation, isolation and characterization. *Journal of Antibiotics*, **38**, 455–9.

Komshian, S. V., Uwaydah, A. K., Sobel, J. D. & Crane, L. R. (1989). Fungemia caused by *Candida* species and *Torulopsis glabrata* in the hospitalized patient: frequency, characteristics, and evaluation of factors influencing outcome. *Reviews of Infectious Diseases*, **11**, 379–87.

Kuranda, M. J. & Robbins, P. W. (1991). Chitinase is required for cell separation during growth of *Saccharomyces cerevisiae*. *Journal of Biological Chemistry*, **266**, 19758-67.

Masubuchi, M., Okuda, T. & Shimada, H. (1993). Antifungal agent, its preparation and microorganism thereof. *European Patent Application*, EP 0537622A.

Matsukuma, S., Ohitsuka, T., Kotaki, H., Shirai, H., Sano, T., Watanabe, K., Nakayama, N., Itezono, Y., Fujiu, M., Shimma, N., Yokose, K. & Okuda, T. (1992). A new series of natural antifungals that inhibit P450 lanosterol C-14 demethylase. 1. Taxonomy, fermentation, isolation and structural elucidation. *Journal of Antibiotics*, **45**, 151–9.

Matsumoto, Y., Matsuda, S. & Tegoshi, T. (1989). Yeast glucan in the cyst wall of *Pneumocystis carinii*. *Journal of Protozoology*, **36**, 21S–2S.

Mercer, E. I. (1991). Morpholine antifungals and their mode of action. *Biochemical Society Transactions*, **19**, 788–93.

Mitani, M. & Inoue, Y. (1968). Antagonists of antifungal substance polyoxin. *Journal of Antibiotics*, **21**, 492–6.

Mizuno, K., Yagi, A., Satoi, S., Takada, M., Hayashi, M., Asano, K. & Matsuda, T. (1977). Studies on aculeacin. I. Isolation and characterization of aculeacin A. *Journal of Antibiotics*, **30**, 297–302.

Moore, K. S., Wehrli, S., Roder, H., Rogers, M., Forrest, J. N., McCrimmon, D. & Zasloff, M. (1993). Squalamine: an aminosterol antibiotic from the shark. *Proceedings of the National Academy of Sciences, USA*, **90**, 1354–8.

Müller, H., Furter, R., Zähner, H. & Rast, D. M. (1981). Metabolic products of microorganisms. 203. Inhibition of chitosomal chitin synthetase and growth of *Mucor rouxii* by nikkomycin Z, nikkomycin X and polyoxin A: a comparison. *Archives of Microbiology*, **130**, 195–7.

Nair, M. G., Mishra, S. K., Putnam, A. R. & Pandey, R. C. (1992). Antifungal anthracycline antibiotics, spartanamycins A and B from *Micromonospora* spp. *Journal of Antibiotics*, **45**, 1738–45.

Nicholas, R. O., Williams, D. W. & Hunter, P. A. (1994). Investigation of the value of β-glucan-specific fluorochromes for predicting the β-glucan content of the cell walls of zoopathogenic fungi. *Mycological Research*, **98**, 694–8.

Nishimoto, Y., Sakuda, S., Takayama, S. & Yamada, Y. (1991). Isolation and characterization of new allosamidins. *Journal of Antibiotics*, **44**, 716–22.

Odds, F. C. (1992). Susceptibility and emerging resistance of pathogenic fungi to current agents. In *The Changing Nature of Fungal Infections – Challenges and Opportunities*, Pub. IBC Technical Services Ltd.

Oki, T., Konishi, M., Tomatsu, K., Tomita, K., Saitoh, K., Tsunakawa, M., Nishio, M., Miyaki, T. & Kawaguchi, H. (1988). Pradimicin, a novel class of potent antifungal antibiotics. *Journal of Antibiotics*, **41**, 1701–4.

Oki, T., Tenmyo, O., Hirano, M., Tomatsu, K. & Kamei, H. (1990). Pradimicins A, B and C: new antifungal antibiotics. II. *In vitro* and *in vivo* biological activities. *Journal of Antibiotics*, **43**, 763–70.

Perfect, J. R., Pickard, W. W., Hunt, D. L., Palmer, B. & Schell, W. A. (1991). The use of Amphotericin B in nosocomial fungal infection. *Reviews of Infectious Diseases*, **13**, 474–9.

Pfaller, M. (1992). The impact of the changing epidemiology of fungal infections in the 1990s. In *The Changing Nature of Fungal Infections – Challenges and Opportunities*, Pub. IBC Technical Services Ltd.

Polak, A. (1990). Amorolfine, RO 14-4767, Loceryl. In *Proceedings of the First International Conference on Antifungal Chemotherapy*, Oiso, Japan. Abstract No. 2-3.

Polak, A. (1988). Mode of action of morpholine derivatives. *Annals of the New York Academy of Sciences*, **544**, 211–8.

Rippon, J. W. (1988). Medical mycology. *The Pathogenic Fungi and the Pathogenic Actinomycetes*. 3rd. edn Pub. W.B. Saunders Company.

Robbins, P. W., Bowen, A. R., Chen-wu, J. L., Momany, M., Szaniszlo, P. J. & Zwicker, J. (1993). The multiple chitin synthase genes of *Candida albicans* and other pathogenic fungi – a review. In *Dimorphic Fungi in Biology and Medicine*, pp. 51–59. (Vanden Bossche, H., Odds, F. & Kerridge, D. eds). Plenum Press.

Ryder, N. S. (1988). Mechanism of action and biochemical selectivity of allylamine antimycotic agents. *Annals of the New York Academy of Sciences*, **544**, 208–20.

Ryder, N. S. (1991). Squalene epoxidation as a target for the allylamines. *Biochemical Society Transactions*, **19**, 774–7.

Sabra, R. & Branch, R. A. (1990). Amphotericin B nephrotoxicity. *Drug Safety*, **5**, 94–108.

Sakuda, S., Nishimoto, Y., Ohi, M., Watanabe M., Takayama, S., Isogai, A. & Yamada, Y. (1990). Effects of demethylallosamidin, a potent yeast chitinase inhibitor, on the cell division of yeast. *Agricultural and Biological Chemistry*, **54**, 1333–5.

San-Blas, G. (1982). The cell wall of fungal human pathogens: its possible role in host–parasite relationships. A review. *Mycopathologia*, **79**, 159–84.

Sarosi, G. A. (1990). Amphotericin B, still the 'gold standard' for antifungal therapy. *Postgraduate Medicine*, **88**, 151–6.

Satoi, S., Yagi, A., Asano, K., Mizuno, K. & Watanabe, T. (1977). Studies on aculeacin. II. Isolation and characterization of aculeacins B, C, D, E, F, and G. *Journal of Antibiotics*, **30**, 303–7.

Sawada, Y., Numata, K., Murakami, T., Tanimichi, H., Yamamoto, S. & Oki, T. (1990). Calcium-dependent anticandidal action of pradimicin A. *Journal of Antibiotics*, **43**, 715–21.

Schmatz, D. M., Abruzzo, G., Powles, M. A., McFadden, D. C., Balkovec, J. M., Black, R. M., Nollstadt, K. & Bartizal, K. (1992). Pneumocandins from *Zalerion arboricola*, IV. Biological evaluation of natural and semisynthetic pneumocandins for activity against *Pneumocystis carinii* and *Candida* species. *Journal of Antibiotics*, **45**, 1886–91.

Schmatz, D. M., Romancheck, M. A., Pittarelli, L. A., Schwartz, R. E., Fromtling, R. A., Nollstadt, K. H., VanMiddlesworth, F. L., Wilson, K. E. & Turner, M. J. (1990). Treatment of *Pneumocystis carinii* pneumonia with 1,3-β-glucan synthesis inhibitors. *Proceedings of the National Academy of Sciences, USA*, **87**, 5950–4.

Speller, D. C. E. (1980). Other antifungal agents. In *Antifungal Chemotherapy*, pp. 183–210. (Speller, D. C. E. ed.). John Wiley & Sons.

Stevens, D. A. (1987). Problems in antifungal chemotherapy. *Infection*, **15**, 87–92.

Stütz, A. (1988). Synthesis and structure-activity correlations within allylamine antimycotics. *Annals of the New York Academy of Sciences*, **544**, 46–62.

Suzuki, S., Isono, K., Nagatsu, J., Mizutani, T., Kawashima, Y. & Mizuno, T. (1965). A new antibiotic, polyoxin A. *Journal of Antibiotics*, **18**, 131.

Takeuchi, T., Hara, T., Naganawa, H., Okada, M., Hamada, M., Umezawa, H., Gomi, S., Sezaki, M. & Kondo, S. (1988). New antifungal antibiotics benanomicins A and B from an *Actinomycete*. *Journal of Antibiotics*, **41**, 807–11.

Terayama, H., Kuzuhara, H., Sakuda, S. & Yamada, Y. (1993). Synthesis of a new allosamidin analog, *N*,*N*-diacetyl-beta-chitobiosyl allosamizoline, and its inhibitory activity against some chitinases. *Biosciences, Biotechnology and Biochemistry*, **57**, 2067–9.

Traber, R., Keller-Juslen, C., Loosli, H. R., Huhn, M. & von Wartburg, A. (1979). Cyclopeptid-antibiotika aus *Aspergillus* arten. Structur der echinocandine C und D. *Helvetica Chimica Acta*, **62**, 1252–67.

Ueki, T., Numata, K., Sawada, Y., Nishio, M., Ohkuma, H., Toda, S., Kamachi, H., Fukagawa, Y. & Oki, T. (1993). Studies on the mode of antifungal action of pradimicin antibiotics. II. D-mannopyranoside-binding site and calcium-binding site. *Journal of Antibiotics*, **46**, 455–64.

VanMiddlesworth, F., Omstead, M. N., Schmatz, D., Bartizal, K., Fromtling, R., Bills, G., Nollstadt, K., Honeycutt, S., Zweerink, M., Garrity, G. & Wilson, K. (1991). L-687,781, a new member of the papulacandin family of β-1,3-D-glucan synthesis inhibitors. I. Fermentation, isolation, and biological activity. *Journal of Antibiotics*, **44**, 45–51.

Walsh, T. J., Jarosinski, P. F. & Fromtling, R. A. (1990). Increasing use of systemic antifungal agents. *Diagnostic Microbiology and Infectious Diseases*. **13**, 37–40.

Warnock, D. W. (1991). Introduction to the management of fungal infections in the compromised patient. In: Warnock, D. W. & Richardson, M. D. eds. *Fungal Infection in the Compromised Patient*, John Wiley & Sons Ltd.

Yadan, J.-C., Gonneau, M., Sarthou, P. & Le Goffic, F. (1984). Sensitivity to nikkomycin Z in *Candida albicans*: the role of peptide permeases. *Journal of Bacteriology*, **160**, 884–8.

Yoshida, M., Ezaki, M., Hashimoto, M., Yamashita, M., Shingematsu, N., Okuhara, M., Kohsaka, M. & Horikoshi, K. (1990). A novel antifungal antibiotic, FR-900848. I. Production, isolation, physico-chemical and biological properties. *Journal of Antibiotics*, **43**, 748–54.

Zhou, Z.-Y., Sakuda, S., Kinoshita, M. & Yamada, Y. (1993). Biosynthetic studies of allosamidin 2. Isolation of didemethylallosamidin, and conversion experiments of ^{14}C-labeled demethylallosamidin, didemethylallosamidin and their related compounds. *Journal of Antibiotics*, **46**, 1582–8.

DISCOVERY AND DEVELOPMENT OF BETA LACTAM ANTIBIOTICS

G. N. ROLINSON

Parkgate House, Newdigate, Dorking, Surrey RH5 5DZ, UK

INTRODUCTION

The beta lactam family of antibiotics presents a large and complex group of compounds which includes the penicillins, cephalosporins, cephamycins, oxacephems, carbapenems and monobactams (Fig. 1). Two penicillins, penicillin G and penicillin V, are produced directly by a fermentation process but the great majority of the beta lactam antibiotics in clinical practice are chemical derivatives of naturally occurring beta lactams. The latter are widely distributed among different groups of micro-organisms. The penicillin and cephalosporin molecules are produced by filamentous fungi, the naturally occurring cephamycins and carbapenems are the products of the actinomycetes while the monobactams are of bacterial origin. In addition, there are two further classes of naturally occurring beta lactams (though not in clinical use as antibiotics) and these are the nocardicins and clavulanic acid.

EARLY HISTORY

In tracing the history of beta lactam antibiotics, it is usual to begin with Alexander Fleming (1929) and his observation that a culture of the genus

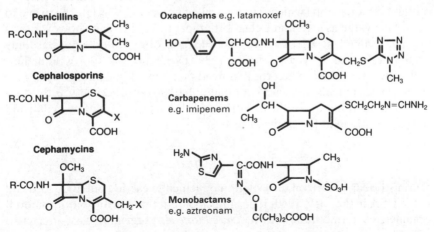

Fig. 1. Structure of penicillins, cephalosporins, cephamycins, oxacephems, carbapenems and monobactams.

Penicillium (originally identified as *Penicillium notatum*) produced an antibiotic which inhibited various bacteria including *Staphylococcus aureus*. Fleming, however, was by no means the first to observe antibiotic production by a culture of *Penicillium*. In the 1870s John Burden Sanderson, Joseph Lister (of antiseptic fame) and William Roberts all reported the production of antibacterial activity by species of *Penicillia* and Lister even went so far as to use the filtrate of his *Penicillium* culture, in 1884, in an attempt to treat an infection by local application. During the previous century, work was also carried out on the isolation of antibacterial agents produced by species of *Penicillium*, an example being the compound we now know as mycophenolic acid, isolated in crystalline form by Gosio in 1896. Later, in 1913, Alsberg and Black isolated another antibiotic from a species of *Penicillium*, also in crystalline form, namely penicillic acid. What the antibiotics were in the case of the observations of Sanderson, Lister and Roberts in the 1870s we cannot say. There is no evidence that any one of these was in fact penicillin, although this possibility cannot be ruled out, and in this connection it is of interest that Lister did refer to abnormal morphological forms of bacteria in his experiments. This, of course, is characteristic of the effect of penicillin, and something Fleming himself was fond of demonstrating. Of particular interest also is the report by Gratia and Dath in 1925 of a *Penicillium* species producing an antibacterial agent showing a bacteriolytic effect. The similarity of this observation with the bacteriolytic effect observed by Fleming is quite striking. It may be therefore that Fleming was not the first to observe the antibiotic action of penicillin. Certainly, the observation of antibiotic production by species of *Penicillia* was not novel. What then made the observation of Fleming so important? The importance lies not in the observation itself because that kind of observation had been made before. The significance lies in the fact that the antibiotic in question (which Fleming called penicillin) was later to prove to be of such value as a systemic chemotherapeutic agent. This, however, was not appreciated at the time of the Fleming observation and was only established some 10 years later in 1939, by Chain and Florey and their colleagues in Oxford, when they showed that penicillin could protect mice from a lethal inoculation with beta-haemolytic streptococci. It was this that provided the stimulus for the isolation and characterisation of penicillin and for its development for clinical use.

THE FIRST PENICILLINS

During the early work on the isolation of penicillin carried out in Oxford and in the USA in the early 1940s it became apparent that the antibiotic material obtained by fermentation with *Penicillium chrysogenum* was, in fact, a mixture of closely related substances, each differing only in the nature of the acyl group R attached to a fused beta lactam thiazolidine ring system (Fig.

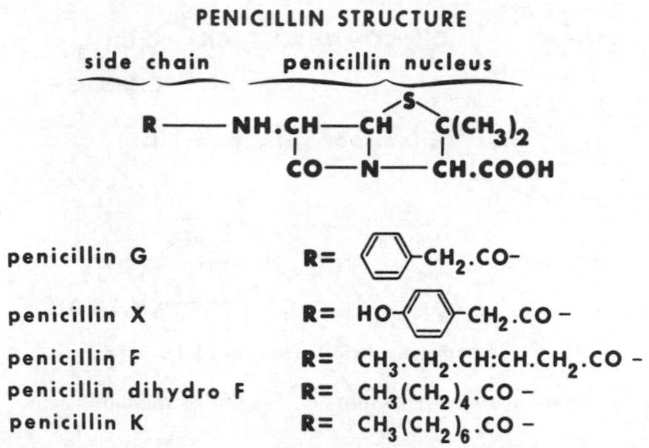

PENICILLIN STRUCTURE

side chain   penicillin nucleus

$$R\text{——}NH.CH\text{——}CH \quad C(CH_3)_2$$
$$CO\text{——}N\text{——}CH.COOH$$

penicillin G   R= ⬡–CH_2.CO–

penicillin X   R= HO–⬡–CH_2.CO –

penicillin F   R= $CH_3.CH_2.CH{:}CH.CH_2.CO$ –

penicillin dihydro F   R= $CH_3(CH_2)_4.CO$ –

penicillin K   R= $CH_3(CH_2)_6.CO$ –

Fig. 2. Structure of naturally occurring penicillins.

2). In other words, penicillin was a family of substances. In the UK each member of the family was designated by numerals and in the USA by the letters X, G, F and K. From this family of penicillins, penicillin G was selected as the preferred member for commercial production and for clinical use. This choice was influenced partly by the biological properties of the different penicillins and partly by the discovery that penicillin G could be produced almost exclusively in the fermentation by the addition of the appropriate side chain structure in the form of a precursor compound, namely, phenylacetic acid.

The existence of a family of penicillins showing some differences in biological properties stimulated interest in the possibility of preparing new penicillins which might display properties superior to those of the naturally occurring compounds. The problem at that time was how such penicillins could be prepared. The total synthesis of the penicillin molecule by chemical means presents a formidable problem even today, and an early approach to the preparation of new members of the family was the addition to the fermentation of precursors which could provide a source of the required side chain. One of the penicillins produced in this way was penicillin V (Behrens *et al.*, 1948) though the important properties of acid stability and absorption following oral administration were not appreciated until much later with the work of Brandl & Margreiter in 1954. Apart from penicillin V, none of the penicillins prepared by the precursor approach appeared sufficiently promising to progress to clinical use and indeed, the diversity of side-chain structure which can be introduced in this way by using precursors is severely limited.

Another approach to the preparation of compounds not otherwise available as naturally occurring substances has been the chemical modification of penicillins which can themselves be obtained by fermentation. Penicillins prepared in this way can thus be said to be semisynthetic. The first penicillins

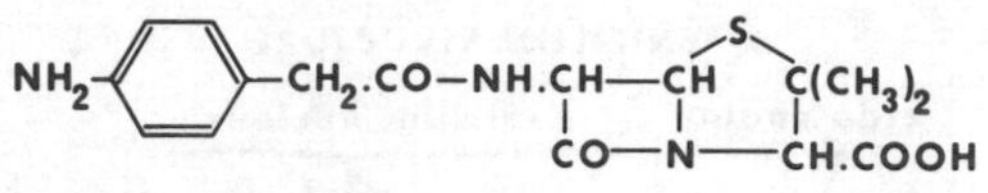

p-aminobenzylpenicillin

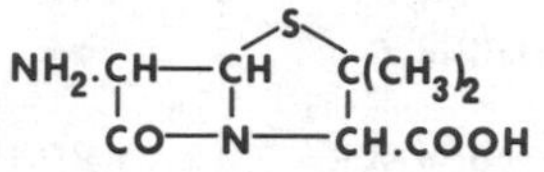

6-aminopenicillanic acid (6-APA)

Fig. 3. Structure of p-aminobenzyl penicillin and 6-aminopenicillanic acid.

of this type were prepared from penicillin X, the p-hydroxy group in the benzene ring of this particular penicillin facilitating substitution by halogens and various azo radicals (Coghill, Stodola & Wachel, 1949). A number of novel penicillins were prepared in this way and some were found to be more potent than penicillin G. However, the spectrum of antibacterial activity was not improved, and none was considered likely to be superior in clinical practice. Some time later (Ballio et al., 1959), a similar approach was adopted for the preparation of novel penicillins, the compounds in this case being prepared by chemical modification of the p-amino group in p-aminobenzylpenicillin (Fig. 3).

THE DISCOVERY OF 6-APA

The p-aminobenzylpenicillin itself was produced by fermentation with Pen. chrysogenum, using p-aminophenylacetic acid as the precursor for the side-chain. During the course of this work it was observed that in control fermentations in which the precursor had not been added, the penicillin titre was extremely low when assayed in terms of antibacterial activity but was very much higher when a chemical assay was used which involved reaction of the beta lactam ring of the penicillin molecule with hydroxylamine and the colorimetric assay of the resulting hydroxamic acid by the addition of ferric salt. These observations led to the recognition that in fermentations carried out in the absence of a side chain precursor the penicillin nucleus, 6-aminopenicillanic acid (6-APA) (Fig. 3) was itself produced as a natural product (Batchelor et al., 1959).

The significance of this compound, 6-aminopenicillanic acid which we can refer to as 6-APA, was that it provided the ideal starting point for the preparation of new penicillins because different side chains could be attached to the nucleus, through the free amino group, by chemical means. A supply of the penicillin nucleus, 6-APA, therefore became of great importance. Although first discovered as a naturally occurring compound, produced by fermentation, a more convenient method of production was found to be the removal of the side chain from penicillin itself. This was first accomplished enzymatically with penicillin V as the substrate, using a deacylase obtained from *Streptomyces* species, and soon after it was found that the side chain of penicillin G could be removed in a similar way with a deacylase of bacterial origin. Later a chemical deacylation process was also developed.

SEMI-SYNTHETIC PENICILLINS

With 6-APA, the way was now open for the preparation of semisynthetic penicillins and two important objectives at once presented themselves. One was the possibility that the spectrum of antibacterial activity might be extended, particularly with respect to the Gram-negative bacilli. The penicillins available at that time, namely penicillins G and V, were well recognised as being highly active against the Gram-positive and Gram-negative cocci but of relatively low activity against the majority of Gram-negative bacilli. Moreover, with certain pathogens, for example *Pseudomonas aeruginosa*, a lack of sensitivity was evident not only to the penicillins but to most of the other antibiotics available as well. Another objective of the work with 6-APA was the preparation of penicillins which might be active against the penicillin-resistant staphylococci. Such strains had appeared within a few years of the widespread clinical use of penicillin G and by 1948 as many as 50% of strains in hospitals were found to be resistant. By the time 6-APA was isolated in 1957, the incidence in many hospitals had risen to as high as 80%. Moreover, not only were such strains resistant to penicillin but they were frequently resistant to most of the alternative antibiotics at that time.

The mechanism of resistance to penicillin in such strains was already known to be the production of the enzyme, penicillinase, which we would now refer to as beta lactamase. This had been established by William Kirby as early as 1944. The staphylococci were resistant, therefore, because the penicillin available at that time, namely penicillin G, was not stable to this enzyme.

Against this background, the isolation of 6-APA clearly offered the possibility of preparing penicillins which would be stable to attack by this enzyme. However, in order to be clinically useful, such penicillins would also have to show an adequate level of antibacterial activity and at the same time possess acceptable pharmacological and toxicological properties. These requirements were met in the compound methicillin (Fig. 4) which at once proved effective in the treatment of infections caused by multi-resistant strains of *Staph.aureus*.

The reason for the stability of methicillin to staphylococcal beta lactamase is that the particular side-chain structure of this penicillin results in very poor

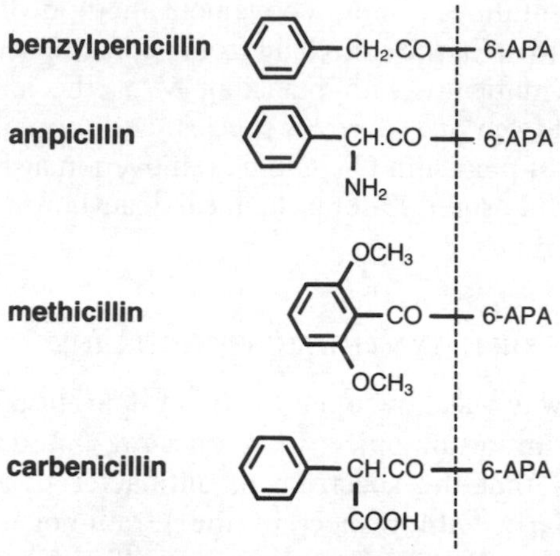

Fig. 4. Structure of benzylpenicillin, ampicillin, methicillin and carbenicillin.

affinity with the enzyme – in other words the K_m value is very high. Unfortunately, the affinity of methicillin for the penicillin target sites within the bacterial cell is also diminished with the result that the level of activity of methicillin against penicillin sensitive staphylococci is much lower than that of penicillin G or V. In addition, methicillin is not absorbed from the gastro-intestinal tract when given by mouth.

Further work with penicillins stable to staphylococcal beta lactamase led to the appearance of compounds which are more active *in vitro* than methicillin and absorbed when given by mouth. Among such compounds can be mentioned nafcillin and the isoxazolyl series, oxacillin, cloxacillin, dicloxacillin and flucoxacillin. Within the isoxazolyl group the serum concentrations reached after oral administration are greatly improved as a result of the introduction of one or more halogens into the side-chain and this results not only in improved absorption but also in a reduction in the extent to which these compounds are metabolized in the body.

Although one of the most important of the compounds to be prepared from 6-APA, methicillin was not the first semisynthetic penicillin to become available for clinical use. The first of these compounds to appear on the market was phenethicillin followed closely by propicillin. These penicillins resemble penicillin V both in chemical structure and in their antibacterial properties but are substantially better absorbed than penicillin V when given by mouth resulting in higher blood levels.

Among the compounds showing a broader spectrum of antibacterial activity, ampicillin (Fig. 4) appeared in 1961 and has remained one of the most widely used of all the semi-synthetic penicillins.

The discovery of ampicillin was also followed by developments in the direction of improved absorption from the gastro-intestinal tract. Certain esters of ampicillin, namely pivampicillin, bacampicillin and talampicillin were introduced in the 1970s which are better absorbed than ampicillin when taken by mouth and then hydrolyse rapidly in the body to give levels of ampicillin superior to those obtained with ampicillin itself. However, it is questionable whether these pro-drugs of ampicillin have any advantage over amoxycillin also introduced in 1970, which shows the same spectrum and potency as ampicillin but is much better absorbed after oral dosage.

Following the introduction of ampicillin in 1961 a further alteration of significance in antibacterial spectrum was not encountered until six years later, when carbenicillin (Fig. 4) was introduced, the first beta lactam antibiotic clinically effective in the treatment of infections caused by *Pseudomonas aeruginosa*. In terms of the MIC, the activity of carbenicillin against *P. aeruginosa* is not high. However, in 1967, when this penicillin was first introduced, the choice of antibacterial agents for the treatment of such infections was severely limited and the appearance of a non-toxic penicillin clinically effective against this pathogen provided a valuable addition to the drugs available, while at the same time opening the way to further advances with other beta lactams active against this particular organism.

Penicillins by definition are *N*-acyl derivatives of 6-APA and all the compounds referred to so far have this structure. However, a different type of derivative of 6-APA appeared in the early 1970s namely amdinocillin (formerly mecillinam) and later the oral pro-drug, pivmecillanam. Mecillinam is a 6-amidinopenicillanic acid. Not only does this compound differ from the penicillins in chemical structure but the spectrum of antibacterial activity also differs from most of the penicillins in that mecillinam is considerably more active against Gram-negative bacilli than it is against the Gram-positive cocci.

The mid-1970s saw the appearance of the ureido penicillins, azlocillin, mezlocillin and piperacillin, and more recently, temocillin. In this compound the nucleus of the penicillin molecule has been substituted with a methoxy group which results in a very high degree of stability to the beta lactamases produced by Gram-negative bacilli.

The synthesis of the new penicillins, using 6-APA as the starting point, has been a major research activity and no fewer than 21 different compounds have appeared on the market in one country or another.

THE CEPHALOSPORINS

In parallel with the penicillins there has also been the development of the cephalosporins. These antibiotics are derived from the naturally occurring compound cephalosporin C (Fig. 5) which is obtained by fermentation using the mould *Cephalosporium*. The structure of cephalosporin C, and its relationship to penicillin, was established (Abraham & Newton, 1961) a few years after the discovery and isolation of 6-APA and at about the same time

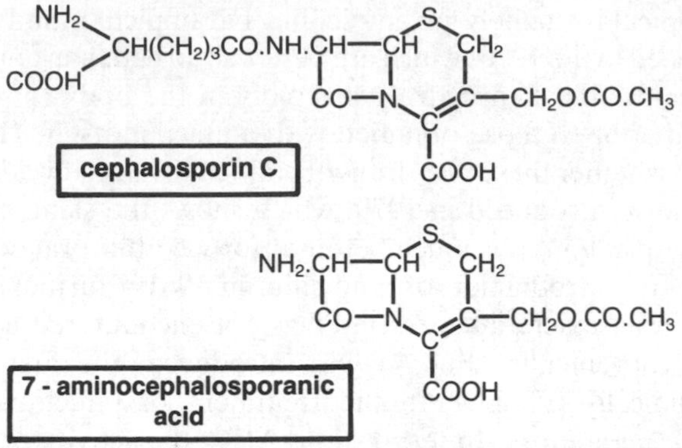

cephalosporin C

7 - aminocephalosporanic acid

Fig. 5. Structure of cephalosporin C and 7-aminocephalosporanic acid.

as the introduction of the first of the semisynthetic penicillins. Removal of the side chain of cephalosporin C yielded a nucleus 7-aminocephalosporanic acid (7-ACA) from which the cephalosporins have been derived as semisynthetic compounds in a manner analogous to the development of the semisynthetic penicillins. However, there is an important difference in the development of these two fields of work. Unlike the penicillin nucleus, the cephalosporin nucleus, 7-ACA, is not a naturally occurring substance. It is only obtained by the removal of the side chain of cephalosporin C. This can be done by a chemical procedure but it is questionable whether there would have been the incentive to attempt this side chain removal had it not been for the knowledge of the corresponding penicillin nucleus, 6-APA, and the demonstration that important derivatives could be prepared from it. It might be argued that on theoretical grounds it could have been envisaged that the new cephalosporins could have been prepared by replacing the side chain of the natural compound with novel structures, but exactly the same could have been said for the penicillin field. The fact is that prior to the isolation of 6-APA no attempt had been made to remove the side chain from penicillin G to obtain such a nucleus from which derivatives could be prepared, despite the interest in varying the side chain structure by the precursor approach and by chemical modification of penicillin X already referred to. Deacylation of penicillin G had simply not been contemplated. Without the isolation of 6-APA, therefore, the deacylation of cephalosporin C might not have been attempted in which case there would not have been the development of cephalosporins.

The first semisynthetic cephalosporin to appear in clinical practice was cephalothin in 1962, followed closely by cephaloridine in 1964. Since that time the cephalosporin group has grown into the largest of the sub-families of beta lactams and these antibiotics are classified as first, second and third generation compounds. At one time, the cephalosporins that can be administered by mouth were confined to the 1st generation but more recently a pro-drug of a second generation cephalosporin has been introduced as well as certain oral third generation compounds.

Progression from first to second to third generation in the cephalosporin family has involved a substantial increase in stability to beta lactamase and a marked increase in antimicrobial activity particularly against *Haemophilus influenzae* and members of the *Enterobacteriaceae*. Certain third generation cephalosporins (but not all) also show important activity against *Pseudomonas aeruginosa*. On the other hand, there has been some loss of potency against *Staph.aureus*. The third generation cephalosporins therefore are not to be thought of as compounds which necessarily replace the first and second generation cephalosporins.

OTHER BETA-LACTAMS

An addition to the cephalosporin family appeared in the early 1970s with the discovery of the cephamycins (Fig. 1) (Nagarajan *et al.*, 1971; Stapley *et al.*, 1972). Although produced by different micro-organisms, namely species of *Streptomyces* (as distinct from filamentous fungi which produce the cephalosporin molecule) the naturally occurring cephamycins have the same side chain as in cephalosporin C, and the same nucleus, except that it is substituted with a methoxy group at position 7. The structural similarity between members of the penicillin, cephalosporin and cephamycin families is quite marked despite their occurrence in different taxonomic groups. From an evolutionary point of view it is also of interest that these compounds share a common biosynthetic pathway.

As with the cephalosporin molecule the naturally occurring cephamycins provide a nucleus from which a number of semi-synthetic cephamycins have been developed. For classification purposes the cephamycins, despite the difference in structure, are generally grouped along with cephalosporins.

In the mid-1970s, the cephalosporins and the cephamycins were joined by another family of beta lactams, namely the carbapenems (Fig. 1). Like the cephamycins these were discovered as naturally occurring compounds produced by species of *Streptomyces* (Kahan *et al.*, 1976) and they now comprise a large and complex family. The carbapenems include the olivanic acids (which were the first naturally occurring beta lactamase inhibitors to be discovered) and also thienamycin, from which imipenem is derived. Regarding the structure of the carbapenems this resembles the penicillins in that the beta lactam ring is fused to a five-membered ring but in the case of the carbapenems this does not contain sulphur but carbon.

In the late 1970s a different kind of development took place in the beta lactam field with the appearance of moxalactam, now named latamoxef (Fig. 1). Latamoxef is also a semisynthetic beta lactam but it is not derived from a naturally occurring nucleus; instead, it is built up almost completely by organic chemistry, starting with penicillin G simply to provide the beta lactam ring. Latamoxef is an oxacephem, having the same nucleus as the cephamycins, with the methoxy group in position 7, but with oxygen in the six-membered ring instead of sulphur.

The sub-family of beta lactams to be discovered most recently, 1980, is the monobactam group (Fig. 1), produced not by fungi or actinomycetes but by bacteria. These are monocyclic beta lactams, hence the name monobactams, and as with the other naturally occurring beta lactams the nucleus of the monobactams provides the starting point for the preparation of semi-synthetic monobactams, for example aztreonam.

THE PROBLEM OF RESISTANCE

Throughout the history of the beta lactam antibiotics, a prime objective has been to develop compounds active against resistant bacterial strains and species. In some cases, resistance is due to diminished permeability of the bacterial cell wall; in other cases to alterations in the target enzymes themselves – the penicillin binding proteins – but in the majority of clinical isolates that are resistant to the beta lactam antibiotics the mechanism of resistance is the production of the enzyme beta lactamase.

At the time of isolation of the penicillin nucleus in 1957, certain pathogens including *Staph.aureus, Klebsiella* species and *Bacteroides* were almost uniformly beta lactamase producers. In other cases, for example *E. coli*, only a proportion of strains showed beta lactamase mediated resistance but over the last 30 years the frequency of resistant strains has increased considerably as a result of selection and genetic transfer of the genome for beta lactamase production. This genetic transfer has also meant that certain pathogens including *Haemophilus influenzae*, the gonococcus, *Moraxella catarrhalis, Enterococcus faecalis* and even the meningococcus now include beta lactamase producing strains whereas at one time all of these pathogens were uniformly non-beta lactamase producers.

Beta-lactamase inhibitors

One approach to the problem of beta lactamase mediated resistance has been the development of beta lactamase inhibitors.

The potential for a beta lactamase inhibitor was recognised as early as the late 1940s and 50s and a number of organic compounds were screened for possible beta lactamase inhibitory activity without leading to a clinically successful agent (Reid *et al.*, 1946, Behrens & Garrison, 1950). In the early 1960s certain of the semisynthetic penicillins were found to act as inhibitors of beta lactamase (Rolinson *et al.*, 1960) but did not find a place in clinical practice in this capacity. Nevertheless the potential for a beta lactamase

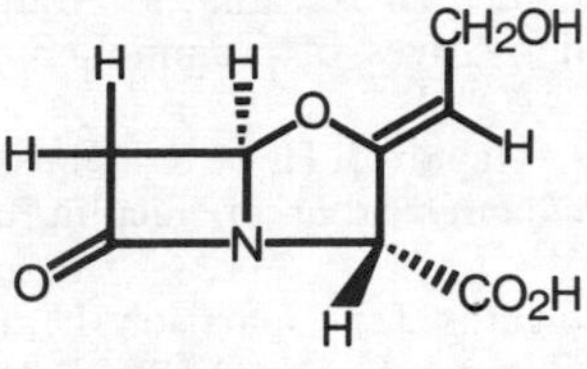

clavulanic acid

Fig. 6. Structure of clavulanic acid.

inhibitor remained clear and in 1967 the author began a programme in which soil micro-organisms were screened for possible production of naturally occurring beta lactamase inhibitors. A suitable test system was devised (Rolinson, 1980) for the detection of beta lactamase inhibitory activity and by the use of this test a family of beta lactamase inhibitors (olivanic acids) was first discovered in certain species of *Streptomyces* and further work led to the discovery of clavulanic acid (Fig. 6) in a strain of *Streptomyces clavuligerus* (Brown *et al.*, 1976). A formulation of clavulanic acid with the semisynthetic penicillin amoxycillin was introduced in 1981 and since then two further beta lactamase inhibitors have appeared, sulbactam (penicillanic acid sulfone) and tazobactam, itself an analogue of sulbactam.

The significance of a combination of a beta lactamase inhibitor with a beta lactam antibiotic, is that the activity of the antibiotic against susceptible organisms is retained but, in addition, there is activity against a high proportion of beta lactamase producing strains and species which otherwise are resistant.

CONCLUSIONS

In reviewing the history and development of the beta lactams over the last 50 years, we can see that this has been a story of the discovery of naturally occurring structures, all related by virtue of containing a beta lactam ring, and the exploitation of these structures by chemical modification. Whether further naturally occurring beta lactams remain to be discovered is debatable but further modifications of the existing members is likely to result in the continued growth of the beta lactam family for some time to come.

REFERENCES

Abraham, E. P. & Newton, G. G. F. (1961). The structure of cephalosporin C. *Biochemical Journal*, **79**, 377–93.
Alsberg, C. L. & Black, O. F. (1913). *US Department of Agriculture Bureau of Plant Industry Bulletin*, 270.

Ballio, A., Chain, E. B., Dentice Di Accadia, F., Batchelor, F. R. & Rolinson, G. N. (1959). Penicillin derivatives of p-aminobenzyl-penicillin. *Nature*, **183**, 180–1.

Batchelor, F. R., Doyle, F. P., Naylor, J. H. C. & Rolinson, G. N. (1959). Synthesis of penicillin: 6-aminopenicillanic acid and penicillin fermentations. *Nature*, **183**, 257–58.

Behrens, O. K., Corse, J., Edwards, J. P., Garrison, L., Jones, R. G., Soper, Q. F., Van Abeele, F. R. & Whitehead, C. W. (1948). Biosynthesis of penicillin. IV. New crystalline biosynthetic penicillins. *Journal of Biological Chemistry*, **175**, 793–809.

Behrens, O. K. & Garrison, L. (1950). Inhibitors for penicillinase. *Archives of Biochemistry*, **27**, 94–98.

Brandl, E. & Margreiter, H. (1954). Ein Saurestabiles biosynthetisches Penicillin. *Oesterreichische Chemiker-Zeitung*, **55**, 11.

Brown, A. G., Butterworth, D., Cole, M., Hanscomb, G., Hood, J. D., Reading, C. & Rolinson, G. N. (1976). Naturally-occurring beta lactamase inhibitors with antibacterial activity. *Journal of Antibiotics*, **29**, 668–69.

Chain, E., Florey, H. W., Gardner, A. D., Heatley, N. G., Jennings, M. A., Orr-Ewing, J. & Sanders, A. G. (1940). *Lancet*, **1**, 1172.

Coghill, R. D., Stodola, F. H. & Wachel, J. L. (1949). *Chemistry of Penicillin* Clarke, H. T., Johnson, J. R. & Robinson, R., eds. Princeton University Press, Princetown, New Jersey, pp. 680–7.

Fleming, A. (1929). On the antibacterial action of cultures of a penicillium with special reference to their use in the isolation of *B. influenzae*. *British Journal of Experimental Pathology*, **10**, 226.

Gosio, B. (1896). Ricerche batteriologische e chimiche sulle alterazioni del mais. Contributo all etiologia della pellagra. (Memoria 2a) *Riv. Igiene Sanit. pubbl.* **7**, 825.

Gratia, A. & Dath, S. (1925). Moississures et microbes bacteriophages. *Comptes Rendus Societé Biologie Paris*, **92**, 461.

Kahan, J. S., Kahan, F. M., Goegelman, R., Currie, S. A., Jackson, M., Stapley E. O., Miller, A. K., Miller, T. W., Hendlin, D., Mochales, S., Hernandez, S. & Woodruff, H. B. (1976). Thienamycin. A new beta lactam antibiotic. I. Discovery and isolation. *16th Interscience Conference on Antimicrobial Agents and Chemotherapy*, Chicago, USA Abstract No. 227.

Kirby, W. M. M. (1944). Extraction of a highly potent penicillin inactivator from penicillin resistant staphylococci. *Science*, **99**, 452–53.

Lister, J. (1971). On bacteria. In *Commonplace Books* Vol I 31-170 Royal College of Surgeons of England, London.

Nagarajan, R., Boeck, L. D., Gorman, M., Hamill, R. L., Higgens, C. E., Hoehn, M. M., Stark, W. M. & Whitney, J. F. (1971). Beta lactam antibiotics from Streptomyces. *Journal of the American Chemical Society*. **93**, 2308–10.

Reid, R. D., Felton, L. C. & Pettroff, M. A. (1946). Prologation of penicillin activity with penicillinase-inhibiting compounds. *Proceedings of the Society of Experimental Biology and Medicine*, **63**, 438–43.

Roberts, W. (1874). Studies of biogenesis. *Philosophical Transactions of the Royal Society of London*, **164**, 457–77.

Rolinson, G. N., Stevens, S., Batchelor, F. R., Cameron-Wood, J. & Chain, E. B. (1960). Bacteriological studies on a new penicillin – BRL 1241. *Lancet*, **ii**, 564–67.

Rolinson, G. N. (1980). The history and background of Augmentin. *Proceedings of the 1st Symposium on Augmentin, Excerpta Medica, International Congress Series*, **544**, 3–7.

Sanderson, J. B. (1871). *13th Report of the Medical Officer of the Privy Council*, Appendix 5, 48–69. Her Majesty's Stationery Office, London.

Stapley, E. P., Jackson, M., Hernandez, S., Zimmerman, S. B., Currie, S. A., Mochales, S., Mata, J. M., Woodruff, H. B. & Hendlin, D. (1972). Cephamycins, a new family of beta lactam antibiotics. I. Production by actinomycetes, including *Streptomyces lactamdurans* sp.n. *Antimicrobial Agents and Chemotherapy*, **2**, 122–31.

THE NEED FOR NEW ANTIBIOTICS: POSSIBLE WAYS FORWARD

HANS ZÄHNER AND HANS-PETER FIEDLER

Institute of Biology II, University of Tübingen, Auf der Morgenstelle 28, D-72076 Tübingen, Germany

THE SITUATION

The opinion frequently proffered, even in some textbooks, that infectious diseases no longer pose a problem thanks to antibiotics and vaccines, is wishful thinking rather than reality. Today, we are further away from mastering infectious diseases than we were 25 years ago (see the alarming publications by Murray, 1991; Neu, 1992; Fox, 1992; ASM News, 1992; Travis, 1994; Kingman, 1994; Berkelman *et al.*, 1994; Davis, 1994; Nikaido, 1994; Spratt, 1994, for example). There are five problem areas which have contributed to this negative assessment of the situation:

1. Changes in the spectrum of pathogens. The best example is the HIV virus; neither a satisfactory vaccine nor a real therapeutic agent is available. However, changes can also be observed in the spectrum of pathogens of bacteria and fungi, such as atypical mycobacteria, aspergilli, *Cryptococcus neoformans, Listeria*, and *Legionella*.
2. Infections are becoming increasingly common following, or concomitant with serious underlying diseases, which wear patients down and reduce their immunity. Examples are HIV as an immunodeficiency itself, bone marrow or organ transplants with the necessary long-lasting immunosuppression achieved with medicines, cystic fibrosis and diabetes patients. Where in the past antibiotics were administered for a few days, for instance, for a lung infection, today they often have to be given for months and, in some cases, even as a prophylactic.
3. Antibiotic-resistant pathogens are on the increase today, particularly in hospitals. Where multi-resistant, Gram-negative bacteria were the problem in the past, it is multi-resistant Gram-positive bacteria that are giving cause for concern today, particularly *Staphylococcus aureus* (MRSA-methicillin-resistant *Staphylococcus aureus*), enterococci and pneumococci. The β-lactam resistance of these strains is due to changes in the penicillin-binding proteins (PBP). In extreme cases the strains are resistant to over ten antibiotics, in many cases the only remaining treatment possible is vancomycin or teicoplanin. Since a number of

vancomycin-resistant strains have already been detected, it should only be a matter of time before multiple resistance, including vancomycin-resistance, develops. These could then be the 'horror bacteria' of the next few years.

4. The deterioration in social conditions in the developing, but increasingly also in the industrialized countries is a crucial contributing factor to the renewed spread of infectious diseases. There is a particularly close correlation between social conditions and the spread of the disease in the case of tuberculosis. The homeless, drug addicts, those with an unbalanced diet or who are undernourished, and those infected with HIV are particularly vulnerable and are increasingly posing a potential threat in terms of the spread of infection.

5. Efforts made to combat infectious diseases have been reduced. For example, programmes on controlling tuberculosis and malaria have been broken off rashly. Infectious disease wards have been reduced in size, and in some cases even closed. In addition to this, the time allotted to medical training for doctors in infectious diseases has been reduced to an unacceptable degree.

Despite the fact that more than 150 antibiotic products are available in German pharmacies today, experts rightly warn that if this trend continues, we will return to the 'pre-antibiotic era'.

REASONS FOR THE DIFFICULTIES

If we take the example of multi-resistant Gram-positive bacteria as an illustration of how we have reached this dreadful state, we see that problems with resistant bacteria come in waves. The first of these was penicillin-resistant staphylococci (β-lactamase forming strains) which, from 1950 onwards, increasingly undermined the success of penicillin. Introduction of β-lactamase stable, semisynthetic penicillins solved this problem. From 1960 onwards, multi-resistant Gram-negative bacteria (*Shigella, Salmonella*) startled experts, initially in Japan but later also in Europe and the USA (for review see Mitsuhashi, 1977). Semi-synthetic β-lactams, this time cephalosporins, also played a large part in the measures taken to overcome this second wave of resistant bacteria.

The third wave facing us now embraces a relatively broad spectrum of Gram-positive bacteria, some of which bear up to ten resistance genes. There is relatively little chance of finding new β-lactams of synthetic or semi-synthetic origin that would be able to inhibit multi-resistant staphylococci or streptococci as well as enterococci and pneumococci. The β-lactam resistance of these strains is due to changes in the PBPs, whereby these changes vary greatly from strain to strain so that it would be difficult to modify the β-lactam structure in such a way that it could overcome all forms of PBP-

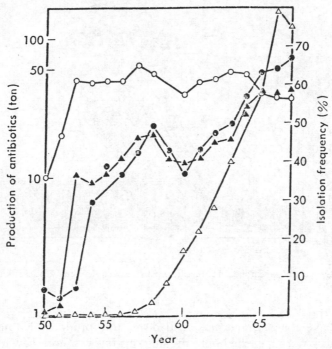

Fig. 1. Production of antibiotics and appearance of antibiotic-resistant Shigella in Japan. ○ streptomycin, ● chlorampenicol, ▲ tetracyclines, △ shigella (Mitsuhashi, 1977).

induced β-lactam resistance at the same time (for review see Georgopapa-dakou, 1993). The number of antibacterial products available today is impressive but deceptive. If instead of looking at the list of products on offer, you focus on the targets in the bacterial cell reached by these substances, you get a different picture; 150 products are indeed available, but they cover only six different target sites.

When an antibiotic is first introduced, the proportion of resistant strains is less than 1%. Between 8 and 12 years later, after widespread use, this level of resistance increases dramatically. Fig. 1 (Mitsuhashi, 1977) verifies this observation using the example of resistant *Shigella* in Japan. The wider the use of an antibiotic, the faster resistant strains spread. This 8–12 year period before a massive development of resistance to sulphonamides and penicillin G, has been confirmed in the case of tetracyclines and aminoglycosides, and is now being repeated with quinolones. Fig. 2 shows the same observation as seen by a cartoonist. A strong correlation exists between the wider use of an antibiotic and the spread of resistant strains, with a delay of 8 to 12 years.

The massive spread of resistant strains and the associated therapeutic problems could be prevented if antibiotics reaching different targets became available every 8–10 years, but this is not the case. In the last decade, the efforts in the pharmaceutical industry, when conducting antibiotic research

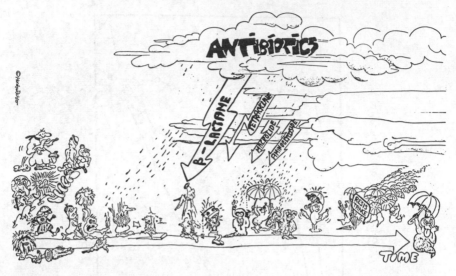

Fig. 2. Lag-phase in appearance of antibiotic resistance bacteria seen by the cartoonist (Monika Richter).

at all, have concentrated on so-called 'me-too' products. No new parent substances were launched onto the market, but yet more new quinolones (over a dozen already) and new β-lactams (already over 100). The field of antibiotics is dominated by β-lactams, which account for around three quarters of the market. The remaining quarter is shared between other 'major' antibiotics (tetracyclines, aminoglycosides, macrolides, sulphonamides, quinolones). From the business viewpoint, focusing on research into further β-lactams makes sense considering the extremely high costs and risks involved in the introduction of a new lead structure, on the one hand, and the dominance of β-lactams on the market, on the other.

Despite this, the call for new lead structures, is justified and priority should be given to this over 'me-too' research. New lead structures with other targets are not only needed against resistant Gram-positive bacteria, but also for use against tuberculosis, mycoses, protozoa and viral infections.

THE SEARCH FOR NEW ANTIBIOTICS

In the search for new pharmaceuticals it should be borne in mind that it takes between 8 and 10 years for a new substance to be introduced. Substances due for launching before the year 2000 must have completed their first clinical trial already. It is not easy to predict in the spring of 1995 which infectious diseases will be untreatable in the year 2005 with antibiotics that are either already available or are under development. Resistant strains are spreading too fast and the change in pathogens is difficult to predict. The speedy advances made in medicine are helping both to spread resistant strains and to change pathogens.

For this reason, antibiotic research will be faced with the difficult task of deciding priorities in 1995 in the hope of having the correct substances available for introduction in the year 2005. This dilemma can only be solved through a relatively broad-based search; there should be a whole range of possible products available at least by the time the first clinical trial takes place.

The financial aspect of such a broad-based search is not the subject of this paper but should be borne in mind when there is a call for more research into antibiotics.

Classic screening for new antibiotics, so successful between 1940 and 1960, has little chance of success. Every year, probably over a million strains of microorganisms are examined for the production of antibiotics, and in most cases identical test methods are applied using microorganisms that are easy to isolate and cultivate (streptomycetes and other actinomycetes, imperfect fungi, *Bacillus, Pseudomonas*, etc.). After over 50 years of screening in this way, it is no surprise that few new substances are selected from an ever-growing mountain of already known ones. Using classic test methods and the usual producer strains for antibiotics that are effective against Gram-positive bacteria, only every 500th activity signal can be attributed to a new substance. The effort required to identify a new substance is great and this is for three reasons:

(a) Even using well-established identification methods, the failure rate remains between 1 and 5%. Thus, despite great efforts, in order to find one new substance, it is necessary to process 5–25 substances which have been declared incorrectly as 'new'.

(b) If a new substance is finally found, it is likely to belong to an already known substance class. Another macrolide or another polyene, for example, has been found, but not the new lead substance desired.

(c) Even if one is lucky enough to discover a new lead structure, there is little probability that it will meet all requirements for development. In most cases, the new lead structure must be 'optimized', potentially a complex affair depending on the structure of the compound involved.

Despite this extremely negative prognosis, we should not give up since there are a whole number of possible ways forward that might lead us to new substances. Some of these ways are outlined here.

Re-evaluation and further development of antibiotics that are 'small' but have already been introduced, including those no longer available on the market

The commercially important antibiotics today are no longer used as un-modified natural substances, they have been optimized in large-scale programmes. Due to the dominance of β-lactams, no or very few 'small' antibiotics have been optimized. Potential candidates for 'optimization' are: phosphomycin, fusidic acid, cycloserine, viomycin, fumagillin, albomycin, novobiocin, coumermycin, amphomycin (or daptomycin to which it is related) and capreomycin. The sites of action of these 'small antibiotics' are clearly different from those of 'major' antibiotics (see vols 3, 5–1 and 5–2 of *Antibiotics*, published by Corcoran & Hahn, 1967–1979; Lambert & O'Grady, 1992).

Falling back on known antibiotics not yet used in man

Of the 6000 and more microbial metabolites described to date, 4000 are antibiotics, some 1% of which have been used in man. There is no doubt that many of those not yet used are unsuitable for medical use, even after intensive derivatization. However, we can assume that they also include substances worth optimising. A candidate for such an optimization programme should meet the following conditions.

- There should be no cross-resistance to substances that are already used in medicine.
- The relationship between activity against microorganisms and toxicity should be favourable.
- The effort required for derivatization or synthesis of the compound should be reasonable.

You could view a programme on further development of 'old' antibiotics as one on 'obsolete stock'. There are two ideas that might be useful when searching for suitable candidates for such a programme.

Idea 1

The first wave of penicillin-resistant staphylococci was the impetus for an intensive search for additional antibiotics. In the ensuing competition between semisynthetic penicillins and new antibiotics, β-lactams were clearly the winners, with the result that work on a whole series of new antibiotics for use in man was then terminated. Among these were:

- orthosomycins: everninomycins (Weinstein *et al.*, 1964; Wagman, Luedermann & Weinstein 1964; Luedermann & Brodsky, 1964; Ganguly & Saksena, 1975; Ganguly & Szulewicz, 1975), avilamycins (Buzetti *et al.*, 1968; Heilman *et al.*, 1979; Keller-Schierlein *et al.*, 1979; Mertz *et al.*, 1986), curamycins (Galmarini & Deulofeu, 1961; Ganguly, Bose & Cappuccino, 1979), flambamycin (Ninet *et al.*, 1974; Ollis *et al.*, 1979 *a,b*).
- lankacidins (also termed bundline) (Gäumann *et al.*, 1960; McFarland *et al.*, 1984; Sawada *et al.*, 1987)
- the sideromycins (Bickel *et al.*, 1960 *a,b*; Nüesch & Knüsel, 1967; Bickel

et al., 1966; Knüsel & Zimmermann, 1975): ferrimycin, succinimycin and danomycin
- pleuromutilin (Högenauer, 1979)
- indolmycin (Rao, 1960; Floss, 1981)

None of these substances shows any cross-resistance to antibiotics used in man, and in each case there is a relatively favourable ratio between antibiotic effect and toxicity. Avilamycin is available as a nutritive antibiotic under the name Maxus; a pleuromutilin derivative (tiamulin) is also used in animal breeding. Since both these products are for special use only and are employed on a very limited scale, there is no reason why they should not be used in man against multiresistant Gram-positive bacteria. The everninomycins are now under re-evaluation (Loebenberg *et al.*, 1993; Mortsell, Hagelberg & Dornbusch 1993; Bauernfeind & Jungwirth 1993; Masoudi *et al.*, 1993; Shlaes *et al.*, 1993; Nakashio *et al.*, 1993; Nishino & Sakurai, 1993; Urban *et al.*, 1993).

Idea 2

Some of the antibiotics not yet used in medicine have a very narrow spectrum of activity that does not embrace any major disease pathogen. Derivatives with a broader spectrum of activity should be possible where poor penetration is the reason for the narrow spectrum. This idea leads us to, amongst others, the large kirromycin group (kirromycin (Wolf & Zähner, 1972), kirrothricin, efrotomycin, aurodox, heneicomycin, factumycin) and pulvomycin. It has been proved clearly for the kirromycin antibiotics that the spectrum is narrow due to lack of penetration and that the target in the cell, the elongation factor Tu, is present in all bacteria (Fischer *et al.*, 1977). Only the elongation factor Tu from *Lactobacillus* is resistant to kirromycin (Wörner, 1982).

Were it possible to overcome the penetration barrier by means of clever derivatization or another trick, we should be a great deal closer to finding a new therapeutic agent. This will have to be proved for some peptide antibiotics that also have a narrow spectrum of activity: stenothricin (Hasenböhler *et al.*, 1974; König *et al.*, 1976); imacidin (Brecht-Fischer, Zähner & Laatsch, 1979; Laatsch, 1982), hormaomycin (Andres *et al.*, 1989; Andres, Wolf & Zähner, 1990) before work on overcoming the penetration barrier for these antibiotics becomes worthwhile.

Search for new antibiotics using new test methods, different microorganisms and varying culture conditions

There are three approaches that can be used to improve our chances of finding new substances: new tests, new organisms, and variation of culture conditions. None of these three options alone guarantees success, and the chances are best if we combine all three.

Options with respect to test methods

Cell-free systems

Classic antibiotic tests with bacteria are based on the decision 'grown/not grown', and over 5000 sites of action may be involved here.

This kind of over-all approach provides a very large number of antibiotics. If you select 100 different actinomycetes at random, and study them with respect to production of antibiotics, you will get about 100 different substances, that is an average of one per strain. Assuming that the probability of finding a substance is the same for each target, limiting the search to one particular site of action, you would have to test 5000 strains before you found a substance. Since this is not enough and experience has shown that at least 3–10 are needed for a suitable lead structure, over 10 000 strains are required to be tested in monotarget screening.

However, a cell-free monotarget test system has its advantages:

- If one uses, say, a single cell-free enzyme for the test, it can be assumed that there are hundreds of thousands of microorganisms in the system that have not yet been investigated and therefore the probability of getting a new substance is relatively high.
- Since there is no penetration barrier, substances can be found that cannot penetrate whole cells. However, the effort required to modify a substance that is only effective in cell-free systems to the extent that it is also effective against whole cells, can be very great.
- Cell-free systems are often characterized by great sensitivity so that even very small concentrations of an active ingredient are detected.

The disadvantages of cell-free test systems should not be forgotten.

- The demands in terms of sample preparation are greater than in a test involving whole cells; the samples should be protease-free, for example.
- The classic plate-diffusion test with bacteria is considerably cheaper than *in vitro* tests and is less likely to meet with problems.

Cyclothialidin (Fig. 3) recently described by Roche Nippon (Nakada *et al.*, 1993; Watanabe *et al.*, 1994; Kamiyama *et al.*, 1994) was fished out of 20 000 strains using an *in vitro* gyrase test. Cyclothialidin shows a very good effect on the cell-free gyrase of Gram-positive and Gram-negative bacteria – an effect that is superior to quinolones. When tested on whole cells it is shown to be a narrow-spectrum antibiotic restricted to only a few *Eubacterium* strains. Cyclothialidin itself is not a potential candidate for antibacterial treatment but should be suitable as a lead structure for the development of new gyrase inhibitors (Goetschi *et al.*, 1993).

The most important of the numerous target sites in the bacterial cell are those that are not present in mammalian cells or are structured differently. This criterion alone does not guarantee the best chance.

The chances should also be reduced if the blocked metabolic chain

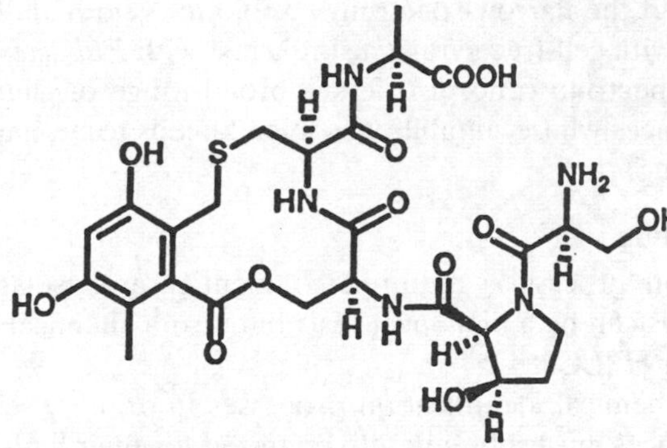

Fig. 3. Cyclothialidin, a new lead structure for gyrase inhibitors.

resulting from the inhibitor can be compensated for by uptake of appropriate modules from the medium or body fluids. For example, the biosynthetic routes to essential amino acids are unlikely to provide suitable target sites since the bacteria can meet their requirement for amino acids from body fluids. The enzymes selected and the target should be genetically conserved. Where the enzymes are already varying greatly, resistance is expected to develop quickly. In our opinion, possible targets that have not been considered yet are:

- early stages of cell membrane synthesis (PEP-transferase, UDP-NAG-enolpyruvate reductase, transglycosylase);
- biosynthesis of teichoic acids in Gram-positive bacteria;
- biosynthesis of KDO and lipid A in Gram-negative bacteria;
- RNA polymerases;
- individual stages of protein biosynthesis not already covered by existing antibiotics, for instance, transformylasis, initiation factors.

More basic research into the comparative biochemistry of eukaryotes and prokaryotes is pressing. If looked at closely, many individual steps for common reactions in eukaryotes and prokaryotes probably differ in detail.

Search for narrow-spectrum antibiotics
More or less the same test strains are used worldwide in the search for new antibiotics; these correlate closely with disease pathogens. If one supplements such a test spectrum with some bacterial strains that have seldom been used as test organisms and then deliberately selects those antibiotics for further processing that specifically inhibit these unusual strains only, there is a good chance that the substances involved are new, perhaps members of novel groups.

These narrow-spectrum antibiotics are not suitable for direct use as therapeutic agents, but can be used as a lead. One could probably just as easily have found the narrow-spectrum antibiotic cyclothialidin, obtained from screening with cell-free gyrase, using a test with *Eubacterium*.

The narrow-spectrum concept offers a broad range of chemically very different substances whose suitability as a lead needs to be clarified before further processing.

Chemical screening

An identification process is required to identify and isolate microbial metabolites. This can be a biological test, but also a chemical process, for example TLC or HPLC.

When using chemical identification processes in primary screening, the resulting substances must subsequently be tested for their biological effect. Most companies still performing antibiotic screening today are not just searching for antibiotic activities. The broader the range of biological activities examined, the more worthwhile chemical screening becomes. Umezawa (1977) introduced the method, Zähner et al. (1988) developed it and several laboratories adopted this way of searching for new metabolites. Nakagawa (1992) has come up with a more recent review, as seen from the Japanese point of view.

Developing this strategy leads to combinations of HPLC with diode-array detection (Fiedler, 1993) or HPLC-MS. The data obtained are stored in a databank so that any new data received can be compared with those already available. Appropriate extension of the databank allows the identification or group allocation at an early stage in screening.

Options with respect to microorganisms

Despite the fact that every year, undoubtedly, more than one million strains are investigated for development of secondary metabolites, in particular antibiotics, and that this number often includes identical strains, this large number accounts for only an extremely small proportion of naturally occurring microorganisms. We assume that less than 10% of all micro-organisms have been examined in culture to date.

A number of problems and observations limit the greater use of the biosynthetic potential of microorganisms:

- In many groups the scientific basis for cultivation of larger numbers of microorganisms is lacking. This applies, for example, to myxomycetes, acrasiomycetes, endosymbiotic microorganisms of animals and plants and sessile bacteria in aquatic systems.
- Even if it is possible to include microorganisms that are difficult to cultivate in screening programmes, these strains – if they produce any interesting substances at all – require great effort to optimize fermen-tation, perform scaling up and improve them genetically.
- A glance at the distribution of those antibiotics described to date shows it to be very uneven. This might, to a small degree, be caused by the fact

that the status of processing is not on the same level, thus the distribution is not really uneven. However, it cannot be disputed that certain groups are less productive despite the fact that they are easy to process, for example, real yeasts, enterobacteriaceae, anaerobic bacteria and halobacteria. The most potent group of microorganisms with respect to the production of secondary metabolites and the one which has been studied the most intensively is actinomycetes, primarily streptomycetes.

What solutions are there or what compromises must we make?

- Studies on the cultivation of organisms which in the past could not be cultivated, have little or no place in a screening programme. Collaboration with college groups will permit access to a small number of such organisms. Co-operation of this kind should also help to stimulate basic research in the isolation and cultivation of previously unexamined organisms.
- In many places, cutbacks have been made in this classic microbiology to the benefit of molecular biology and microbial genetics. This situation should be rectified; however, not at the cost of molecular biology.
- Past observations have shown there to be a positive correlation between the diversity of macroflora or macrofauna and the diversity of associated microorganisms. By selecting the biological material for initial isolation of microorganisms from areas with isolated macroflora, there is a greater chance of isolating strains that will also provide new products. Regions with original flora of interest are, for instance parts of Australia, New Zealand, Ceylon, Madagascar, Malaysia, and so on.
- Botanists know so-called gene centres for a number of plant families. Today, soil samples from the rhizosphere of plants from gene centres remain a source of a considerable proportion of strains previously unexamined, even those from easily accessible groups such as actinomycetes or imperfect fungi.
- Soil samples, when collected are often not used for the isolation of strains until months later. Direct isolation of microorganisms from the sample locally is desirable. Storage and transport reduce the diversity of microflora more than the total bacterial count of samples.

Spreading the variation of secondary metabolites with genetic methods

Advances in the genetics of secondary metabolism make it possible today to 'construct' hybrid antibiotics. In the polyketide series for example, the polyketide synthesis genes or erythromycin, tetracenomycin and actinorhodin are available for such constructions. Soon it will be possible to construct hybrid antibiotics in the area of non-ribosomal polypeptide synthesis.

In the work carried out with polyketide synthesis genes to date, hybrid constructs were obtained by chance, rather than by design. With erythromycins, the first examples of directed modification through gene inactivation

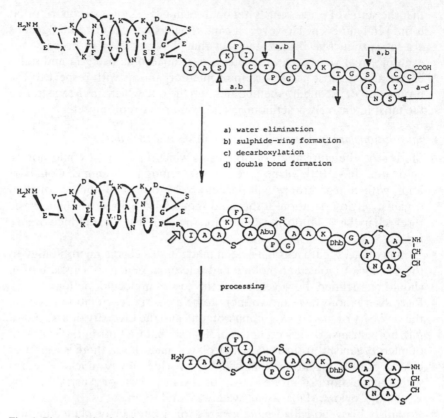

Fig. 4. Biosynthesis of epidermin by Staphylococcus epidermidis (Schnell *et al.*, 1988).

have been realized. Since it is not yet possible, despite considerable effort, to predict which parts of the erythromycin molecule must be modified, in order to obtain new and better properties, we still have a long way to go – despite the great advances made – until rational, genetic improvement of macrolides is possible.

We are further ahead with ribosomally synthesized polypeptide antibiotics, the lantibiotics. Three teams at the University of Tübingen: Prof. Jung (Organic Chemistry), Prof. Götz (Bacterial Genetics), and Prof. Zähner (Antibiotics), have developed the lantibiotics epidermin and gallidermin (Allgaier *et al.*, 1985; Schnell *et al.*, 1988) to such an extent that directed modification using genetics has been possible (Götz, 1994, personal communication). Fig. 4 shows the biosynthesis of epidermin and Fig. 5 summarizes the results obtained to date on structure/activity relationships. The exchanges of amino acids that led to an increase in activity are marked with + and ++, those that led to a decrease in activity with − and −−.

Lantibiotics are a rapidly growing group of substances the efficacy of

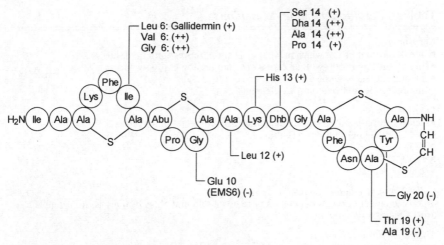

Fig. 5. Exchanges of amino acids in epidermin by genetical methods. Effects of antibacterial activity. (Götz, 1994, personal communication).

which in animals would suggest that we are not so very far away from their possible use in man.

Options with respect to fermentation

Variation of culture conditions is an area that has been rarely exploited. A comparison of the possible variations, as summarized in Table 1, with the variations actually used at the screen level, would suggest that only a small proportion of the biosynthesis potential has been exploited to date. Too often, one accepts conditions that permit optimum growth although it is generally assumed that optimum growth and evolution of the secondary metabolism are mutually exclusive.

FINAL COMMENTS: TWO THESES

Thesis 1

• Long-term basic microbiological research must be developed.

A number of ways of doing this which may help us out of this resistance dilemma have been illustrated here. Several ways should be used when trying to improve our chances. Regardless of the methods chosen, more basic microbiological research is desirable or a prerequisite. It should cover the following areas:

1. Methods of isolating and cultivating microorganisms that have not yet been accessed or only with great difficulty.
2. Studies on the transportation of antibiotics into the bacterial cell.
3. Comparative biochemistry of prokaryotes and eukaryotes.

Table 1. *Variation of culture conditions*

Nutrient broth
- use of nutrient broths which prevent C, N and P repression, if possible
- use of nutrient broths with an excess of C and limited N supply and vice versa
- medium supplementing towards the end of the growth phase
- nutrient broth deficient in trace elements

Physical conditions
- temperature, possible shift in temperature towards end of log phase
- pH including pH shift during culture
- pO_2
- pCO_2
- osmotic values

Collection of metabolites using adsorber resin
or ion exchanger during fermentation or removal of metabolites by dialysis, for example using a membrane fermenter

Creation of stress conditions
- osmotic stress
- stress through heavy metals
- stress through inhibitor supplement

Microbial conversions of certain parent substances
- incorporation of modified modules using deliberate fermentation
- incorporation of modified modules using mutasynthesis
- microbial transformation of biologically active parent substances

4. Mode of action of antibiotics.
5. Pathogenicity factors.

Thesis II

- Long-term strategies to prevent the development and spread of resistant bacteria must be developed.

The numerous ways of obtaining new antibiotics outlined here may help to improve the current situation with respect to resistance. If, however, we continue to use new antibiotics in the same way as we have done in the past, the next dilemma is only a matter of time. Questions related to resistance management could not be dealt with in this overview. However, in the long term, management of the development of resistance is likely to become a more important issue than the search for new antibiotics.

REFERENCES

Allgaier, H., Jung, G., Werner, R. G., Schneider, U. & Zähner, H. (1985). Elucidation of the structure of epidermin, a ribosomally synthesized tetracyclic heterodetic polypeptide antibiotic. *Angewandte Chemische, International Edition, English*, **24**, 1051–3.
Andres, N., Wolf, H. & Zähner, H. (1990). Hormaomycin, a new peptide lactone

antibiotic effective in inducing cytodifferentiation and antibiotic biosynthesis in some *Streptomyces* spp. *Zeitschrift fur Naturforschung*, **45c**, 851–5.

Andres, N., Wolf, H., Zähner, H., Rössner, E., Zeeck, A., König, W. A. & Sinnwell, V. (1989). Hormaomycin, ein neues Peptid-lacton mit morphogener Aktivität auf Streptomyceten. *Helvetica Chimica Acta*, **72**, 426–37.

ASM News. (1992). Anticipating antibiotic resistance in nosocomical infections. *American Society for Microbiology News*, **58**(3), 124–5.

Bauernfeind, A. & Jungwirth, M. (1983). *In vitro* activity of everninomicin (EVE) in comparison with vancomycin (Van). *Proceedings of the 33rd Interscience Conference on Antimicrobial Agents and Chemotherapy*, New Orleans, USA. Abstract N. 458.

Berkelman, R. L., Bryan, R. T., Osterholm, M. T., LeCuc, J. W. & Hughes, J. M. (1994). Infectious disease surveillance a crumbling foundation. *Science*, **264**, 368–70.

Bickel, H., Gäumann, E., Keller-Schierlein, W., Prelog, V., Vischer, E., Wettstein, A. & Zähner, H. (1960a). Über eisenhaltige Wachstumsfaktoren, die Sideramine, und ihre Antagonisten, die eisenhaltigen Antibioticka, Sideromycine. *Experientia*, **16**, 129–33.

Bickel, H., Güumann, E., Nussberger, G., Reusser, P., Vischer, E., Voser, W., Wettstein, A. & Zähner, H. (1960*b*). Über die Isolierung und Charakterisierung der Ferrimycine A1 und A2, neuer Antibiotica der Sideromycin Gruppe. *Helvetica Chimica Acta*, **43**, 2105–18.

Bickel, H., Mertens, P., Prelog, V., Seibl, J. & Walser, A. (1966). Über die Konstitution von Ferrimycin A1. *Tetrahedron* Suppl. Part 1, 171–9.

Brecht-Fischer, A., Zähner, H. & Laatsch, H. (1979). Imacidin, ein neues Acylpeptidantibioticum aus *Streptomyces olivaceus*. *Archives of Microbiology*, **122**, 219–229.

Buzetti, F., Eisenberg, F., Grant, N. H., Keller-Schierlein, W., Voser, W. & Zähner, H. (1968). Avilamycin. *Experientia*, **24**, 320–3.

Corcoran, J. W. & Hahn, F. E. (eds.). (1975). *Antibiotics, Vol. III. Mechanism of Action of Antimicrobial and Antitumor Agents*. Springer-Verlag, Berlin, Heidelberg, New York.

Davis, J. (1994). Inactivation of antibiotics and the dissemination of resistance genes. *Science*, **264**, 375–82.

Fiedler, H. P. (1993). Biosynthetic capacities of actinomycetes. 1. Screening for secondary metabolites by HPLC and UV–visible absorbance spectral libraries. *Natural Product Letters*, **2**, 119–28.

Fischer, E., Wolf, H., Handtke, K., Parmeggiani, A. (1977). Elongation factor Tu resistance to kirromycin in an *Escherichia coli* mutant altered in both tuf genes. *Proceedings of the National Academy of Sciences, USA*, **74**, 4341–5.

Floss, H. G. (1981). Biosynthesis of some aromatic antibiotics. In: Corcoran J. W., *Antibiotics Vol. IV.*, pp. 236–61, Springer Verlag, Berlin, Heidelberg, New York.

Fox, J.L. (1992). Coalition reacts to surge drug resistance TB. *American Society for Microbiology News*, March 1992, 135–9.

Galmarini, O. L. & Deulofeu (1961). Curamycin. I. Isolation and characterization of some hydrolysis products. *Tetrahedron*, **15**, 76–86.

Ganguly, A. K., Bose, A. K. & Cappuccino, N. F. (1979). Structural studies on curamycins. *Journal of Antibiotics*, **32**, 1213–16.

Ganguly, A. K. & Saksena, A. K. (1975). Structure of everninomicin B. *Journal of Antibiotics*, **28**, 707–9.

Ganguly, A. K. & Szulewicz, S. (1975). Structure of everninomicin C. *Journal of Antibiotics*, **28**, 710–12.

Gäumann, E., Hütter, R., Keller-Schierlein, W., Neipp, L. Prelog, V. & Zähner, H. (1960). Lankamycin und Lankacidin. *Helvetica Chimica Acta*, **43**, 601–6.

Georgopapadakou, N. H. (1993). Penicillin-binding proteins and bacterial resistance to β-lactams. *Antimicrobial Agents and Chemotherapy*, **37**, 2045–53.

Goetschi, E., Angehrn, P., Gmuender, H., Hebiesen, P., Link, H., Masciardi, R. & Neilsen, J. (1993). Cyclothialidin and its cogeners: a new class of DNA gyrase inhibitors. *Pharmaceutical Therapy*, **60**, 367–80.

Hasenböhler, A., Kneifel, H., König, W. A., Zähner, H. & Zeiler, H. J. (1974). Stenothricin, ein neuer Hemmstoff der bakteriellen Zellwandsynthese. *Archives of Microbiology*, **99**, 307–21.

Heilmann, W., Kupfer, E., Keller-Schierlein, W., Zähner, H., Wolf, H. & Peter, H. (1979). Avilamycin C. *Helvetica Chimica Acta*, **62**, 1–6.

Högenauer, G. (1979). Tiamulin und Pleuromutilin. In: Hahn, F. E., ed., *Antibiotics Vol. V/Part 1*, Springer Verlag, Berlin, New York, pp. 344–360.

Kamiyama, T., Shimma, N., Ohtsuka, T., Nakayma, N., Itozono, Y., Nakada, N., Watanabe, J. & Yokose, K. (1994). Cyclothialidin, a novel E.coli DNA gyrase inhibitor.II. Isolation, characterization and structure elucidation. *Journal of Antibiotics*, **47**, 37–45.

Keller-Schierlein, W., Heilman, W., Ollis, W. D., Smith, C. (1979). Die Avilamycine A und C: Chemischer Abbau und spektroskopische Untersuchungen. *Helvetica Chimica Acta*, **62**, 7–20.

Kingman, S. (1994). Resistance a European problem, too. *Science*, **264**, 363–5.

Knüsel, F. Zimmermann, W. (1975). Sideromycins. In: Corcoran, J. W. & Hahn, F. E. eds *Antibiotics Vol III.*, Springer Verlag, pp. 653–67.

König, W. A., Engelfried, C., Hagenmaier, H. & Kneifel, H. (1976). Struktur des Peptidantibioticums Stenothricin. *Liebigs Annals of Chemistry*, 2011.

Laatsch, H. (1982). Immacidin. *Liebigs Annals of Chemistry*, 28–40.

Lambert, H. P. & O'Grady. (1992). *Antibiotics and Chemotherapy*, Churchill, Livingstone 1992, pp. 1–561.

Loebenberg, D., Cacciapuotti, A., Napples, L., Moss, E. L., Menzel, F., Hare, R. S. & Miller, G. H. (1993). Biological activity of SCH 27899 (EvE), an everninomicin antibiotic. *Proceedings of the 33rd Interscience Conference of Antimicrobial Agents and Chemotherapy*, New Orleans, USA. Abstract No. 456.

Luedermann, G. M. & Brodsky, B. (1964). *Micromonospora carbonacea* sp., an everninomicin-producing organism. *Antimicrobial Agents and Chemotherapy*, 47–52.

McFarland, J. W., Pirie, D. K., Retsuma, J. A. & English, A. R. (1984). Side chain modification in lankacidin group antibiotics. *Antimicrobial Agents & Chemotherapy*, **25**, 266–33.

Masoudi, A., Callion, J., Mazeau, C., Minozzi, C., Miller, G. & Bismuth, R. (1993). *In vitro* antibacterial activity of everninomicin (SCH 27899) compared with vancomycin and taicoplanin against clinical isolates of staphylococci. *Proceedings of the 33rd Interscience Conference on Antimicrobial Agents and Chemotherapy*, New Orleans, USA. Abstract No. 459.

Mertz, J. L., Peloso, J. S., Barker B. J., Babitt, G. E., Occolowitz, V. L., Simson, V. L. & Kline, R. M. (1986). Isolation and structural identification of nine avilamycins. *Journal of Antibiotics*, **39**, 877–87.

Mitsuhashi, S. (1977). *R-Factor. Drug Resistance Plasmid*. University Park Press, London, Tokyo, pp. 1–45.

Mortsell, E., Hagelberg, A. & Dornbusch, K. (1993). *In vitro* activity of SCH 27899 on *Clostrium difficile* at various pHs. *Proceedings of the 33rd Interscience Confer-*

ence on Antimicrobial Agents and Chemotherapy, New Orleans, USA. Abstract No. 461.

Murray, B. E. (1991). New aspects of antimicrobial resistance and the resulting therapeutic dilemmas. *Journal of Infectious Diseases*, **163**, 1185–94.

Nakada, N., Shimada, H., Hirata, T., Aoki, Y., Kamiyayma, T., Watanabe, J. & Arisawa, M. (1993). Biological characterization of cyclothialidine, a new DNA-gyrase inhibitor. *Antimicrobial Agents and Chemotherapy*, **37**, 1656–61.

Nakagawa, A. (1992). Chemical Screening. In: Omura, S., ed. *The Search for Bioactive compounds from Microorganisms*, Springer Verlag, pp. 263–80.

Nakashio, S., Iwasawa, H., Hori, S. & Shimada, J. (1993). Comparative anti-microbial activity of everninomicin (SCH 27899) a new oligosaccharide antibiotic. *Proceedings of the 33rd Interscience Conference on Antimicrobial Agents and Chemotherapy*, New Orleans, USA. Abstract No. 461.

Neu, H. C. (1992). Crisis in antibiotic resistance. *Science*, **257**, 1064–73.

Nikaido, H. (1994). Prevention of drug access to bacterial targets. Permeability barriers and active efflux. *Science*, **264**, 382–7.

Ninet, L., Benazet, F., Charpentie, Y., Dubest, N., Florent, J., Lunel, L., Mancy, D. & Preud'homme, J. (1974). Flambamycin, a new antibiotic from *Streptomyces hygroscopicus*. *Experientia*, **30**, 1270–3.

Nishino, T. & Sakurai, M. (1993). *In vitro* activity of everninomicin (SCH 27899). *Proceedings of the 33rd Interscience Conference on Antimicrobials and Chemo-therapy*, Abstract No. 462.

Nüesch, J. & Knüsel, F. (1967). Sideromycins. In: Gottlieb, D. & Shaw, P.D., eds, *Antibiotics I. Mechanism of Action*. Springer Verlag, Berlin, New York, pp. 499–541.

Ollis, W. D., Jones, S., Smith, Ch. & Wright, D. E. (1979a). The orthosomycin family antibiotics. III. Mass spectral studies of flambamycin and its degradation products. *Tetrahedron*, **35**, 1003.

Ollis, W., Sutherland, I. O., Taylor, B. F., Smith, Ch. & Wright, D. E. (1979b). The orthosomycin family antibiotics. II. The ^{13}C NMR spectra of flambamycin and its derivatives. *Tetrahedron*, **35**, 993.

Rao, K. V. (1960). PA155H: A new antibiotic. *Antibiotics and Chemotherapy*, **10**, 312–15.

Sawada, H., Suzuki, T., Akiyama, S. & Nakao, Y. (1987). Stimulatory effect in cyclodexdrins on the production of lankacidin group antibiotics by *Streptomyces* spp *Applied Microbiology and Biotechnology*, **26**, 522–6.

Schnell, N., Entian, K. D., Schneider, U., Götz, F., Zähner, H., Kellner, R. & Jung, G. (1988). Prepeptide sequence of epidermin, a ribosomally synthesized antibiotic with four sulphide-rings. *Nature*, **333**, 276–8.

Shlaes, D. M., Shlaes, J. H., Etter, L., Hare, R. S. & Miller, G. H. (1993). SCH 27899 (EVE) an everninomycin active against multiply-resistant Enterococci and Staphylococci. *Proceedings of the 33rd Interscience Conference on Antimicrobial Agents and Chemotherapy*, New Orleans, USA, Abstract No. 460.

Spratt, B. G. (1994). Resistance to antibiotics mediated by target alterations. *Science*, **264**, 388–93.

Travis, J. (1994). Reviving the antibiotic miracle? *Science*, **264**, 360–2.

Umezawa, H. (1977). In Takeuchi, T., ed. *Institute of Microbial Chemistry 1962–1977*. Microbial Chemistry Research Association pp. 1–200.

Urban, C., Maiano, N., Mosinka-Snipas, K., Wadee, C., Chahrour, T. & Rahal, J. J. (1993). Comparative *in vitro* activity of SCH 27899, a novel everninomicin. *Proceedings of the 33rd Interscience Conference on Antimicrobial Agents and Chemotherapy*, New Orleans, USA, Abstract No. 463.

Wagman, G. H., Luedermann, G. M. & Weinstein, M. J. (1964). Fermentation and isolation of everninomicin. *Antimicrobial Agents and Chemotherapy*, 33–37.

Watanabe, J., Nakada, N., Swairi, S., Shimada, H., Ohshima, S., Kamiyama, T. & Arisawa, M. (1994). Cyclothialidine, a novel DNA gyrase inhibitor. I. Screening, taxonomy, fermentation and biological activity. *Journal of Antibiotics*, **47**, 32–6.

Weinstein, M. J., Luedermann, G. M., Oden, E. M. & Wagman, G. M. (1964). Everninomicin, a new antibiotic complex from *Micromonospora carbonacea*. *Antimicrobial Agents and Chemotherapy*, 24–32.

Wolf, H. & Zähner, H. (1972). Kirromycin. *Archives of Microbiology*, **83**, 147–54.

Wörner, W. (1982). Heterogenität der Elongationsfaktoren bei Eubakterien – eine strukturelle und funktionelle Charakterisierung. Dissertation, Tübingen 1982.

Zähner, H., Drautz, H., Fiedler, H.-P., Grote, R., Keller-Schierlein, W., König, W. A. & Zeeck, A. (1988). Ways to new metabolites from Actinomycetes. In Okami, Y., Beppu, T. & Ogawara, H., eds *Biology of Actinomycetes '88*. Japan Scientific Soc. Press, Tokyo, pp. 171–77.

CONTROL OF FUNGI PATHOGENIC TO PLANTS

PHILIP E. RUSSELL, RICHARD J. MILLING and KEN WRIGHT

AgrEvo UK Ltd, Chesterford Park, Saffron Walden, Essex CB10 1XL, UK

INTRODUCTION

About 50 years ago, the plant protection industry was emerging from a period dominated by inorganic fungicides such as Bordeaux mixture and organo-mercury substances, and beginning to experiment with compounds derived from more modern organic synthesis research. In this review, we will delve into the history of the chemical control of plant disease, describe the properties and the biochemical modes of action of some of the compounds available today and glance into the future by reviewing the requirements for new fungicides and the compounds which are emerging to meet these needs. Last, we will touch upon approaches which we expect to complement chemical control of fungi to provide the protection that our crops need to allow us to feed an ever-expanding world population.

PLANT DISEASE, PAST AND PRESENT

The realization that fungal diseases are a problem

It is a reasonable assumption that fungal diseases have afflicted plants since they first evolved. However, very early records of disease are not always clear, as the true nature of the causal agents was not known and the disorders of plants tended to be attributed to natural phenomena, or to the wrath of the gods. The reports that exist leave no doubt, however, that diseases were taken very seriously. An extract from a Sumerian farm almanac written about 1700 BC refers to Samara disease of barley and calls upon the goddess Ninkilim for help (Krämer, 1963). This state of affairs remained for many years and it is unclear when man realized that he could intervene and provide some practical protection for his crops rather than rely upon appeasing the elements through incantations and even sacrifice.

By the eighteenth century, ideas were being modified. Particularly in India, plants were thought to suffer from human ailments such as bile, phlegm, wind, indigestion and tumours. Attempts to provide cures were based on human medication and included application of various concoctions made from herbs, honey, milk, ghee and even cowdung (Raychaudhuri,

Table 1. *Notable plant disease epidemics*

Disease	Cause	Location

Potato blight
  Irish famine
  German famine
	Phytophthora infestans	USA, Europe
		1845–46
		1916–17

Wheat black stem rust	*Puccinia graminis*	USA 1904, 1916, 1935 Europe 1932
Hop downy mildew	*Pseudoperonospora humuli*	Europe 1924–26
Dutch Elm disease	*Ceratocystis ulmi*	Europe, USA 1920–40 1970–80
Rice brown spot Bengal famine	*Helminthosporium oryzae*	India 1943
Tobacco blue mould	*Peronospora tabacina*	USA, Europe 1960–61

From Waldron (1935), Klinkowski (1970).

1964), although still frequently reinforced by incantation. As science developed through the eighteenth and nineteenth centuries, more disorders of plants were recognized, and in the latter part of this era an association was made between the incidence of disease and the presence of microorganisms. The conclusion at the time, however, was that the microorganisms were the result of disease rather than the cause, and it took some pioneering work in the mid-1800s to show the true relationships. The history is fascinating and the reader is referred to Schumann (1991) and Large (1940) for full accounts.

Although scientific investigations in that era were exploring a vast array of fungal, bacterial and viral disorders, the diseases considered most important were those that had an immediate, visible and direct effect on the crop, such as powdery and downy mildews, potato blight and diseases of fruit. All of these could devastate crops very quickly. In this respect, the capacity of plant disease to cause major problems should not be underestimated. An early biblical reference to effects can be found in Kings (1) 8:32. 'If there be in a land famine, if there be pestilence, blasting, mildew, locust or if there are caterpillars . . .'. In more modern times, the effects of the Irish potato famines of 1845/46, when some 1.5 million Irish emigrated to the USA, are well documented (Large, 1940) although it should be appreciated that such epidemics were also ravaging most of Europe and the USA. Remember, too, that this was at a time when the cause of blight was not known. The fungus, called initially, *Botrytis infestans* and later renamed *Phytophthora infestans*, was known to associate with the rotting plants but had not been shown to be the cause. Since then, other epidemics have wreaked havoc at various times in different parts of the world (Table 1). It can easily be argued

Table 2. *Key dates in fungicide discovery*

Date	Fungicide	Main use
60	Wine	Seed treatment
1637	Brine	Seed treatment
1755	Arsenic	Seed treatment
1802	Lime sulfur	Fruit
1847/50	Sulfur dusts	Vines: Powdery mildews
1885	Bordeaux mixture	Vines: Downy mildews, potato blight[a]
1913	Organo Hg	Seed
1934	Dithiocarbamates (thiram, ferbam, maneb, zineb, mancozeb)	General
1952	Captan (captafol, folpet)	General/Vines
1964	Dichlofluanid	General
	Chlorothalonil	General
	Benzimidazoles	General First systemics
1965	Rice antibiotics	Rice
1967	Pyrimidines	Seed
1968	Organo-phosphorus	General
	Triazoles/imidazoles	Cereals/Fruit/Vine: Powdery mildews and leaf spots
1969	Morpholines	General
1974	Dicarboximides	Vines: *Botrytis*
1976	Cymoxanil	Downy mildew
1977	Phenylamides	Downy mildew
1977	Fosetyl-Al	Downy mildew
1988	Dimethomorph	Downy mildew
1990	Anilino-pyrimidines	General
1992	Methoxyacrylates	General

[a]Since their early introduction, copper fungicides have found general uses to control a vast array of fungal leaf spot and bacterial diseases.

that, without the benefit of modern crop protection methods for agriculture, the list would be longer, with subsequent devastating effects on the world economy.

Early disease control by fungicides

The first conscious effort to control plant disease by chemical means, rather than by medicaments, was the use of lime sulfur and sulfur dusts to control powdery mildews of fruit and vines in the early nineteenth century, although cereal seed treatments, including brine and arsenic, had been used earlier. Table 2 shows key landmarks in fungicide discovery up to the present day.

At the time of the potato famines, and spurred on by successes with lime sulfur in controlling other diseases, an attempt was made in 1845 by Morren, then Director of the School of Agriculture at Liège, Belgium, to use a mixture of lime, salt and copper sulfate as a soil drench to control tuber rot

Table 3. *Fungicide recipes from the early twentieth century*

Bordeaux mixture	Copper sulfate	2 lb
	Fresh burned lime	2 lb
	Water	10 gallons
Woburn Bordeaux Emulsion	Copper sulphate	10 oz
	Lime water	8½ gallons
	Water to make up to . . .	10 gallons
	Paraffin (solar distillate)	22½ oz
Ammoniacal copper carbonate (especially for tomatoes)	Carbonate of copper	1 oz
	Carbonate of ammonia	5 oz
	Soft water	16 gallons
Violet fungicide (especially for violet, pansy, viola)	Copper sulfate	3 lb 4 oz
	Copper carbonate	5 lb 8 oz
	Permanganate of potash	1 oz
	Water	22 gallons
Lime sulfur spray	Flowers of sulfur	1 lb
	Quicklime	15 lb
	Water	50 gallons

From Sanders (1910).

from blight. The experiment failed; if only the mixture had been applied to the foliage! In practice, it was not until some 40 years later in 1885 that Bordeaux mixture, a product of the reaction between copper sulfate and lime, was used as a fungicide. The mixture, originating as 'Médoc Mixture' was used initially to prevent pilfering of grapes along pathways. Millardet, Professor of Botany at the Science Faculty of Bordeaux University, noticed the effect on downy mildew of vines and then worked with the Chateau Dauzac and Beaucaillou to perfect the Bordeaux mixture which has subsequently become a household name (Viennot-Bourqin & Lafon, 1985). By this time, the true causes of disease were well known and much research was being conducted on the effect of copper and sulfur preparations on pathogenic fungi. Bordeaux mixture and its variants as well as lime sulfur and sulfur alone are, of course, still widely used today.

Early in the twentieth century, disease control practice was very different from today. Recipes for home production of fungicides were common (Table 3), virtually all based on Bordeaux mixture, sulfur, lime and other inorganics. Safety standards and instructions as we now know them were absent. This can be illustrated by reference to the preparation of an iron sulfate solution to wash the walls of a greenhouse in which diseased tomatoes had been grown: '. . . the (sulfuric) acid (1 pint) should first be poured over the iron sulfate (25 lb) placed in a wooden, not a metal, vessel and then the water (50 gallons) be gradually and very cautiously added' (Sanders, 1910).

The crops considered worth treating were limited to orchards, ornamentals (especially roses), vegetables and vines. All these are high value crops upon which the effect of disease was very obvious, ranging from total destruction to poor quality produce. Treatment was also relatively easy with the equipment available. Cereal seed was also treated to prevent various disorders, primarily smuts and bunts, which have a clearly debilitating effect on the crop.

At the beginning of this era, no thought was given to foliar treatment of cereals. They were a low value crop, the need for disease control was not perceived and the technology for fungicide application was not developed.

Rules and regulations governing fungicide use were negligible, although a reference to control of lettuce downy mildew did indicate that fungicides should not be used due to their poisonous nature (Sanders, 1910).

Fungicide use and manufacture could thus be regarded as a home industry supported by a limited number of manufacturers supplying fungicide concentrates and raw ingredients.

Fifty years ago . . .

After the discovery of Bordeaux mixture, there was a period of 50 years while sulfur products and various mixtures based on Bordeaux mixture dominated disease control, a period only punctuated by the development of organo-mercury cereal seed treatments (Riehm, 1913). The end of this period coincides with an important historical milestone in the context of this book, about 50 years ago, with the discovery in the mid-1930s of the dithiocarbamate fungicides, compounds initially used as rubber accelerators.

The development of these compounds and relatives during the late 1930s set the scene for fungicide research but it was not until the discovery of captan ('Kittleson's Killer') by Kittleson (1952) and the related phthalimides, captafol and folpet (Fig. 1), followed by a rush of fungicides in the early 1960s, that fungicide science was really born. Many of these latter fungicides possessed a very desirable property, systemicity. Epitomized by the benzimidazoles, such compounds were able to exert an effect at sites remote from the point of application and were heralded as revolutionizing crop protection. Their introduction also extended the scope of fungicide

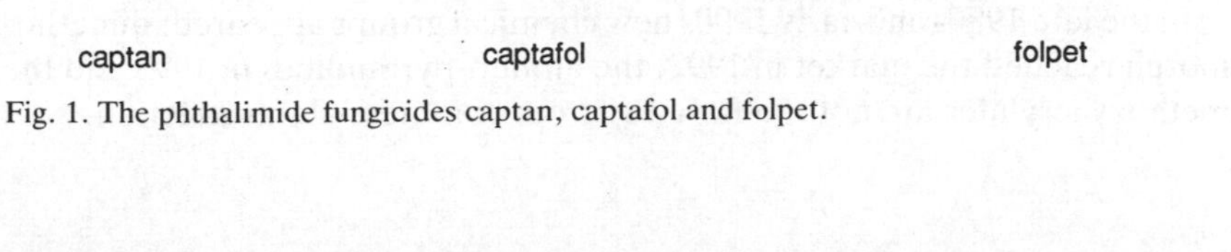

Fig. 1. The phthalimide fungicides captan, captafol and folpet.

use, because the new active molecules made it possible to determine more accurately the damage that plant pathogens could inflict. They also provided more flexibility, in that spray intervals were extended from the normal 7–10 days (or less) for protectant materials up to 14 days or longer. Gone were the days when the damage had to be obvious to be regarded as important. It was now possible to state that the hitherto innocuous or overlooked diseases that caused no apparent harm were in fact yield-limiting. The developing pesticide industry thus realized the potential financial benefit of further fungicide discovery. The advent of new molecules, combined with subsequent increases in plant pathology research, and the recognition of the value of the cereals market led to increased experimentation on cereals which, in turn, allowed the damaging effects of cereal diseases to be understood.

At the same time, regulations were being introduced to ensure the safety of the new molecules to the user, the consumer and the environment. The financial costs of research and development began to rise and gradually the research programmes became market driven and targeted towards diseases whose control could give a financial yield which would justify the development of a new fungicide. The aims of fungicide discovery moved from satisfying numerous small diverse markets to producing molecules for use on a scale large enough to provide good profits. The crops involved became cereals, vines, potatoes, rice and apples. Many vegetables, ornamentals and other crops became 'add-ons', only considered when a market position had been established with a major crop.

Through the 1970s, new compound types appeared at regular intervals as fungicide use and market values expanded. At this time there can be no doubt that the azole fungicides were the most significant discovery, being broad spectrum, often systemic products with very high levels of activity. However, the impact of the dicarboximides in controlling *Botrytis* and the phenylamides in providing systemic control of *Phycomycetes* must not be forgotten.

The 1980s were largely a time of consolidation. New azoles and phenylamides were introduced, and already established compounds found extended uses in the expanding market. Not surprisingly, the extensive and sometimes exclusive use of many compounds had by now led to problems of resistance and a realization that the agrochemical industry must adopt strict anti-resistance strategies in order to safeguard its products for the mutual benefit of the industry, the farmer and the consumer. This period was the 'me-too' era, as companies 'patent-busted' areas of known chemical activity to produce small improvements in the form of their own molecules for particular markets. There was little true innovation.

In the late 1980s and early 1990s new chemical groups appeared: dimethomorph reached the market in 1992, the anilino-pyrimidines in 1993 and the methoxyacrylates are not too far away from commercial introduction.

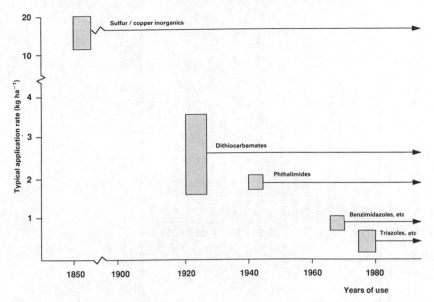

Fig. 2. Declining dose rates of fungicide classes by year of first introduction. The vertical bar represents the typical range of application rates for compounds of each class (kg ha^{-1}) and the horizontal axis the years of use with the left-hand edge of the bar indicating the year of first introduction.

During all this time the economics of fungicide use, the practicality of use and the environmental concern over use have all increased. All are to some extent influenced by the amount of active ingredient required to control the disease. The advances made in the discovery of new fungicides with a decreased impact can be seen by reference to Fig. 2 which shows how dose rates have declined over the years. Total reliance on fungicides applied at 3–20 kg ha^{-1} of the active ingredient has now advanced to improved control from materials applied at 125 g ha^{-1} or less.

WHAT COMPOUNDS ARE CURRENTLY AVAILABLE TO CONTROL FUNGAL DISEASES OF CROPS?

Table 4 shows the major groups of compounds available today. Some have been in use for many years, and it is interesting to note that these generally have a non-specific mechanism of action. As the amount of research which went into the discovery of new fungicides was increased, a trend developed to more specific modes of action which parallels, and may well be connected with, the change towards lower dose rates.

The sterol biosynthesis inhibitors have almost 20% of the world fungicide market, closely followed by the dithiocarbamates at almost 16% and the methylbenzimidazole carbamate (MBC) fungicides with 10%. In view of the importance of these groups, each will be reviewed in turn below. Meanwhile

Table 4. *Major classes of fungicides currently used in the main crops*

Class	Year first reported[a]	Value[b] (US $ million)	Number of compounds	Mode of action[c,d]
Inorganics	1802[e]	610	7	Non-specific
Dithiocarbamates	1934[f]	875	9	Non-specific/cellular thiol groups
Phthalimides	1952	410	7	Non-specific/cellular thiol groups
Guanidines	1957	+	3	Non-specific/membrane effects
MBCs/N-phenylcarbamates	1964/1983[g]	575	5/1	β-tubulin
Antibiotics	1965	+	4	Various, including chitin synthase, trehalase and protein synthesis
Carboxanilides	1966	+	6	Succinate dehydrogenase
Pyrimidines/Azoles	1967[h]/1968[h]	1005	29	Sterol 14 α demethylase
Organophosphorus	1968	365	5	Phosphatidylethanolamine methyl transferase[i]
Hydroxypyrimidines	1968	+	3	Adenosine deaminase
Morpholines	1969	310	3	Sterol Δ^{14} reductase/$\Delta^8 \rightarrow \Delta^7$ isomerase
Dicarboximides	1974	+	4	Unknown/lipid peroxidation
Melanin Biosynthesis Inhibitors	1976	+	2	Tetra/trihydroxynaphthalene reductase
Phenylamides	1977	+	4	RNA polymerase 1
Phenylpyrroles	1988	+	2	Unknown/glucose phosphorylation (?)[j]
Anilino-pyrimidines	1990	+	3	Enzyme secretion[k]
Others		+	8	Various
Combined value of those marked (+)		1410		

[a]From Worthing & Hance (1991) unless otherwise stated.
[b]County NatWest Wood Mackenzie (1991).
[c]Corbett *et al.* (1984).
[d]Köller (1992).
[e]Forsyth (1802).
[f]Tisdale & Williams (1934).
[g]Noguchi *et al.* (1983).
[h]Krämer (1983).
[i]Not the mode of action of pyrazophos.
[j]Jespers *et al.* (1993).
[k]Milling *et al.* (1993).

it is interesting to note the large number of pyrimidine/azoles available to control fungi through inhibition of the sterol 14α-demethylase. This reflects the fact that the target enzyme is able to accommodate a wide range of structural variants, so that it has been possible to develop many compounds which combine the attributes of efficacy at the target site and the right combination of physico-chemical properties and metabolic stability to gain access to it. This may well be less true of some of the other sites, although the patent situation, the size of the market which is being targeted and the relative biological efficacy of different classes of compound all also play critical roles.

Dithiocarbamates

It will be apparent from our brief account of the history of fungicide development that, until the 1930s, disease control programmes relied virtually exclusively upon copper and sulfur. The first major success of a subsequent search for new materials was the discovery of the dithiocarbamate fungicides patented 60 years ago by Tisdale and Williams (1934). Since the 1930s, several fungicides of this group have appeared, for example, thiram, ferbam, maneb and zineb, culminating in the introduction of the most successful of all, mancozeb in 1961 (Fig. 3). When launched, they offered immediate advantages over copper and sulfur in that they were safer to the plants, more convenient to handle and, above all, were more effective. They are, however, surface-acting protectant materials which still have to be applied at frequent intervals. Their spectrum of activity encompassed much of that covered by both copper and sulfur. Today, the most widely used compounds are mancozeb and thiram with the other derivatives being less important. Mancozeb is considered as a cheap commodity product but, while its activity may have been surpassed by more modern materials, it is still a valuable compound for cheap, effective disease control. It is also regarded as an essential mixture component for use in resistance management strategies designed to protect modern, 'at-risk' fungicides such as the phenylamides. Thiram has found a similar niche but is also used as a seed

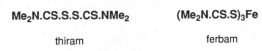

Me$_2$N.CS.S.S.CS.NMe$_2$ (Me$_2$N.CS.S)$_3$Fe

thiram   ferbam

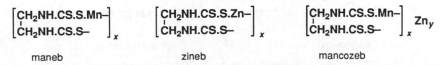

maneb   zineb   mancozeb

Fig. 3. The dithiocarbamate fungicides thiram, ferbam, maneb, zineb and mancozeb.

treatment, primarily for broad-leaf crops, as well as being a component of foliar spray mixtures.

The dithiocarbamates include both the dialkyldithiocarbamates and the alkylene bisdithiocarbamates (the most important of which are thiram and mancozeb respectively) which are thought to differ subtly in their modes of action (Krämer, 1983; Corbett et al., 1984). The dialkyldithiocarbamates are copper chelators, and it is thought that their ability to carry copper into the fungal cell and subsequently interfere with the function of lipoamide-containing dehydrogenases, at least in part explains their mechanism of action. The alkylene bisdithiocarbamates react with cellular thiol groups generally, with no enzyme being especially sensitive. Both groups are therefore non-specific in their mode of action. This 'multi-site' activity, as it is often called, would be expected to lessen the chances of resistance developing, at least by modification of the target site; certainly no problems of resistance are known.

Methylbenzimidazole carbamates (MBCs)

The so-called MBC fungicides are closely related to methylbenzimidazole carbamate, MBC, more properly known as carbendazim, and include the related benomyl and thiophanate methyl which are precursors for carbendazim, and thiophanate itself which is converted to the corresponding ethyl ester (Fig. 4). These materials were introduced in the 1960s and 1970s (Delp 1987) into a market accustomed to protectants such as copper formulations and dithiocarbamates which had to be applied at frequent intervals. The MBCs were systemic, able to penetrate the plant and available for trans-

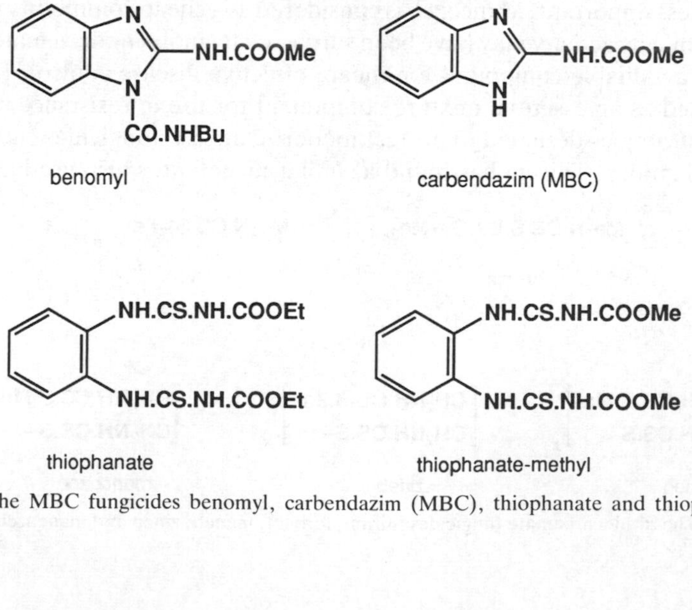

benomyl

carbendazim (MBC)

thiophanate

thiophanate-methyl

Fig. 4. The MBC fungicides benomyl, carbendazim (MBC), thiophanate and thiophanate methyl.

location via the xylem to areas remote from the point of application. This, combined with their excellent activity, made them more flexible than the protectants and helped to extend the time interval between applications. Their activity spectrum covered the Deuteromycetes, Ascomycetes and some Basidiomycetes, groups that encompass the world's most important pathogens including powdery mildews and leaf spots of cereals, fruit, nuts, vegetables and ornamental crops. MBC fungicides could be used as seed treatments or root drenches to control root and foliar diseases, as foliar sprays, injection treatments and as post-harvest applications to prevent storage rots. Because of their superior activity and lower dose rates, they quickly dominated fungicide use programmes in many crops, frequently to the total exclusion of other materials.

Studies during the early 1970s by Davidse and others showed that inhibition of nuclear division by the MBCs was primarily responsible for their fungicidal activities (for reviews, see Corbett *et al.*, 1984; Davidse, 1986; 1987). This was subsequently shown to result from interference with the assembly of tubulin into microtubules. Binding experiments with $[^{14}C]$-carbendazim showed that the fungicide formed a complex with fungal β-tubulin, and that the affinity of binding was positively correlated with fungal sensitivity to the compound (Davidse & Flach, 1977). Inhibition of microtubule assembly by carbendazim was shown *in vitro* with yeast β-tubulin (Kilmartin, 1981) and *in vivo* in hyphal tips of *Fusarium acuminatum* (Howard & Aist, 1980).

With the emergence of MBC-resistant strains, it was demonstrated that resistance was always associated with decreased $[^{14}C]$-carbendazim binding to β-tubulin (for review see Ishii, 1992). MBC-resistant *Botrytis cinerea* and *Penicillium expansum* strains were found to be highly sensitive to *N*-phenylcarbamates (NPCs), and negative cross-resistance was demonstrated between the two classes of chemicals (Leroux & Gredt, 1982). Thus carbendazim is now often used in mixtures with the NPC diethofencarb (Fig. 5) to control resistant strains. The β-tubulin gene was studied in strains of *Neurospora crassa* displaying a range of MBC- and NPC-sensitive phenotypes (Fujimura *et al.*, 1992). The work revealed that mutations in the β-tubulin gene conferring varying degrees of MBC- and/or NPC-sensitivity resulted in single amino acid changes at position 198 of the protein. Different amino acids in this position conferred greater sensitivity to one or other chemical class, with glutamate present in the MBC-sensitive wild-type

EtO—NH.COOPr^i

EtO

Fig. 5. The *N*-phenylcarbamate diethofencarb.

strain, and glycine resulting in the most extreme of NPC-sensitivity. Thus, after 20 years of research into the mode of action of antimicrotubule fungicides, the precise binding site on the molecular target has been identified. Paradoxically, this has been aided to a large extent by the emergence of pathogens resistant to the fungicides.

As a consequence of the single site of action of MBC fungicides and the apparent fitness of fungal strains with modifications at this site, this class of compounds is now seen as a high resistance risk group for many pathogens. However, resistance management strategies have been devised, based on the use of mixtures or alternating spray programmes with compounds of a different mode of action. Where MBC fungicides are used as recommended, and not relied upon exclusively, they are still very valuable, effective compounds.

Azoles

The introduction of the first azole fungicides to agriculture in the early 1970s set a milestone in disease control. Just as the limitations of much earlier fungicides had been realized and solved by introduction of the dithiocarbamates and MBC fungicides some years previously, so in the late 1960s and early 1970s the limitations of the more recent introductions were becoming clear. MBC fungicides began to suffer resistance problems, while the search for compounds active at lower dose rates continued in an attempt to improve on the dithiocarbamates.

The first triazoles, triadimefon and propiconazole, which combined systemic action via xylem translocation with activity at much lower dose rates (125 g ha^{-1} of the active ingredient for cereals), were thus very well received. They, and subsequent introductions such as the imidazole prochloraz, found immediate widescale use as cereal fungicides (see Fig. 6 for structures). The spectrum of activity of the early triazoles was, however, limited and more modern derivatives have activity against a wider range of fungi on more crops and, often, at lower dose rates. The activity spectra cover Ascomycetes, Deuteromycetes and Basidiomycetes. Major differences do occur in activity spectra between different azoles, and the perfect azole with a spectrum covering all of the fungi able to be controlled by the azoles has not yet been found. Because of their high levels of activity and broad spectrum, they can be regarded as the most important group of modern fungicides and new molecules such as epoxiconazole (Ammermann et al., 1990), fluquinconazole (Russell et al., 1992) and triticonazole (Gaulliard et al., 1994) continue to be introduced (Fig. 7).

Prompted by the development of resistance in other fungicide groups, much effort has been devoted to evaluating the resistance risk of azoles and to developing anti-resistance strategies. Initial studies suggested that the development of resistance to azoles was under multigene control (Van Tuyl,

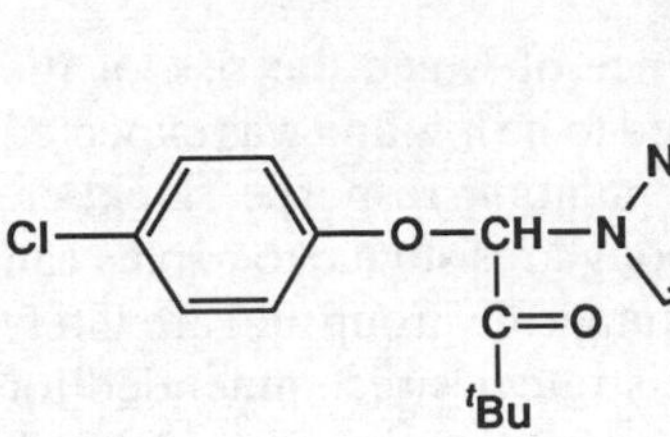

triadimefon

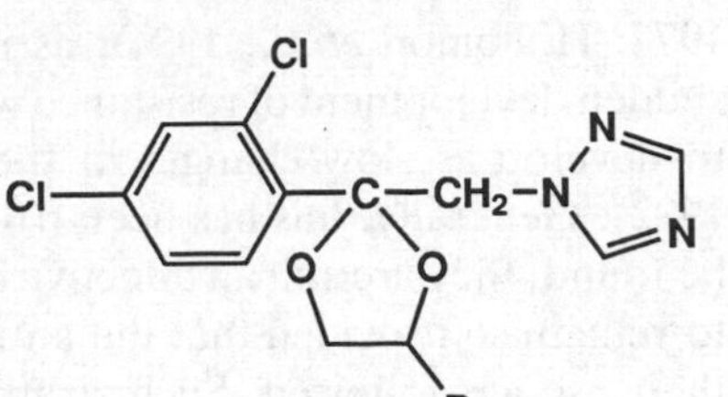

propiconazole

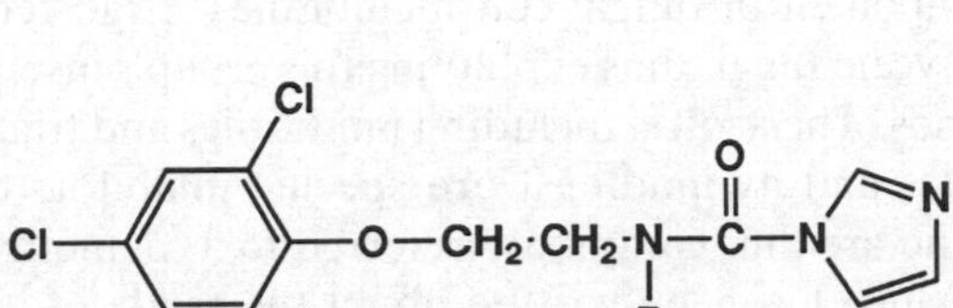

prochloraz

Fig. 6. The triazoles triadimefon and propiconazole, and the imidazole prochloraz.

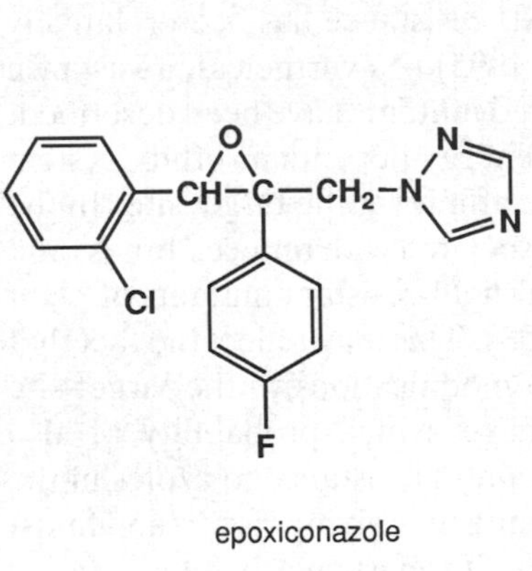

epoxiconazole

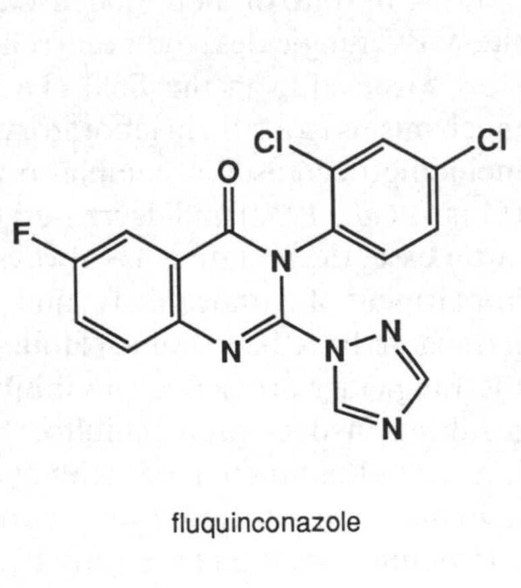

fluquinconazole

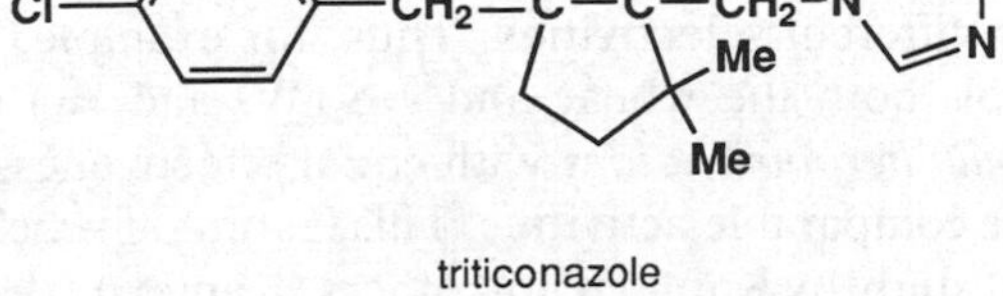

triticonazole

Fig. 7. The new azoles epoxiconazole, fluquinconazole and triticonazole.

1977; Hollomon *et al.*, 1984), as a consequence of which the risk of the sudden development of resistance was perceived to be low and was expected to develop as slow changes in the fungal population response (Dekker, 1982). Generally, this has been true and, although resistance to azoles can be found, they are still an effective and major fungicide group and are likely to remain so provided that the anti-resistance strategies recommended for their use are followed. Such strategies have been summarised by Russell (1993, 1994).

The azole fungicides act by inhibiting the biosynthesis of ergosterol, the major sterol component of fungal cell membranes. Ergosterol is not synthesized by Oomycete fungi, thus explaining this group's insensitivity to this class of compounds. The azoles, including imidazoles and triazoles together with the pyridines and pyrimidines, are specific inhibitors of sterol 14α-demethylation and are thus commonly referred to as demethylation inhibitors, or DMIs. Much has been written about the mode of action of these fungicides, and we will therefore not cover the matter extensively here; for recent reviews, see Köller (1992) and Vanden Bossche (1988). However, two particular aspects which may have contributed to some extent to the widescale use of such large numbers of compounds (Table 4) will be discussed.

First, in spite of their mode of action as single-site inhibitors, and unlike the MBC fungicides, the occurrence of DMI resistance has been relatively slow to develop in the field (Hollomon, 1993). Nevertheless, resistance mechanisms to azoles in laboratory-generated mutants have been described, including decreased accumulation via an energy-dependent efflux system (Guan *et al.*, 1992) and decreased fungicide affinity at the target site. In the latter case, the mutation has always been associated with reduced fitness and impairment of pathogenicity, and no similar field-resistant mutants of plant pathogens have been found (Hollomon, 1993). This may reflect the fact that the fungicides are active-site inhibitors, so modifications to the target site resulting in decreased inhibitor binding have a high probability of also impairing the function of the enzyme. Certainly, resistance to azoles in the field has been associated with many different genes giving rise to an almost continuous variation in sensitivities to DMIs. It is particularly interesting to note the apparent lack of cross-resistance between triazoles and imidazoles (Hollomon, 1993).

A second interesting aspect of azole fungicides is that although, as a group, they offer a very broad spectrum of activity, individual compounds have significantly different selectivities. Thus, for example, the imidazole prochloraz controls both the wheat and rye (W- and R-) pathotypes of *Pseudocercosporella herpotrichoides* with equal effectiveness, whilst other azoles do not have comparable activities. This feature of prochloraz was the subject of a recent study by Kapteyn and others (Kapteyn, 1993; Kapteyn *et al.*, 1992, 1994).

Table 5. *Worldwide crop losses due to plant disease*

	Loss in 1987		1994 value
Crop	%	10^6 tonnes	US $ billions
Wheat	9.1	64.7	12.5
Rice	8.9	77.1	18.1
Potatoes	21.8	93.4	10.4
Sugarbeet/cane	16.5	406.7	11.7
Vegetables	10.1	64.8	20.6
Fruit, grapes	16.4	74.8	22.0
Others	14.0	138.6	35.4
Total		920.1	130.7

From James *et al.* (1991).

The accumulation of prochloraz in W- and R-type strains of *Pseudocercosporella herpotrichoides* and other fungi was compared with that of a triazole. No general correlation was found between the sensitivities of different fungi and the amounts of fungicide accumulated. Furthermore, although prochloraz was accumulated to somewhat higher levels than the triazole in both W- and R-types, accumulation of the triazole was greatest in the R-type which was less sensitive to the triazole than was the W-type (Kapteyn *et al.*, 1992).

Cell-free sterol biosynthesis assays for two different fungi were developed and were shown to be the most useful means of evaluating the relative potencies of DMIs *in vitro*. With *Botrytis cinerea*, data for the potencies of prochloraz and the triazole *in vitro* correlated well with their relative fungitoxicities, but with *Penicillium italicum* there was no such correlation (Kapteyn *et al.*, 1994). It was concluded that differences in affinity at the target site were only one factor affecting the selective fungitoxicities of DMI fungicides. Furthermore, since differential metabolism could not be demonstrated, other mechanisms such as delivery to the target site may be involved.

More studies such as these will be needed if we are to understand the differences in efficacies of individual compounds. Also the behaviour of the compound on the crop, and the physiology of the different fungi in the field will have a major impact on the potential for disease control.

THE NEED FOR NEW FUNGICIDES

Despite all the advances in fungicide chemistry and subsequent improvements in disease control, the world still loses approximately 12% of its food production to plant diseases (Table 5). A further 20% is estimated to be lost

Table 6. *Estimated losses in the USA if crop protection chemicals were not used*

Crop	Percentage loss
Wheat, Great Plains	70
Soybeans, South	50
Corn, Corn Belt	60
Apples, North	100
Potatoes, Northeast	100
Melons, California	45
Lettuce, California	96
Strawberries, California	94
Cole Crops, California	95
Tomatoes, California	70
Sorghum, Texas	50

From Office of Technology Assessment: *Pest Management Strategies in Crop Protection*, Vol. 1, October 1979.

through weed competition, insect damage and natural disasters. It is a sobering realization that losses were estimated at 920 million tonnes in 1987, an amount worth about 130 billion US dollars at today's values. On a world basis, 30% of the losses occur in Asia, especially in the developing countries, followed by 25% in Europe where intensive agriculture presents pathogens with excellent conditions for development and spread (James, 1981). Losses in the USA were estimated at 16% of the world total. Against this figure, it is interesting to note the possible effects of not using crop protection chemicals. Such estimates have been produced for the USA (OTA, 1979) (Table 6).

These current, actual, losses must now be considered in the context of an ever-expanding world population. It has been estimated that more people died from hunger between 1983 and 1985 than died as a result of World War I and II combined. The current world population of over 5 billion is expected to reach about 8 billion by 2025 and possibly 12 billion by 2045. This will require approximately 300% more food to be produced. We in Europe are accustomed to thinking of surpluses, but a loss of 2.2% in world production could result in starvation of 264 million people (Schumann, 1991).

Efficient disease control agents will be required to maximise the yield potential of crops grown to feed the increasing population. The need for such disease control is not often questioned, but the need for new chemicals is; why, if modern materials are effective, do we need more?

The reasons are not always linked with intrinsic activity against the pathogen but with environmental concerns, the economics of disease control and with the prevention of fungicide resistance. Concerns over the environment and safety to the user and consumer can never be ignored and despite the fact that modern fungicides are tested to very high standards it is a valid argument that safety can always be improved. This can be done by discovering new molecules active at lower dose rates and with even lower toxicity to non-target organisms combined with better physico-chemical properties to minimise the environmental impact. New fungicides are also required that are cheaper and easier to apply, thus making the use of more efficient crop protection measures more amenable to low input agriculture. Compounds able to be used at longer spray intervals and as components of integrated pest management programmes aided by disease prediction schemes will also help contribute to these requirements.

Resistance management is also a critical issue. Experience has shown that fungicide resistance can jeopardize the usefulness of modern fungicides and that careful management is required to maintain their value to the user. Strategies for ensuring this, depend upon limiting the selection pressure exerted by the 'at risk' fungicide on the pathogen. This is generally achieved by mixing, or alternating the use of, the 'at risk' fungicide with a compound having a different biochemical mode of action or mechanism of resistance. At present, such strategies depend heavily on the availability of older commodity fungicides, especially dithiocarbamates, and should such products be withdrawn from the market for any reason, there would be serious effects on disease control programmes. New compounds with new modes of action, preferably presenting a low resistance risk themselves, are needed.

CURRENT OBJECTIVES

Markets

The search for new fungicides is intrinsically linked to the requirement for subsequent sales of product to provide a profit over the considerable research and development expenditure needed to put one new molecule on the market. These costs are currently estimated at up to £60 million. Table 7 shows the principal fungicide markets targeted by the agrochemical industry and their value in 1989. Within these, the major crop sectors are cereals, vines, potatoes, rice and fruit. The markets for vegetables, plantation crops and ornamentals are high in total, but comprise many individual market sectors on a multitude of crops which complicates the introduction of new products. Such markets are therefore usually considered once a new product has shown promise in one of the larger single crop sectors.

Chemical properties

Chemical properties can be divided into two areas: properties that concern the environment, user and consumer and those that relate to the biological performance against the target pathogen. Properties relating to the environment, user and consumer are defined by the requirements for compounds to

Table 7. *Principal fungicide markets targeted by agrochemical companies*

Pathogen	Market value (US $ million)[a]	Main crops
Powdery mildews	697	Cereals, fruit, vegetables
Downy mildews	494	Vines, potatoes, vegetables
Leaf spots	847	Cereals, fruit, plantation crops, vegetables, ornamentals
Rice blast	393	Rice
Apple scab	138	Apples
Rusts	172	Cereals, coffee
Botrytis and related	258	Vines, fruit, vegetables
Unaccounted	1101	

[a]Adapted from *AGROW* (1992) **153**, 16.

satisfy various safety regulations in force at the time and will not be discussed here. Biological performance properties, however, are largely a function of user requirements. High activity at minimum cost is a frequent criterion. This should be combined with a high degree of reliability and flexibility in use. To a large extent, these are attributes of the molecule itself but they can be influenced by formulation. Extensive studies on the mode of action of fungicides and the infection and colonisation processes of fungi have now shown that systemicity, once thought essential to the success of a chemical synthesis programme, is not critical. Emphasis is being placed on optimization of biological activity, based on improved knowledge of how and where in the fungal infection process the fungicide acts. Depending on the type of compound, it may be more appropriate for a fungicide to remain on the leaf surface or to penetrate the cuticle but not be translocated away from the point of application.

HOW TO GET WHAT WE WANT?

This is not the place for a detailed description of the discovery and development process for a fungicide, but it may be valuable to the reader to outline the process.

When an organic synthesis chemist prepares a new candidate fungicide, the idea for which compound to make will have its origin in one of four approaches:

(i) *New chemistry;* compounds will be made and tested with no specific reason as to why they might be fungicidal. The chemist might, for instance, be seeking to exploit newly available intermediates or methodology.

(ii) *Analogue chemistry;* a compound will be prepared on the basis of its similarity to a structure of known fungicidal potency, in the hope that the closely related new compound will have improved properties.

(iii) *Natural product lead following;* with this approach, clues are taken from nature by synthesising chemicals related to compounds of natural origin and proven activity.

(iv) *Biochemical design;* the idea here is to choose a biochemical process believed or proven to be essential to the target fungus and design effective inhibitors for fungicide evaluation. This approach is of course attractive from a scientific point of view but, regrettably, no commercial compound has yet been discovered by this route.

Whatever the inspiration behind the synthesis of a compound, it will need to be evaluated *in vivo* in a biological screen carefully designed to reveal any activity of potential commercial interest. If it is active in the first stage, it will be followed up in a stepwise fashion by more extensive tests, culminating in field trials in relevant parts of the world. Perhaps eight years later, it will emerge, if it turns out to be the one compound in about 30 000 which combines novelty, efficacy, and safety to the user, the consumer and the environment with the opportunity to make a profit, as a new product.

Particularly in the fourth approach, and usually in all cases where the ultimate biochemical target of a class of chemicals is known, compounds will be tested at the biochemical (enzyme/receptor) level at an early stage. Biochemical results can be faster than those generated *in vivo* and when examined alongside results *in vivo*, can form the basis for ideas about why a potent compound has poor activity *in vivo*. Biochemical experiments at the enzyme/receptor and/or cell/tissue level can then be conducted to investigate a discrepancy which could be caused, for instance, by lack of uptake or the occurrence of metabolic degradation.

Full-scale biochemical screening, in the sense of high-throughput screens without any pre-selection of compounds is carried out within the agrochemical industry, but to a lesser extent than within pharmaceutical companies. The main reason for this is that it is still easy and cheap in the agrochemical industry to screen against the ultimate target organism whereas this is obviously not true in human healthcare research, in which model systems are obligatory rather than supplementary.

EMERGING COMPOUND CLASSES

In recent years, a number of fungicide groups with novel chemistry has been developed (Fig. 8). Fenpiclonil, introduced in 1988, was the first of the phenylpyrroles to be launched, and a second analogue, fludioxonil, has now been brought to the market. Both are related to the natural antibiotic pyrrolnitrin and are used in particular for cereal seed treatments (Nyfeler &

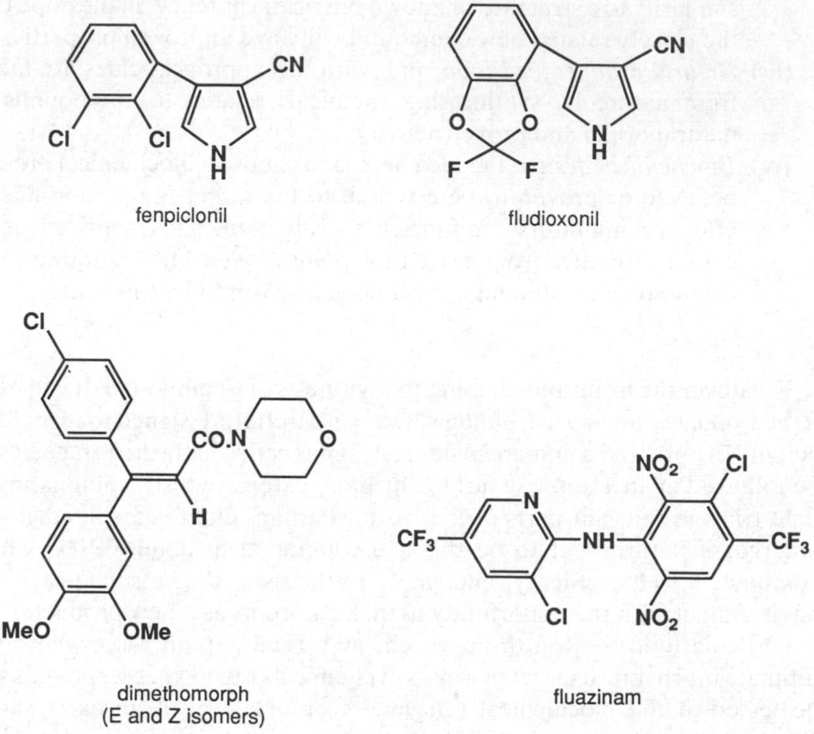

Fig. 8. The recently introduced fungicides fenpiclonil, fludioxonil, dimethomorph and fluazinam.

Ackermann, 1992). Dimethomorph is a new cinnamic acid derivative for use against *Plasmopara viticola* on vines and *Phytophthora infestans* on tomatoes and potatoes (Albert *et al.*, 1988). Preliminary studies point to effects at the level of the cell wall (Kuhn *et al.*, 1991), but most importantly it appears to have a novel mode of action and it is not cross-resistant to phenylamide fungicides. Fluazinam is another new Oomycete fungicide for use in vines and potatoes, but its spectrum of activity also includes *Botrytis cinerea* (Anema *et al.*, 1992). Fluazinam has an interesting biochemical mode of action, being a potent uncoupler of oxidative phosphorylation (Guo *et al.*, 1991), a property which might normally be of concern from an animal toxicity point of view. It is, however, rapidly metabolized in mammalian mitochondria to a product without uncoupling activity (Guo *et al.*, 1991).

The anilino-pyrimidines, represented by pyrimethanil (Neumann *et al.*, 1992) and cyprodinil (Bocquet *et al.*, 1994), represent the latest new class of fungicides to be commercialized (Fig. 9). Pyrimethanil is a product of our own research and has demonstrated excellent efficacy against *Botrytis cinerea* on vine, fruits, vegetables and ornamentals, and against *Venturia*

pyrimethanil   cyprodinil

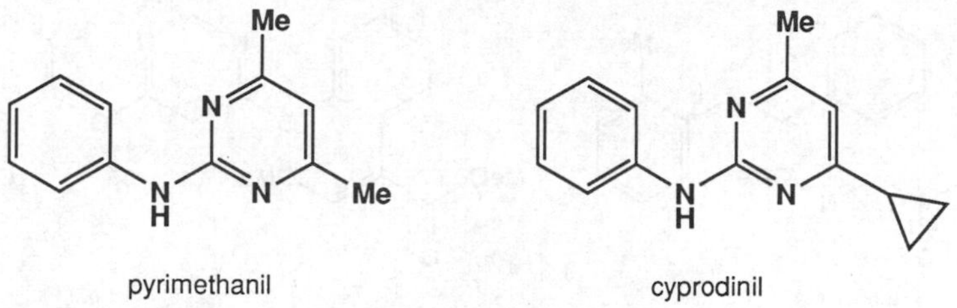

Fig. 9. The anilino-pyrimidines pyrimethanil and cyprodinil.

spp. in apples and pears. Pyrimethanil has protective, translaminar and root systemic activity against *Botrytis*. For *Venturia* control it is used preventatively, but it also has 3 days curative efficacy. Pyrimethanil has a different mode of action to all current *Botrytis* and *Venturia* fungicides, and does not show cross-resistance to other compound classes. It inhibits the secretion of fungal enzymes, and thus for example was shown to prevent the digestion and lysis of host cells that occurs at the site of *Botrytis* infection (Milling *et al.*, 1993). Protein synthesis is not inhibited, and as yet the precise site of action of pyrimethanil resulting in inhibition of protein secretion still remains to be elucidated.

A new group of fungicides which is showing great potential for the future is the β-methoxyacrylates. These are synthetic derivatives of the antifungal antibiotics, the strobilurins (Anke *et al.*, 1977). There is currently intense interest in this area of chemistry, with patent applications from a dozen or more companies, and several compounds undergoing development (Ammermann *et al.*, 1992; Godwin *et al.*, 1992). The mode of action of the strobilurins and the related natural products oudemansin and myxothiazol was first described by Becker and co-workers (Becker *et al.*, 1981). They inhibit respiratory electron transport between cytochrome b and cytochrome c_1 of ubiquinol cytochrome c reductase. It is now known that they bind to a specific site on cytochrome b, distinct from that of antimycin (Brandt *et al.*, 1988). The natural product-leads suffered from poor photochemical stability and excessive volatility, but the E-β-methoxyacrylate toxophore was identified as a useful starting point from which to synthesize new molecules with improved biological and physical properties (Beautement *et al.*, 1991). The resulting fungicides, exemplified so far by BAS 490F and ICIA 5504 (Fig. 10) are especially broad spectrum in their activities, are useful both as protectant or curative treatments, and have good persistency (Ammermann *et al.*, 1992; Godwin *et al.*, 1992). Although some years from commercialization, this story of innovation bodes well for our ability to combat fungal diseases of crops into the next century. By continued research and development of new chemical classes with novel modes of action and still greater biological utility, we will be able to maintain and probably improve disease control efficiency, overcome the spectre of fungicide

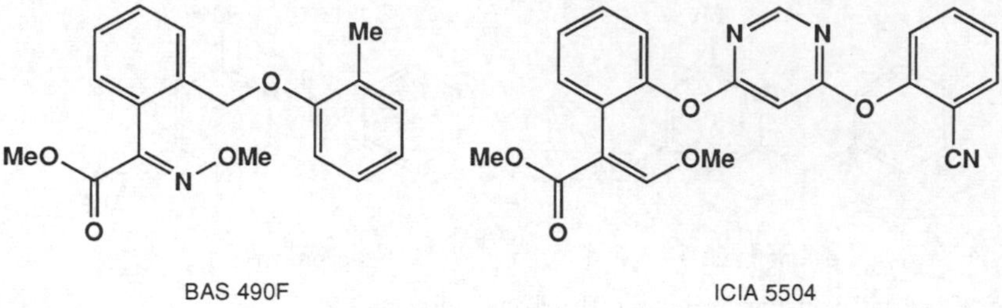

BAS 490F   ICIA 5504

Fig. 10. The recent fungicides BAS 490F and ICIA 5504, inspired by the naturally occurring strobilurins.

resistance, and avoid the devastating yield losses due to fungal diseases which were commonly experienced in agriculture only 50 years ago.

WHAT DOES THE FUTURE HOLD?

The past 50 years have seen great advances in the control of plant diseases by fungicides derived from synthetic chemistry. There is no doubt that such research will continue, but it is also clear that other routes to discovery and disease control will rise in importance.

Maybe the most promising route, and one that until recently had not received much attention, is the potential use of natural products as fungicides. Micro-organisms are capable of producing a vast array of chemicals as exudates or metabolites, and great interest is now being shown in searching for fungicides or fungicide chemistry-leads from amongst these substances.

There is also great interest in using chemicals to stimulate host plant defence mechanisms, a process termed 'induced resistance'. It is expected that such an approach could utilize compounds which are intrinsically less toxic as they would act to 'switch on' the plant's natural defence system.

The concept of using genetically modified crops has links to several disease control possibilities. Introduction of foreign genes could transform a susceptible plant into a resistant one with a distinct possibility that the resistance could be more robust and longer lasting than resistance introduced as a result of conventional breeding programmes. Similarly, genetic modification could raise the efficiency of a plant's natural defence mechanism.

In itself, none of these approaches is likely to provide the ultimate answer. They should all be regarded as potential components of an integrated approach to disease control programmes.

In order to aid such programmes, the use of disease diagnosis and forecasting systems will provide means of directing the use of fungicides more effectively. This will have the dual benefits of increasing the efficiency of treatment and of avoiding unnecessary applications, to the benefit of the environment.

Fungicide resistance is rising in importance and the threat it poses to the agrochemical industry is taken very seriously. New molecules to be introduced to the market are now required to have an anti-resistance strategy available before product introduction. In this way it is hoped to avoid the loss to the user of essential tools used in the fight against plant disease.

In conclusion, fungicide science and associated means of disease control are expanding and must continue to do so if the main objectives of disease control are to be achieved, namely the provision and protection of an adequate food supply for all.

REFERENCES

Albert, G., Curtze, J. & Drandarevski, Ch. A. (1988). Dimethomorph (CME 151), a novel curative fungicide. In *Brighton Crop Protection Conference, Pests and Diseases*, Vol. 1, pp. 17–24. British Crop Protection Council, Farnham, UK.

Ammermann, E., Loecher, F., Lorenz, G., Janssen, B., Karbach, S. & Meyer, N. (1990). BAS 490F – a new broad spectrum fungicide. In *Brighton Crop Protection Conference, Pests and Diseases*, Vol. 2, pp. 407–14. British Crop Protection Council, Farnham, UK.

Ammermann, E., Lorenz, G., Schelberger, K., Wenderoth, B., Sauter, H. & Rentzea, C. (1992). BAS 490F – a broad spectrum fungicide with a new mode of action. In *Brighton Crop Protection Conference, Pests and Diseases*, Vol. 1, pp. 403–10. British Crop Protection Council, Farnham, UK.

Anema, B. P., Bouwman, J. J., Komyoji, T. & Suzuki, K. (1992). Fluazinam: a novel fungicide for use against *Phytophthora infestans* in potatoes. In *Brighton Crop Protection Conference, Pests and Diseases*, Vol. 1, pp. 663–8. British Crop Protection Council, Farnham, UK.

Anke, T., Oberwinkler, F., Steglich, W. & Schramm, G. (1977). The strobilurins – new antifungal antibiotics from the basidiomycete *Strobilurus tenacellus. Journal of Antibiotics*, **30**, 806–10.

Anon (1992). World fungicide market spurred by sales in Western Europe and Far East. *AGROW* **153**, 16.

Beautement, K., Clough, J. M., de Fraine, P. J. & Godfrey, C.R.A. (1991). Fungicidal β-methoxyacrylates: from natural products to novel synthetic agricultural fungicides. *Pesticide Science*, **31**, 499–519.

Becker, W. F., von Jagow, G., Anke, T. & Steglich, W. (1981). Oudemansin, strobilurin A, strobilurin B and myxothiazol: new inhibitors of the bc$_1$ segment of the respiratory chain with an E-β-methoxyacrylate system as common structural element. *FEBS Letters*, **132**, 329–33.

Bocquet, G., Sylvestre, M. & Speich, J. (1994). Le cyprodinil. Fongicide céréales. *Phytoma*, **458**, 53-5.

Brandt, U., Schägger, H. & von Jagow, G. (1988). Characterisation of binding of the methoxyacrylate inhibitors to mitochondrial cytochrome c reductase. *European Journal of Biochemistry*, **173**, 499–506.

Corbett, J. R., Wright, K. & Baillie, A. C. (1984). *The Biochemical Mode of Action of Pesticides*. Academic Press, London.

County NatWest Wood Mackenzie (1991). Agrochemical service database reference vol. 1 & 2.

Davidse, L. C. (1986). Benzimidazole fungicides: mechanism of action and biological impact. *Annual Review of Phytopathology*, **24**, 43–65.

Davidse, L. C. (1987). Biochemical aspects of benzimidazole fungicides – action and resistance. In *Modern Selective Fungicides – Properties, Applications, Mechanisms of Action* (Lyr, H., ed.). pp. 245–57. Longman Group UK Ltd, London, and VEB Gustav Fischer Verlag, Jena.

Davidse, L. C. & Flach, W. (1977). Differential binding of methyl benzimidazol-2-yl carbamate to fungal tubulin as a mechanism of resistance to this antimitotic agent in mutant strains of *Aspergillus nidulans. Journal of Cell Biology*, **72**, 174–93.

Dekker, J. (1982) Counter measures for avoiding fungicide resistance. In *Fungicide Resistance in Crop Protection* (Dekker, J. and Georgopoulos, S. G., eds.). pp. 172–86. Pudoc, Wageningen.

Delp, C. J. (1987) Benzimidazole and related fungicides. In *Modern Selective Fungicides – Properties, Applications, Mechanisms of Action* (Lyr, H., ed.). pp. 233–44. Longman Group UK Ltd, London, and VEB Gustav Fischer Verlag, Jena.

Forsyth, W. (1802). *A Treatise on the Culture and Management of Fruit Trees.* Nichols, London.

Fujimura, M., Kamakura, T., Inoue, H., Inoue, S. & Yamaguchi, I. (1992). Sensitivity of *Neurospora crassa* to benzimidazoles and N-phenylcarbamates: Effect of amino acid substitutions at position 198 in β-tubulin. *Pesticide Biochemistry and Physiology*, **44**, 165–73.

Gaulliard, J. M., Chazalet, M., Saillard, A. & Gouot, J. M. (1994). Efficacy of triticonazole seed treatment against common eyespot in wheat. In *Seed Treatment: Progress and Prospects.* British Crop Protection Council Monograph, **57**, 79–84.

Godwin, J. R., Anthony, V. M., Clough, J. M. & Godfrey, C. R. A. (1992). ICIA 5504: A novel, broad spectrum, systemic β-methoxyacrylate fungicide. In *Brighton Crop Protection Conference, Pests and Diseases*, Vol. 1, pp. 435–42. British Crop Protection Council, Farnham, UK.

Guan, J., Kapteyn, J. C., Kerkenaar, A. & de Waard, M. A. (1992). Characterisation of energy-dependent efflux of imazalil and fenarimol in isolates of *Penicillium italicum* with a low, medium and high degree of resistance to DMI-fungicides. *Netherlands Journal of Plant Pathology*, **98**, 313–24.

Guo, Z., Miyoshi, H., Komyoji, T., Haga, T. & Fujita, T. (1991). Uncoupling activity of a newly developed fungicide, fluazinam [3-chloro-N-(3-chloro-2,6-dinitro-4-trifluoro-methylphenyl)-5-trifluoromethyl-2-pyridinamine]. *Biochimica et Biophysica Acta*, **1056**, 89–92.

Hollomon, D. W. (1993). Resistance to azole fungicides in the field. *Biochemical Society (UK) Transactions*, **21**, 1047–51.

Hollomon, D. W., Butters, J. & Clark, J. (1984). Genetic control of triadimenol resistance in barley powdery mildew. In *British Crop Protection Conference, Pests and Diseases*, Vol. 2, pp. 477–82. British Crop Protection Council, Farnham, UK.

Howard, R. J. & Aist, J. R. (1980). Cytoplasmic microtubules and fungal morphogenesis: ultrastructural effects of methyl benzimidazol-2-yl carbamate determined by freeze-substitution of hyphal tip cells. *Journal of Cell Biology*, **87**, 55–64.

Ishii, H. (1992). Target sites of tubulin-binding fungicides. In *Target Sites of Fungicide Action* (Köller, W., ed.). pp. 43–52. CRC Press, Boca Raton, USA.

James, W. C. (1981). Estimated losses of crops from plant pathogens. In *CRC Handbook of Pest Management in Agriculture*, (Hanson, A. A., ed.). Vol. 1, pp. 79–94.

James, W. C., Teng, P. S. & Nutter, F. W. (1991). Estimated losses of crops from plant pathogens. In *CRC Handbook of Pest Management in Agriculture* (Hanson, A. A., ed.). Vol. 1, pp. 15–51.

Jespers, A. B. K., Davidse, L. C. & de Waard, M. A. (1993). Biochemical effects of the phenylpyrrole fungicide fenpiclonil in *Fusarium sulphureum* (Schlect). *Pesticide Biochemistry and Physiology*, **45**, 116–29.

Kapetyn, J. C. (1993). Biochemical mechanisms involved in selective fungitoxicity of fungicides which inhibit sterol 14α-demethylation. PhD Thesis, University of Wageningen.

Kapetyn, J. C., Pillmoor, J. B. & De Waard, M. A. (1992). Biochemical mechanisms involved in selective fungitoxicity of two sterol 14α-demethylation inhibitors, prochloraz and quinconazole: accumulation and metabolism studies. *Pesticide Science*, **36**, 85–93.

Kapetyn, J. C., Milling, R. J., Simpson, D. J. & De Waard, M. A. (1994). Inhibition of sterol biosynthesis in cell-free extracts of *Botrytis cinerea* by prochloraz and prochloraz analogues. *Pesticide Science*, **40**, 313–19.

Kilmartin, J. V. (1981). Purification of yeast tubulin by self-assembly *in vitro*. *Biochemistry*, **20**, 3629–33.

Kittleson, A. R. (1952). A new class of organic fungicides. *Science*, **115**, 84–96.

Klinkowski, M. (1970). Catastrophic plant diseases. *Annual Review of Phytopathology*, **8**, 37–60.

Köller, W., ed., (1992). *Target Sites of Fungicide Action*. CRC Press, Boca Raton, USA.

Krämer, S. N. (1963). *The Sumerians*, pp. 340–42. University of Chicago Press, Chicago, USA.

Krämer, W. (1983). Fungicides and bactericides. In *Chemistry of Pesticides* (Büchel, K. H., ed.). pp. 227–321. John Wiley & Sons, New York.

Kuhn, P. J., Pitt, D., Lee, S. A., Wakley, G. & Sheppard, A. N. (1991). Effects of dimethomorph on the morphology and ultrastructure of *Phytophthora*. *Mycological Research*, **95**, 333–40.

Large, E. C. (1940). *The Advance of the Fungi*. Dover Publications, New York.

Leroux, P. & Gredt, M. (1982). Negatively correlated cross resistance between benzimidazole fungicides and *N*-phenylcarbamate herbicides. In *Systemische Fungizide und Antifungale Verbindungen. 6th International Symposium*. (Lyr, H. & Polter, C. eds.). Vol. 1, pp. 297–301. Akademie-Verlag, Berlin.

Milling, R. J., Richardson, C. J. & Daniels, A. (1993). Pyrimethanil inhibits hydrolytic enzyme secretion by *Botrytis* spp. and prevents lysis of host cells. In *Sixth International Congress of Plant Pathology*, *(Abstracts)*. National Research Council of Canada, Ottawa.

Neumann, G. L., Winter, E. H. & Pittis, J. E. (1992). Pyrimethanil: a new fungicide. In *Brighton Crop Protection Conference, Pests and Diseases*, Vol. 1, pp. 395–402. British Crop Protection Council, Farnham, UK.

Noguchi, H., Kato, T., Takahashi, J., Ishiguri, Y., Yamamoto, S. & Kamoshita, K. (1983). Alkyl *N*-3,4-dialkoxy-phenyl-carbamate(s) and analogues useful as broad spectrum fungicides with low phytotoxicity. *European Patent* 078 663 A2.

Nyfeler, R. & Ackermann, P. (1992). Phenylpyrroles, a new class of agricultural fungicides related to the natural antibiotic pyrrolnitrin. In *Synthesis and Chemistry of Agrochemicals III* (Baker, D. R., Fenyes, J. G. & Steffens, J. J., eds.). ACS Symposium Series **504**.

OTA (1979). Office of Technology Assessment: Pest Management Strategies in Crop Protection, Vol. 1, October 1979.

Raychaudhuri, S. P. ed., (1964). *Agriculture in Ancient India*. pp. 84–98. Indian Council Agricultural Research, New Delhi.

Riehm, E. (1913). Prufung einiger Mittel zur Bekämpfung des Steinbrandes. *Mitteilungen der Kaiserlich Biologischen Anstalt fur Land-u-Forstwirtschaft*, **14**, 8–9.

Russell, P. E., Percival, A., Coltman, P. M. & Green, D. E. (1992). Fluquincona-zole, a novel broad-spectrum fungicide for foliar application. In *Brighton Crop Protection Conference, Pests and Diseases*, Vol. 1, pp. 411–18. British Crop Protection Council, Farnham, UK.

Russell, P. E. (1993). Fungicide Resistance Action Committee 1993. *Pesticide Outlook*, **4**, 21–3.

Russell, P. E. (1994). Fungicide Resistance Action Committee. In *GIFAP Resistance Newsletter No. 2*. GIFAP, Brussels.

Sanders, T. W. (1910). *Garden Foes*. Collingridge, London.

Schumann, G. L. (1991). *Plant Diseases, their Biology and Social Impact*. APS Press, St. Paul, Minnesota, USA.

Tisdale, W. H. & Williams, I. (1934). Disinfectant and fungicide. US Patent 1 972 961.

Vanden Bossche, H. (1988). Mode of action of pyridine, pyrimidine and azole antifungals. In *Sterol Biosynthesis Inhibitors, Pharmaceutical and Agrochemical Aspects* (Berg, D. & Plempel, M., eds.). pp. 79–119. Ellis Horwood Ltd, Chichester, UK.

Van Tuyl, J. M. (1977). Genetics of fungal resistance to systemic fungicides. In *Mededelingen Landbouwhogeschool Wageningen*, Vol. 77-2, pp. 1–136.

Viennot-Bourqin, G. & Lafon, R. (1985). La naissance de la Bouillie Bordelaise. In *Fungicides for Crop Protection, 100 Years of Progress. BCPC Monograph*, **31**, 3–10.

Waldron, L. R. (1935). Stem rust epidemics and wheat breeding. *North Dakota Agriculture Experiment Station Circular*, **57**, 12.

Worthing, C. R. & Hance, R. J. (1991). *The Pesticide Manual*, 9th edn. British Crop Protection Council, Farnham, UK.

QUINOLONES: SYNTHETIC ANTIBACTERIAL AGENTS

DANIEL T. W. CHU and LINUS L. SHEN

Anti-Infective Research Division, Abbott Laboratories, Abbott Park, Illinois 60064-3500, USA

INTRODUCTION

The discovery and development of a useful antimicrobial agent depends to a great extent on the search for analogues of a lead compound. Systematic intellectual use of this approach is the foundation of modern medicinal chemistry. The term 'structure–activity relationship' can be found in many medicinal scientific publications. The application of analogue preparation has dominated drug research, especially in the antibacterial field, for many years. For example, in the β-lactam antibiotic area, introduction of a new side-chain on to the β-lactam molecule produced many beneficial effects such as improved potency, activity spectrum and pharmacokinetic properties, and enhanced stability to acid or β-lactamases. Further chemical modifications have produced more than 5000 semi-synthetic β-lactam molecules, including the development of many penicillins, cephalosporins, carbapenems and monobactams for clinical uses against bacterial infections.

For many years, new antibacterial lead structures have originated from microbiological fermentations or isolation of natural products and plant extracts. The early antibiotic β-lactams, macrolides, aminoglycosides and tetracyclines were isolated from moulds or fungi. The identification of new lead compounds from natural sources dominated most of the antibacterial discovery effort. Synthetic antibacterial agents play a minor role in combating bacterial infections. Recently, with the discovery of quinolone antibacterial agents, the emphasis in antibacterial research has shifted to the utilization of totally synthetic compounds as leads. In fact, the use of synthetic chemicals for antimicrobial chemotherapy is not new. Sulphonamides were the first synthetic agents effective in the treatment of bacterial infections, signalling the beginning of the present chemotherapeutic era. Prontosil, the earliest of the sulphonamides, discovered by Domagk (1935) is a diazo compound having sulphanilamide as its active moiety. Among the newer totally synthetic antibacterial agents, the fluoroquinolones, because of their excellent antibacterial spectrum and potency, provide the greatest impact for the treatment of bacterial infections and fuel the continuing use of synthetic compounds. Many excellent reviews on microbiological and clinical aspects (Fernandes, 1988; Høiby, 1986; Hooper & Wolfson, 1985; Janknegt, 1986; Maple, Brumfitt & Hamilton-Miller, 1990; Neu, 1987, 1989; Paton & Reeves, 1988; Walker & Wright 1991; Wolfson & Hooper, 1985) and structure–activity relationships (Chu & Fernandes, 1989; Chu *et al.*, 1985; Cornett & Wentland, 1986; Domagala, 1994; Mitscher *et al.*, 1988; Mitscher, Devasthale & Zavod, 1990; Radl, 1990; Schentag & Domagala, 1985) of quinolones have been published over the past several years. Recently, several chapters of books on the chemistry, microbiology, modes of action, pharmacology and other clinical aspects of quinolones have also been written (Andriole, 1988; Chu & Fernandes, 1991; Crumplin, 1980; Fernandes, 1989; Rosen, 1990; Siporin, Heifetz & Domagala, 1990; Wolfson & Hooper, 1989*a,b*). Hence, in this review it is not our intention to give a comprehensive coverage of the topic. We should like, however, to provide a general overview of this class of antibacterial agents with particular emphasis on recent developments. Since many quinolone derivatives having potent antibacterial activities have been described in the literature, it may be difficult to mention all of them in this presentation. Hence, only those quinolones that have been approved for clinical use with significant value and those under commercial development having major therapeutic potential as well as those that may have an impact on structure–activity relationship are described here.

OLDER QUINOLONES

The research on quinolone antibacterial agents began more than 20 years ago with the isolation of 6-chloro-1-ethyl-1,4-dihydro-4-oxoquinoline-3-carboxylic acid (Fig. 1). It was isolated as a byproduct from the mother liquors in the commercial manufacture of chloroquine, an antimalarial drug, and was found to be active against Gram-negative bacteria. Although the biological activity was low, because of its structural simplicity, the medicinal chemist George Lesher was able to prepare many more potent analogues and identified nalidixic acid (Fig. 1) for clinical development (Lesher *et al.*, 1962). It was introduced into clinical practice in the early 1960s for the treatment of urinary tract infections as it gave a high concentration in urine upon oral administration. However, there were needs for further improvement because of its moderate activity against Gram-negative bacteria, lack of activity against Gram-positives and pseudomonads, as well as anaerobes. Its low blood level prevents it being used for systemic infections.

Nalidixic acid is the prototype of the highly potent class of antibacterial agents called quinolones, a new class of totally synthetic broad-spectrum antibacterial agents having at least a bicyclic structure. These agents generally consist of a 1-substituted-1,4-dihydro-4-oxopyridine-3-carboxylic

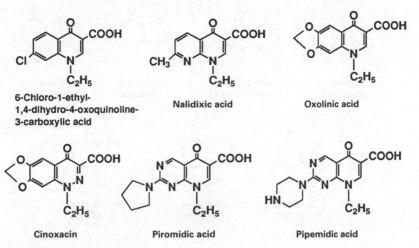

Fig. 1. First-generation quinolones.

acid moiety combined with an aromatic or heteroaromatic ring fused at the 5- and 6-positions as shown in Fig. 2. Although there are many different ring systems associated with more than 5000 reported quinolones, those having significant antibacterial activity can be represented by nine basic ring skeletons. The complexity of the systematic nomenclatures of these series of compounds prompted Smith (1984) to propose a simple solution for the naming of these ring skeletons. 4-Quinolone is used as a generic name for the 4-oxo-1,4-dihydro-quinoline skeleton. Thus, nalidixic acid is a 8-aza-4-quinolone. Derivatives with an extra ring or rings attached to the basic skeleton at positions other than 2 and 3 are considered as 4-quinolones. Fig. 3 illustrates the use of the Smith method in naming these nine ring skeletons. Many clinicians and microbiologists, however, would prefer to name all of them as quinolones. Because the newer very potent quinolones possess a fluorine atom at the 6-position, they may also be referred to as fluoroquinolones.

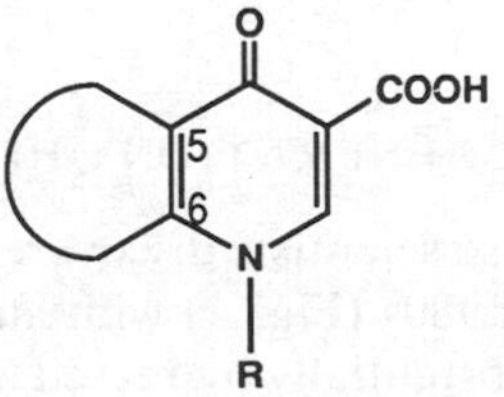

Fig. 2. General structure of quinolones.

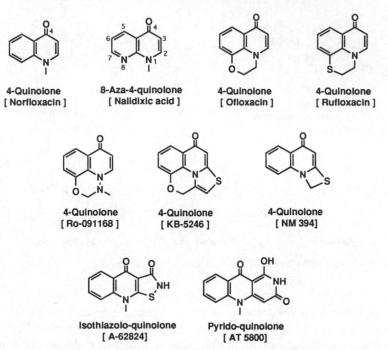

Fig. 3. Different ring skeletons of quinolones. An example of an antibacterial compound is given in brackets for each ring skeleton.

Nalidixic acid, and congeners such as oxolinic acid, cinoxacin, piromidic acid and pipemidic acid, with which it is used clinically, are considered as first generation quinolones (Fig. 1). As a class, they possess moderate activity against susceptible bacteria, but do not have favourable pharmaco-kinetic properties, providing unsatisfactory blood levels for the treatment of systemic infections such as pneumonia or skin infections. They are often used for the treatment of urinary tract infections caused by susceptible organisms. Side-effects on the central nervous system and gastrointestinal intolerance are common when they are given at high doses to patients. Bacterial resistance may develop fairly rapidly during therapy. All these factors make these early quinolones unattractive for extensive clinical use.

SECOND-GENERATION QUINOLONES

In the early 1980s, Japanese scientists discovered the fluorinated quinolone norfloxacin (Koga *et al.*, 1980) (Fig. 4) with antibacterial activity, both in potency and spectrum, substantially increased and expanded over the first-generation quinolones. In addition to having excellent activity against

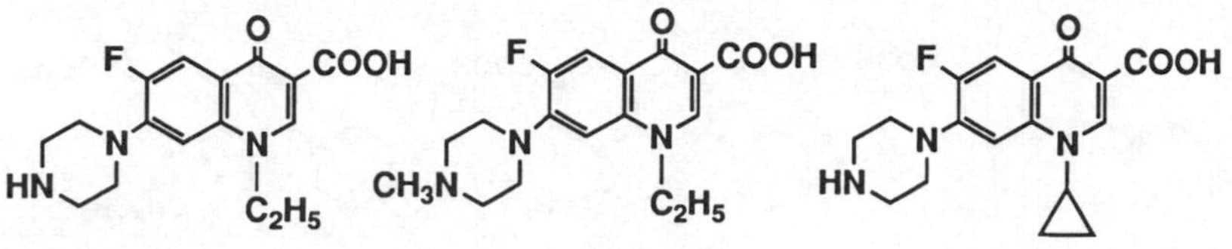

Fig. 4. Second-generation quinolones.

Gram-negative bacteria, it also has significant activity against Gram-positives. This compound represents a major therapeutic advance having very good clinical efficacy for urinary tract infections. However, its pharmacokinetic properties are not good enough for systemic use, especially against lower respiratory tract infections.

Pefloxacin (Fig. 4) was prepared in an attempt to improve the pharmacokinetic profile of norfloxacin (Goueffon et al., 1981). It is converted to norfloxacin to some extent in vivo, although it is slightly less biologically active than norfloxacin. Since the discovery of norfloxacin, there have been intense research efforts to find more potent and useful quinolones for clinical development, and considerable advances in their therapeutic use have been made. The initial success in treating urinary tract infections has been expanded to include various systemic infections such as those of the lower and upper respiratory tracts, skin and soft tissue, eye and bone, as well as sexually transmitted diseases, and ophthalmological infections (Paton & Reeves, 1988; Shah, 1987; Siporin et al., 1990; Wolfson & Hooper, 1985, 1989b). Because of the increase in clinical uses and expanded antibacterial spectrum, these newer quinolones are generally grouped together as second-generation quinolones.

Other than norfloxacin and pefloxacin, the second-generation quinolones currently in clinical use include ciprofloxacin (Wise, Andrews & Edwards, 1983), enoxacin (Matsumoto et al., 1984), ofloxacin (Sato et al., 1982), fleroxacin (Chin, Brittain & Neu, 1986), lomefloxacin (Hokuriku Pharmaceutical Co., 1986), tosufloxacin (Chu et al., 1986; Soejima & Shimada, 1989) and sparfloxacin (Miyamoto et al., 1990) (Fig. 5). These newer compounds have a fluorine atom at the C-6 position of the quinolone, naphthyridine or benzoxazine ring system. Their expanded antibacterial spectrum also includes activity against Mycobacterium tuberculosis and anaerobic bacteria. They possess superior pharmacokinetic profiles over the older quinolones (for example, higher oral bioavailability and long serum half-life), and they can be administered both orally and parenterally. Because of their high potency and efficacy in oral mode of administration,

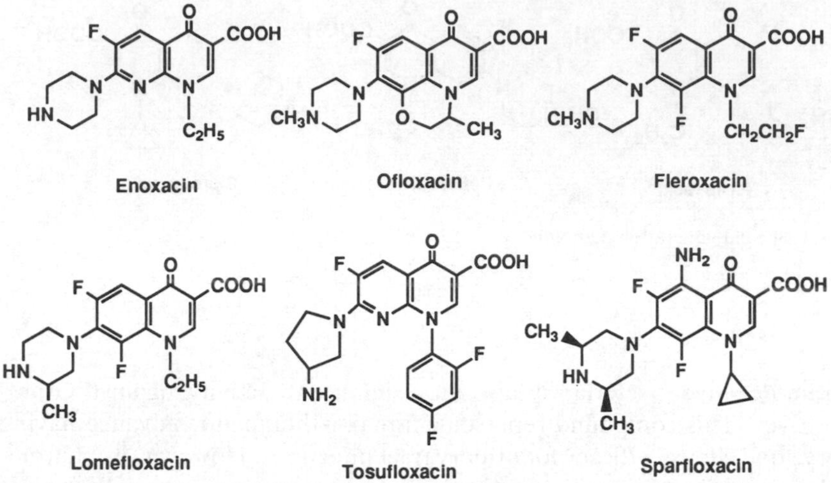

Enoxacin   Ofloxacin   Fleroxacin

Lomefloxacin   Tosufloxacin   Sparfloxacin

Fig. 5. Fluorinated second-generation quinolones.

they compare very favourably with the third- and fourth-generation cephalosporins. Currently, they are well accepted in clinical practice as one of the dominant classes of antibacterial agents against a number of bacterial infections.

Antibacterial activity

The second-generation quinolones possess a broad spectrum of antibacterial activity against a wide range of clinically important pathogens. In addition to being active against Gram-negative and Gram-positive aerobes, a few quinolones of this class also possess antibacterial activity against anaerobic organisms. They are rapidly bactericidal with a minimum bactericidal concentration (MBC) that is either the same as or within two-fold of the minimum inhibitory concentration (MIC) (Kenny *et al.*, 1989; Krausse & Ullmann, 1988; Tjiam *et al.*, 1986). They are extremely active against many common Gram-negative bacteria, especially enterobacteriaceae, enterogens and respiratory pathogens; they also have reasonable activity against staphylococci and streptococci, and their activity against pseudomonads is acceptable but not exceptional. Among the quinolones used clinically, only tosufloxacin and sparfloxacin possess reasonable activity against anaerobes. Of the common genital pathogens, the second-generation quinolones are very active against *Neisseria gonorrhoeae* and *Haemophilus ducreyi*, but are only moderately active against mycoplasma, ureaplasma and chlamydia. They are found to have activity against *Rickettsia conovic, Mycobacterium tuberculosis* and *M. leprae*, as well as *M. avium* complex. The microbiologi-

Table 1. In vitro *Antibacterial activity of several fluoroquinolones*

Organism	MIC$_{90}$ (μg/ml)						
	CIP	SPA	TOSU	OFL	ENO	NOR	LOM
Gram-positive aerobes							
S. aureus	0.5	0.06	0.016	0.5	2	2	2
Coagulase-neg. Staph.	0.5	0.12	0.03	0.5	1	2	1
S. pneumoniae	2	0.5	0.06	4	16	8	8
Gram-negative aerobes							
E. coli	0.06	0.03	0.03	0.125	0.25	0.125	1
Klebsiella spp.	0.125	0.25	0.125	0.5	1	0.5	1
E. cloacae	0.125	0.13	0.20	0.5	0.5	0.25	0.5
Serratia spp.	0.25	2		2	4	2	2
P. mirabilis	0.06	0.5	0.25	0.5	0.5	0.25	1
H. influenzae	0.06	0.06	0.03	0.12	0.5	5	0.12
M. catarrhalis	0.5	0.03	0.03	1	0.25		0.25
P. aeruginosa	1	4	1	8	8	8	12.5
Legionella spp.	2	0.06		3	8	25	
C. trachomatis	1		0.25	0.5	8	16	
M. hominis	0.5	0.03		2		8	
Anaerobes							
B. fragilis	8	4	2	8	32	32	8
Peptoco. peptostrep.	2	1	2	4	8	16	
Clostridium spp.	8	4	0.5	8		32	

Abbreviations: MIC$_{90}$, minimum drug concentration inhibiting growth of 90% of the strains tested; CIP, Ciprofloxacin: SPA, sparfloxacin: TOSU, tosufloxacin; OFL, ofloxacin; ENO, enoxacin; NOR, norfloxacin; LOM, lomefloxacin. Compiled from published data in Barry & Fuchs, 1991; Fernandes *et al.*, 1986; Hardy *et al.*, 1987; Hooper, 1986; Kenny *et al.*, 1989; Krausse & Ullmann, 1988; Nicolle *et al.*, 1988; Pernet, 1989; Tjiam *et al.*, 1986; Venezia *et al.*, 1988; and Wolfson & Hooper, 1989*b*.

cal activity of a few representative quinolones including ciprofloxacin is shown in Table 1.

Pharmacological and clinical aspects

The second-generation quinolones possess favourable pharmacokinetics over the older quinolones (Andriole, 1988; Leigh *et al.*, 1988; Moellering & Neu, 1989; Mouton & Leroy, 1991; Shah, 1991; Siporin, 1989; Speller & Wise, 1988; Wolfson & Hooper, 1989*b*). Good oral bioavailability of the fluoroquinolones is their primary advantage over other antibacterial agents, ranging from 30% for norfloxacin to nearly 100% for ofloxacin, pefloxacin, fleroxacin, lomefloxacin and enoxacin. Peak serum concentrations are normally achieved within two hours after oral dosing. Although tosufloxacin provides 1 μg/ml as the peak serum level, most of the second-generation fluoroquinolones give much higher peak serum concentrations. For example, a peak serum concentration of 4.36 μg/ml for fleroxacin with an

Table 2. *Comparative pharmacokinetics of several fluoroquinolones*

Fluoroquinolones	Oral dose (mg)	C_{max} (μg/h/ml)	AUC (mg/h/l)	$T_{1/2}$ (h)	Urinary recovery (%/24h)
Ciprofloxacin	500	2.3	9.9	3.9	30.6
Sparfloxacin	400	1.6	32.3	17.6	8.8
Tosufloxacin	300	0.8		3.9	25.0
Ofloxacin	600	10.7	57.5	7.0	73.0
Enoxacin	600	3.7	28.8	6.2	61.2
Norfloxacin	400	1.45	5.4	3.2	27.0
Lomefloxacin	200	1.7	12.9	8.5	79.5
Fleroxacin	400	4.4	78.3	12.0	42.5

Abbreviations: C_{max}, maximum drug concentration in serum; AUC, area under the curve; $T_{1/2}$, half-life.

oral dose of 400 mg was obtained (Karabalut & Drusano, 1993). Variable oral absorption has been observed with quinolones having low water solubility. While ofloxacin is highly water soluble, the solubility for ciprofloxacin at physiological pH is low giving a high degree of variability in absorption. The comparative pharmacokinetic profiles of several clinically important quinolones are given in Table 2.

The quinolones have low protein binding to human serum albumin, ranging from 13.8% for norfloxacin (Bergan, 1988) to 32% for enoxacin (Pocidalo, Vachon & Regnier, 1985). The serum half-lives of the second-generation quinolones are relatively long, ranging from 4 h for norfloxacin to 17 h for sparfloxacin, so that an oral clinical dosage regimen of 1–2 times daily is possible. In general, quinolones with a methyl-substituted piperazine-1-yl group at the C-7 position are long-acting antibacterial agents having high serum level and long half-lives (Granneman *et al.*, 1991; Shiba *et al.*, 1988; Wise & Lockley, 1988; Wolfson & Hooper, 1989*b*). Biotransformation rates of various quinolones are quite different: for example, there is more than 80% metabolism of pefloxacin, whereas <10% of ofloxacin and lomefloxacin are metabolized. While most of the quinolones are excreted by the kidney, a few others such as enoxacin, tosufloxacin and sparfloxacin can also be excreted in the bile to a small extent. Pefloxacin and the animal antibacterial drug difloxacin have been shown to be excreted in the bile with subsequent re-absorption from the gut.

Quinolones, once absorbed, are readily distributed to the tissues and penetrate into cells and body fluids. The penetrative capacities of quinolones into lung tissue, tonsils, prostatic tissue, bones, blister fluid and sinus secretions are excellent, ranging from 1-4 times the serum concentration. Thus, their concentrations in serum and tissues are generally sufficiently high enough above the MIC_{90} for many Gram-negative and Gram-positive aerobes, giving good clinical efficacy against simple, serious and chronic bacterial systemic infections.

Intravenous formulations for several fluoroquinolones have been developed for ciprofloxacin, pefloxacin, and ofloxacin. The successful development of intravenous dosage for each individual quinolone depends on their water solubilities.

The second-generation quinolones, in contrast to the older quinolones, are found to be highly effective for the treatment of both complicated and uncomplicated urinary tract infections. They are considered as the drug of choice for the management of gastrointestinal tract infections, due to their potent activity against Gram-negative bacteria, particularly multiply-resistant *Shigella* and *Salmonella* spp.. With the exception of sparfloxacin and tosufloxacin, the second-generation quinolones have moderate activity *in vitro* against *Streptococcus pneumoniae* and enterococci, providing borderline efficacy against infections with these bacteria. Since the concentration of quinolones in the prostatic tissue and fluid can be maintained at sufficiently high levels, they are highly effective for both acute and chronic prostatitis. For pelvic inflammatory disease, the second-generation quinolones (with the possible exception of tosufloxacin and sparfloxacin) should not be used as a single agent for the treatment of this kind of infection. They are very effective for the treatment of acute pyelonephritis due to the fact that high concentrations can be achieved in both the kidney and urine. They have also been demonstrated to be highly efficacious for the treatment of sexually-transmitted diseases and can be used in the single-dose treatment of uncomplicated gonococcal urethritis and cervicitis. Multiple doses are needed for the effective treatment of genital infections with *Chlamydia trachomatis* or *Urealplasma urealyticum* and single-dose treatment for these infections is not recommended.

The clinical and bacteriological cure rates for quinolone therapy of respiratory tract infections are as high as 90%. Yet, pneumonia due to *Streptococcus pneumoniae* is not effectively treated with quinolones owing to their less potent activity *in vitro* against this organism. Fluoroquinolones are the favourable drugs for the treatment of Gram-negative bacillary osteomyelitis due to their high penetration into bone as well as their ease of administration. They have also been shown to be effective for the treatment of skin and skin-structure infections.

A number of side-effects with varying frequency and importance have been found to be associated with quinolone therapy. A few cases of crystalluria are found to occur with those quinolones having poor water solubility. With the exception of lomefloxacin and ofloxacin, many clinically-used quinolones are found to interact with the anti-asthmatic compound theophylline and with caffeine by interfering with their clearance. Enoxacin and a few other quinolones interact with non-steroid anti-inflammatory drugs (NSAIDs) which has lead to a few cases of convulsions.

Patients on quinolone therapy may experience CNS side-effects such as dizziness and insomnia to varying degrees. The incidence of gastrointestinal disturbances, such as nausea and diarrhoea, occur with a frequency that is similar to that of other antibiotic therapy. Interactions of quinolones with antacids are common, leading to reduced absorption of the drug. More importantly, however, co-administration of cimetidine or other related compounds with quinolones does not affect the absorption, peak serum-level and other pharmacokinetic parameters of the fluoroquinolones. An area of concern with quinolones is the cartilage changes seen in experimental young animals on quinolone treatment. Because of the potential to cause cartilage damage, fluoroquinolones are restricted for routine use in paediatrics and they are not recommended for use in pregnant women. Nowadays, fluoroquinolones are well accepted in clinical practice for a broad variety of clinical indications, including infections of the respiratory tract, urinary tract, bone, skin and skin structure as well as sexually-transmitted diseases.

DNA GYRASE – TARGET ENZYME OF QUINOLONES

The discovery of DNA gyrase in 1976 (Gellert *et al.*, 1976) marked the beginning of a new era in quinolone development. The identification of this enzyme as the target of nalidixic acid and oxolinic acid (Gellert *et al.*, 1977; Sugino *et al.*, 1977) generated elevated interest not only in the structure and activity of DNA gyrase and DNA topology but also in the molecular mode of action of quinolones.

DNA gyrase is an essential type II DNA topoisomerase which exists only in bacteria (Cozzarelli, 1980; Gellert, 1981; Reece & Maxwell, 1991; Sutcliffe, Gootz & Barrett, 1989; Wang, 1982). The enzyme has a tetrameric structure composed of two A-subunits, the 97 kilodalton proteins encoded by the *gyrA* gene, and two B-subunits, the 90 kilodalton proteins encoded by the *gyrB* gene. The functions of the A- and B-subunits are the DNA cutting/resealing and the ATP-driven energy-transduction processes, respectively. The enzyme catalyses the supercoiling reaction of the covalently closed-circular (ccc) DNA through a concerted strand breaking–passing–resealing process. These sequential enzyme-catalysed reactions result in an underwound DNA molecule which spontaneously adopts a negatively supercoiled form. Cozzarelli and his coworkers (Brown & Cozzarelli, 1979; Morrison & Cozzarelli, 1981) proposed a DNA supercoiling model consisting of the following four steps: (i) binding of gyrase to DNA substrate to stabilize a positive DNA node; (ii) cleavage of DNA at 4-base pair staggered sites at the node, forming covalent linkages between a tyrosine group on the gyrase A-subunit and the 5′-end of the DNA chain; (iii) passage of the intact DNA segment at the node through the opened DNA gate, thus inverting the sign of the node; and (iv) resealing the DNA break and completing a supercoiling run. Genes of both subunits of the *E. coli* (Hallett *et al.*, 1990) and *S. aureus*

Table 3. *Antibacterial activity and the gyrase supercoiling inhibition activity of several selected quinolones*

Compound number	Quinolone	[a]MIC (μg/ml)	[b]IC$_{50}$ (μg/ml)
1	Ciprofloxacin	0.01	0.9
2	Du-6859	0.013	0.18
3	Temafloxacin	0.02	1.1
4	Sparfloxacin	0.025	0.27
5	Norfloxacin	0.05	1
6	Pefloxacin	0.05	1.3
7	Ofloxacin	0.1	2
8	Difloxacin	0.1	3
9	Oxolinic acid	0.15	3.1
10	Nalidixic acid	1.56	26

[a]Minimum inhibitory concentration for the growth of *E. coli* H560 cells, determined by the twofold agar dilution method. Data were determined at Abbott.
[b]IC$_{50}$ is defined as the drug concentration causing 50% inhibition of the supercoiling activity of *E. coli* DNA gyrase. Data for DU-6859 and sparfloxacin are obtained from published data (Hoshino *et al.*, 1992). Others are taken from Shen *et al.* (1989c).

(Brockbank & Barth, 1993) DNA gyrase have been cloned and overproduced. So far, the only part of the enzyme structure solved by X-ray crystallography is the 43 kD N-terminal portion of the B-subunit (Wigley *et al.*, 1991).

Quinolone-type drugs such as oxolinic acid trap the catalytic intermediate (that is, the DNA gate) by stabilizing the enzyme–DNA complex, thus preventing enzyme turnover. Moreover, the stabilized complex, termed a 'cleavable complex', was believed to trigger a hitherto unknown cellular process leading to cell death (Drlica, 1984; Kreuzer & Cozzarelli, 1979). The stabilized intermediate complex has both DNA strands cleaved, and this phenomenon can be demonstrated by the addition of a protein denaturant that converts the circular DNA to the linear form with the A-subunit attached covalently to the revealed 5'-ends (Morrison & Cozzarelli, 1979). These observations provided key evidence that the target of the drug is the A-subunit of DNA gyrase.

Quinolones are known to be extremely potent bactericidal agents with MIC values that are normally one or two orders of magnitude lower than the inhibitory activities against the purified enzyme. Table 3 and Fig. 6 show the correlation of the antibacterial activity and the DNA gyrase inhibition activity of ten quinolones. As clearly seen in Fig. 6, the two activities showed

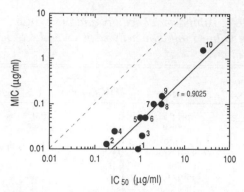

Fig. 6. Correlation of the antibacterial activity and the gyrase supercoiling inhibition activity of ten quinolones. Data used for the plot are listed in Table 3, which also gives the definitions of MIC and IC_{50}. The solid line is the first-order regression line with the correlation coefficient (r value) shown. The dashed line is the 1:1 relationship diagonal line. The numbers near each symbol are the compound numbers listed in Table 3.

good correlation, with the antibacterial concentrations on the average about 20-fold lower than the anti-gyrase concentrations.

At least three factors may contribute to this phenomenon: (1) the drug is concentrated in the cell by an active transport process, (2) a small signal produced at sub-IC_{50} concentration is sufficient and can be amplified by some unknown cellular process leading to the cell killing, and (3) there are secondary targets in the cell responsible for the lethality. For each quinolone, one or more of the above factors may contribute to its bactericidal action against different bacterial species including resistance mutants.

Molecular mechanism of quinolone inhibition of DNA gyrase

The mode of inhibition mechanism of quinolone to DNA gyrase at molecular level has not been elucidated completely. In 1989, Shen and his coworkers proposed the cooperative quinolone–DNA binding model as illustrated in Fig. 7 (Shen *et al.*, 1989e). This model proposed that the drug-binding site is induced during the gate-opening step of the supercoiling process. The separated short DNA strands between the 4-base-pair staggered cuts may form a denatured DNA bubble that is a preferred site for the drug to bind in a multi-dimensional fashion. Quinolones, with their simple molecular structure and few functional binding groups, acquire high binding affinity via self-association of the drug molecules. Two types of interactions have been proposed on the basis of nalidixic acid crystal structures: the π–π stacking between the quinolone rings and the tail-to-tail hydrophobic interactions between the N-1 substitution groups (Fig. 7, right panel). Such interactions not only enhance the binding strength but also provide flexible adjustment in fitting the receptor conformation in the binding pocket.

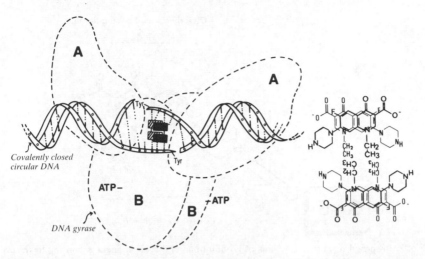

Fig. 7. Quinolone-DNA co-operative binding model for the inhibition of DNA gyrase. Quinolone molecules (filled and slashed rectangles at the centre of the diagram) bind to a gyrase-induced DNA site during the intermediate gate-opening step of the DNA supercoiling process. Binding of the drug is through hydrogen-bonding (indicated by dotted lines) to the unpaired bases. The stacking of the rectangles mimics the mode of drug self-association shown at the right. The tyrosine-122 residues of the gyrase A-subunits are covalently linked transiently to the 5'-end of the DNA chain (Horowitz & Wang, 1987) during the supercoiling catalysis. Subsequent opening of the DNA chains, between the four base-pair staggered cuts, results in a locally denatured DNA bubble which is proposed to be the site for drug binding. The binding of the drug molecules thus stabilizes the enzyme–DNA intermediate. When relaxed DNA substrate (represented by the short DNA segment in the diagram) is used, ATP is required for the induction of the drug binding site. Dashed curves denote the shape of the DNA gyrase, a tetrameric topoisomerase composed of two A-subunits and two B-subunits, as revealed by the electron microscopic image of the *M. Luteus* enzyme (Kirchhausen *et al.*, 1985). Reprinted partly from Shen *et al.* (1989e) with permission.

The proposed co-operative binding model was based on the radioligand ultrafiltration binding studies carried out using pure gyrase, pure DNA and homopolymers (Shen & Pernet, 1985; Shen *et al.*, 1989*b,d*). Two unique characteristics of quinolone action at the molecular level are its lack of binding to the target enzyme and its capability to bind preferentially to single-stranded rather than double-stranded DNA. From this point of view, quinolones are essentially DNA-targeted drugs. However, the participation of the enzyme in creating the binding site and the capability of the drug to saturate the binding pocket in a co-operative manner constitute the basis of the drug's specificity and superior potency.

The DNA-binding property of quinolones has raised concerns about possible side-effects or toxicity. At higher drug concentrations, non-specific drug binding does occur. In an earlier study, we showed that this binding may be antagonized by cellular DNA-binding components such as polyamines or histones (Shen *et al.*, 1989*a,c*). The antagonizing effect is particu-

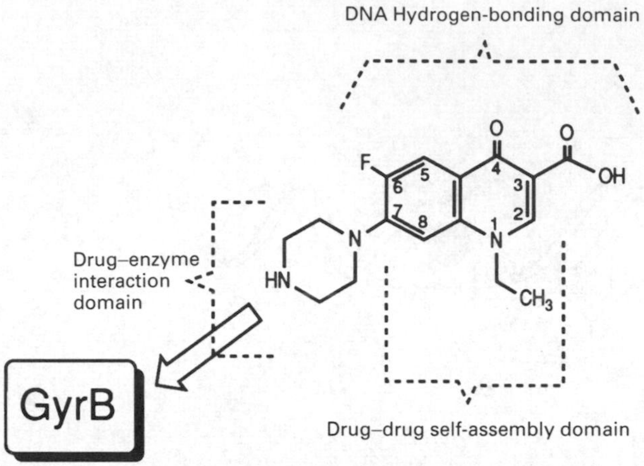

Fig. 8. Proposed functional domains of a quinolone antibacterial agent according to the co-operative drug–DNA binding model (Shen *et al.*, 1989*e*). The structure of norfloxacin is shown as an example. The C-7 substituents are believed to interact with the gyrase-B-subunit for further strengthening the binding (see text for explanation).

larly effective with spermine which exists in mammalian but not in bacterial cells. These studies implied that the DNA-unwinding effect observed *in vitro* may not take place in host cells.

The interaction between DNA and the drug may be mediated by magnesium ions (Palu' *et al.*, 1992). In this mode of drug binding, the quinolone binds to sites on DNA via magnesium-bridges formed between the phosphate group on DNA and the carbonyl and carboxyl groups of norfloxacin. Stacking the DNA base in a single-stranded region with the quinolone ring provide additional binding strength. The role of metal ions in nalidixic acid binding to DNA has been proposed to be catalytic (Crumplin, Midgley & Smith, 1980). This model was based on the finding that copper ions were required for nalidixic acid binding to DNA but were not retained in the drug–DNA complex and therefore not involved in the direct binding. Whether magnesium ions play a similar catalytic role in norfloxacin binding is unknown.

The co-operative quinolone–DNA binding model agrees in general with the current structure–activity relationships of quinolones (Chu & Fernandes, 1989; Mitscher, Zavod & Sharma, 1989, 1990; Mitscher, Devasthale & Zavod, 1990; Rosen, 1990; Wentland, 1990) and has provided a general guideline for efforts to synthesize active quinolones with novel structures (Chu *et al.*, 1988; Hubschwerlen *et al.*, 1992; Mitscher *et al.*, 1991). The model suggests that there are three functional domains on the quinolone molecule (Fig. 8): the DNA binding domain, the drug self-association domain and the drug–enzyme interaction domain. The hypothetical quinolone–enzyme interaction domain, namely the space occupied by the substituents at C-7, was proposed to be a possible location of functional groups for further strengthening drug binding to the complex, and this proposal is consistent with the observed potency increase of new amphoteric quinolones with piperazinyl and other basic substituents at this position.

RECENT DEVELOPMENTS

Resistance

The development of bacterial resistance to naturally-derived antibiotics is thought to arise by the acquisition of resistance genes from an exogenous source. These resistance genes encode enzymes that may chemically modify and inactivate the antibiotic. They may also modify the antibiotic target either by generating a new functional enzyme that may have low affinity for the antibacterial agent or by increased production of the original target enzyme. Resistant bacteria may develop an efflux system to get rid of the antibiotic. The origins of these resistance genes found in soil bacteria are thought to have evolved over hundreds of millions of years ago as a natural defence by the bacteria against antibiotics produced by other soil bacteria or by themselves. However, for synthetic antibacterial agents, such resistance genes are not believed to be present naturally in soil bacteria. It is unlikely that soil bacteria over the course of millions of years may have needed to produce a gene product to combat a novel class of non-existent chemicals. At present, enzymes that can inactivate or modify synthetic antibacterial agents such as fluoroquinolones and sulphonamides have not been identified or isolated. Resistance against synthetic antibacterial agents is believed to arise by modification of the target enzyme as described above (Maxwell, 1992). Without the availability of existing exogenous genes in soil bacteria, whose products either inactivate or modify the antibacterial compound, pathogens may be expected to develop resistance to synthetic antibacterial agents by mutational or recombinational events resulting in the development of an altered target.

Emergence of quinolone resistance during therapy for organisms having MIC values close to the achievable drug concentration has recently been documented. Mechanisms of quinolone resistance in the clinical setting may be mediated by altered DNA gyrase, decreased drug permeation or increased drug efflux. Wide-scale resistance to quinolones is expected to develop if existing quinolones are used improperly. For example, the extensive use of enoxacin, fleroxacin, and lomefloxacin for pneumonia or bacterial bronchitis when *S. pneumoniae* is the aetiological agent will certainly increase the development of quinolone-resistant strains. With the wide and careless use of quinolones, resistance is likely to become a major problem in the future. A recent susceptibility surveillance study of ciprofloxacin indicated a substantial increase in resistant *S. aureus, S. epidermidis, Acinetobacter* sp., *E. faecalis, E. faecium, P. stuartii,* and *P. cepacia* (Thornsberry *et al.,* 1993). Resistance to fluoroquinolones is mediated via mutations giving altered amino acids in the quinolone resistance-determining region of DNA gyrase. Clinical isolates of methicillin-resistant *S. aureus* (MRSA) have a single mutation at Ser^{84} or Glu^{88} of the A-subunit of DNA gyrase providing high-level resistance. Increased activity of an efflux system can also give rise to resistance in some isolates (Sreedharan *et al.,* 1990).

A hypothetical model of DNA gyrase resistance to quinolones

The co-operative drug–DNA binding model suggests an indirect interaction between quinolone and the enzyme. Consequently, questions have been raised concerning how such a model can explain quinolone-resistance mutations that were mapped on the structural genes of both A- and B-subunits of DNA gyrase (Yamagishi *et al.,* 1986; Yoshida *et al.,* 1990, 1991). The co-operative binding model provides a plausible explanation to this question. In fact, it was predicted at the time of the proposal of the model in 1989, that any mutation, in either *gyrA* or *gyrB,* which alters the geometry of the enzyme-induced drug-binding pocket on DNA could lead to drug resistance (Shen *et al.,* 1989*e*). Such a mutation is more likely to occur in *gyrA* since the A-subunit is directly involved in forming the DNA gate, that is the proposed drug-binding site. The gyrase B-subunit, the energy-transduction machinery that provides a conformational change to assist the DNA gate-opening and strand-passing, obviously is also important to the conformation of such a site.

As mentioned earlier, the C-7 substituents were proposed to be the domain that interacts with the enzyme for further strengthening drug binding (Fig. 8). Recent results with quinolone-resistant mutants suggested the possibility of the C-7 substituent interaction site being the B-subunit, as judged from the differential activity response of quinolones with charged and non-charged C-7 substituents to two different types of *gyrB* mutations that have altered charge properties (Yoshida *et al.,* 1991). Such a direct interaction would not, however, be expected to provide sufficient bond strength to account for the drug binding *per se.* This type of interaction, if any, is more likely to play a subordinate role.

Using a gel-filtration method, the binding of [^{3}H]enoxacin to the complex formed between pBR322 DNA and several mutant gyrase preparations was studied recently (Yoshida *et al.,* 1993). This technique measured tightly-bound ligands that were not dissociated during the filtration step. The results of this study favoured the notion that quinolone binds tightly only to the enzyme–DNA complex, and that both A- and B-subunits were involved with the binding.

The high-level quinolone resistance of gyrA Ser[83] E. coli mutants and the weaker binding of quinolone to the complex formed between DNA and the mutant gyrase compared with that between DNA and the wild-type gyrase, have prompted the proposal that Ser[83] is a direct quinolone-binding site mediated through a hydrogen bond (Maxwell, 1992). The role of this type of interaction in the overall binding of the drug needs to be further elucidated.

Potential secondary target: bacterial topoisomerase IV

Recently, a new topoisomerase (topoisomerase IV) was identified and purified from E. coli (Kato et al., 1990, 1992; Peng & Marians, 1993b). The enzyme is composed of two subunits encoded by the parC and parE genes. Both topoisomerase IV and gyrase are type II enzymes and they share some similarity in their structures and functions. For example, the active form of the topoisomerase IV has a tetrameric $(ParC)_2(ParE)_2$ structure in solution (Peng & Marians, 1993b), and the subunits are highly homologous to the subunits of DNA gyrase (that is, 36% between parC and gyrA, and 40% between parE and gyrB) (Kato et al., 1990). Gyrase acts primarily to solve topological problems during DNA replication and to relieve the positive supercoiling accumulated during DNA transcription. Genetic studies indicate that topoisomerase IV is required at the terminal stages of DNA replication, and mutants in topoisomerase IV genes are deficient in chromosomal partitioning (Adams et al., 1992; Peng & Marians, 1993b). The enzyme is essential for cell viability and is sensitive to both types of gyrase inhibitors, the coumerins and quinolones (Kato & Ikeda, 1993; Peng & Marians, 1993b), but not to the mammalian topoisomerase II inhibitors m-AMSA and VP-16. Interestingly, the inactive isomer o-AMSA was found to inhibit the relaxation activity of topoisomerase IV (Kato & Ikeda, 1993). The possibility of topoisomerase IV being a secondary target of quinolone is currently under investigation. Preliminary results have revealed that some quinolones are active against both DNA gyrase and topoisomerase IV (Peng & Marians, 1993b), but they are relatively more potent against the gyrase than the topoisomerase IV, especially in bacterial cells (Zechiedrich et al., 1994).

THIRD-GENERATION QUINOLONES

Although the second-generation quinolones enjoy their dominance as major chemotherapeutic agents against various bacterial infections, they have limited activities against a number of clinically-important Gram-positive bacteria such as Streptococcus pneumoniae, Streptococcus pyogenes, Staphylococcus aureus, and enterococci. Ciprofloxacin, the market leader of the second-generation quinolones, has low potency against anaerobes, and resistance to this compound is increasing rapidly. To overcome this

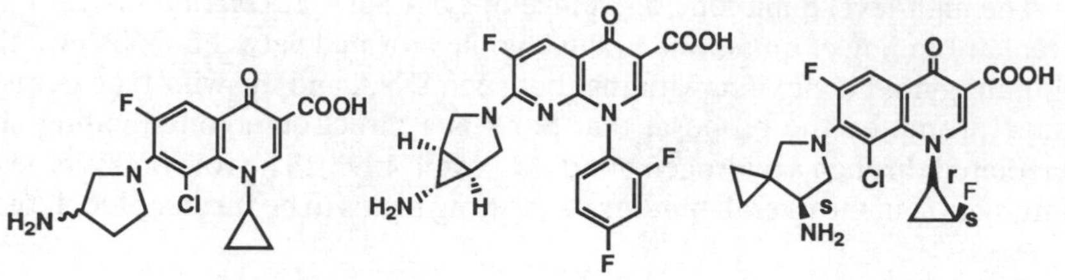

Fig. 9. Third-generation quinolones.

deficiency, several novel fluoroquinolones are now at various stages of development. These third-generation quinolones are characterized by having substantially enhanced activities against streptococci, staphylococci and enterococci; they also have excellent activity against anaerobes. Among third-generation quinolones the most promising compounds include clinafloxacin (Barrett *et al.*, 1991), CP-99,219 (Gootz *et al.*, 1992) and DU-6859a (Sato *et al.*, 1992) (Fig. 9). Their microbiological activities *in vitro* are summarized in Table 4.

Table 4. *Antibacterial activity* in vitro *of several fluoroquinolones under development*

Organism	[a]MIC$_{90}$ (μg/ml)		
	Clinafloxacin	CP-99,219	DU-6859a
Gram-positive aerobes			
S. aureus	0.06	0.06	0.03
S. epidermidis	0.25	0.25	0.03
S. pneumoniae	0.25	0.25	0.12
Enterococcus faecalis	0.5	2	1
Enterococcus faecium	1.0	4	4
Gram-negative aerobes			
E. coli	0.016	0.125	0.06
Klebsiella spp.	0.06	0.25	0.06
E. cloacae	0.03	0.5	0.1
S. marcescens	0.25	4	1.56
P. mirabilis	0.03	1	0.03
M. influenzae	0.03	0.03	0.05
M. catarrhalis	0.03	0.06	0.025
P. aeruginosa	1	2	2
L. pneumophila	0.015	0.06	0.0005
Anaerobes			
B. fragilis	0.5	0.39	0.5
Clostridium spp.	2		0.2

[a]MIC$_{90}$ is the minimum inhibitory concentration causing inhibition of growth of 90% of the strains tested.

These quinolones have potential advantages over the second-generation quinolones and may extend their clinical uses to cover other indications that are not recommended previously. For example, because of their high potency against lower respiratory tract pathogens, their acceptance for use for this indication is anticipated. However, although these quinolones hold great promise, their lack of good activity against quinolone- and methicillin-resistant *Staphylococcus aureus, Enterococcus faecium*, and ciprofloxacin-resistant *Pseudomonas aeruginosa* may preclude their more widespread use.

PERSPECTIVES

The quinolone antibacterial agents have emerged as one of the dominant classes of chemotherapeutic drugs for the treatment of various bacterial infections. Agents now under development having improved activity against Gram-positive pathogens may have potential to further expand the clinical uses of this class of drug. However, there are several areas of concern related to their safety and the development of resistance.

Quinolones have been shown to induce arthropathy in juvenile animals, and because of the potential of having a similar effect in man, quinolones are contraindicated in children, adolescents and pregnant women. Clinical experience with these drugs in prepubertal patients in recent years indicated the incidence (although infrequent) of reversible arthralgia (Schaad, 1992). In a retrospective evaluation of 634 children and adolescents treated with ciprofloxacin, arthralgia was found in eight (1.3%) children (Chysky *et al.*, 1991). Although the concern over arthropathy with quinolone use seems justifiable, the risk/benefit analysis may continue to permit quinolone treatment in children with severe and life-threatening infective diseases (Stahlmann, Forster & Van Sickle, 1993).

Quinolones have been shown to inhibit mammalian enzymes such as DNA polymerase alpha, and topoisomerases I and II at very high concentrations (Hussy *et al.*, 1986; Riesbeck, Bredberg & Forsgren, 1990). These data, together with the fact that they interact with DNA, do not give us much comfort on their long-term safety. Future quinolone research should be focused on the discovery of new quinolones with low genotoxic potential having minimal activity on mammalian topoisomerase II. An alternative approach is to search for extremely potent fluoroquinolones so that a very low concentration of the drug is needed to give a good clinical efficacy, thereby minimizing the toxicity potential. Because of the emergence of resistant bacteria, especially multiple antibiotic methicillin-resistant *Staphylococcus aureus* (MRSA), enterococci and ciprofloxacin-resistant *S. aureus, S. epidermidis* and *P. aeruginosa*, the search for DNA gyrase inhibitors with novel structures will no doubt be important since the inhibition of the enzyme may be at a different site as for the currently used quinolones and therefore not be expected to have cross resistance. The 2-pyridone A-86719 is a DNA gyrase inhibitor of novel structural class and has been identified with the above criteria in mind. It is extremely potent against MRSA and ciprofloxacin-resistant *S. aureus* and *S. epidermidis*, as well as enterococci (Chu *et al.*, 1994). Quinolone research is also focused on minimizing the side-effects on the CNS and gastrointestinal tract. Other goals include improved pharmacokinetics to cover meningitis and endocarditis, and enhanced solubility. Although the search for quinolones for the treatment of cancer (Chu *et al.*, 1992; Kohlbrenner *et al.*, 1992), viral infections (Ikeda, Yazawa & Nishimura, 1987; Kreuzer, 1989; Maschera *et al.*, 1993) and parasitic diseases (Sarma, 1989; Wyler, 1989) is in its infancy, major advances in quinolone research are expected to be found in these chemotherapeutic areas in the not too distant future.

REFERENCES

Adams, D. E., Shekhtman, E. M., Zechiedrich, E. L., Schmid, M. B. & Cozzarelli, N. R. (1992). The role of topoisomerase IV in partitioning bacterial replicons and the structure of catenated intermediates in DNA replication. *Cell*, **71**, 277–88.

Andriole, V. T. (1988). *The Quinolones*. Academic Press, London.

Barrett, M. S., Jones, R. N., Erwin, M. E., Johnson, D. M. & Briggs, B. M. (1991). Antimicrobial activity evaluations of two new quinolones, PD 127391 (CCI-960 and AM-1091) and PD 131628. *Diagnosis of Microbiological Infectious Diseases*, **14**, 389–401.

Barry, A. L. & Fuchs, P. C. (1991). *In vitro* activities of sparfloxacin, tosufloxacin, ciprofloxacin, and fleroxacin. *Antimicrobial Agents and Chemotherapy*, **35**, 955–60.

Bergan, T. (1987). Quinolones. In *Antimicrobial Agents Annual* (Peterson P. K. & Verhoef, J., eds). pp. 169–183. New York, Elsevier.

Brockbank, S. M. & Barth, P. T. (1993). Cloning, sequencing, and expression of the DNA gyrase genes from *Staphylococcus aureus*. *Journal of Bacteriology*, **175**, 3269–77.

Brown, P. O. & Cozzarelli, N. R. (1979). A sign inversion mechanism for enzymatic supercoiling of DNA. *Science*, **206**, 1081–3.

Chin, N., Brittain, D. & Neu, H. (1986). *In vitro* activity of Ro 23-6240, a new fluorinated 4-quinolone. *Antimicrobial Agents and Chemotherapy*, **29**, 675–80.

Chu, D. T. & Fernandes, P. B. (1989). Structure–activity relationships of the fluoroquinolones. *Antimicrobial Agents and Chemotherapy*, **33**, 131–5.

Chu, D. T., Fernandes, P. B., Claiborne, A. K., Pihuleac, E., Nordeen, C. W., Maleczka, R. E. & Pernet, A. G. (1985). Synthesis and structure–activity relationships of novel arylfluoroquinolone antibacterial agents. *Journal of Medicinal Chemistry*, **28**, 1558–64.

Chu, D. T. W. & Fernandes, P. B. (1991). Recent developments in the field of quinolone antibacterial agents. *Advances in Drug Research*, **21**, 39–144.

Chu, D. T. W., Fernandes, P. B., Claiborne, A. K., Gracey, H. E. & Pernet, A. G. (1986). Synthesis and structure–activity relationships of new arylfluoronaphthyridine antibacterial agents. *Journal of Medicinal Chemistry*, **29**, 2363–69.

Chu, D. T. W., Fernandes, P. B., Claiborne, A. K., Shen, L. L. & Pernet, A. G. (1988). Structure–activity relationships in quinolone antibacterials: design, synthesis and biological activities of novel isothiazoloquinolones. *Drugs Experimental Clinical Research*, **14**, 379–83.

Chu, D. T. W., Hallas, R., Clement, J. J., Alder, J., McDonald, E. & Plattner, J. J. (1992). Synthesis and antibacterial activities of quinolone antineoplastic agents. *Drugs in Experimental and Clinical Research*, **18**, 275–82.

Chu, D. T. W., Li, Q., Claiborne, A., Raye-Passarelli, Cooper, C., Fung, A., Lee, C., Tanaka, K., Shen, L. L., Donner, P., Armiger, Y. & Plattner, J. J. (1994). Synthesis and antibacterial activity of A-86719.1 and related 2-pyridones: a novel series of potent DNA gyrase inhibitors. *Abstract of the 34th Interscience Conference on Antimicrobial Agents and Chemotherapy*, Orlando, Florida, Abstract F41.

Chysky, V., Kapila, K., Hullmann, R., Arcieri, G., Schacht, P. & Echols, R. (1991). Safety of ciprofloxacin in children: worldwide clinical experience based on compassionate use. Emphasis on joint evaluation. *Infection*, **19**, 289–96.

Cornett, J. B. & Wentland, M. P. (1986). Quinolone antibacterial agents. In *Annual Report Medicinal Chemistry* (Bailey, D. M. & Sciavolino, F. C. eds). pp. 139–148. Academic Press, Orlando.

Cozzarelli, N. R. (1980). DNA gyrase and the supercoiling of DNA. *Science*, **207**, 953–60.

Crumplin, G. C. (1980). *The 4-Quinolones: Antibacterial Agents in vitro*. Springer-Verlag, London.

Crumplin, G. C. Midgley, J. M. & Smith, J. T. (1980). Mechanisms of action of nalidixic acid and its congeners. *Antibiotics Chemistry*, **8**, 13–35.

Domagala, J. M. (1994). Structure–activity and structure–side-effect relationships for the quinolone antibacterials. *Journal of Antimicrobial Chemotherapy*, **33**, 685–706.

Domagk, G. (1935). Ein betrag zur chemotherapie der bakteriellen infektionen. *Deutsche Medizinische Wochenschrift*, **61**, 250–3.

Drlica, K. (1984). Biology of bacterial deoxyribonucleic acid topoisomerases. *Microbiological Reviews*, **48**, 273–89.

Fernandes, P. B. (1988). Mode of action, and *in vitro* and *in vivo* activities of the fluoroquinolones. *Journal of Clinical Pharmacology*, **28**, 156–68.

Fernandes, P. B. (1989). Editor. *International Telesymposium on Quinolones*. J. R. Prous Scientific Publishers, Barcelona.

Fernandes, P. B., Shipkowitz, N., Bower, R. R., Jarvis, K. P., Wiesz, J. & Chu, D. T. (1986). *In vitro* and *in vivo* potency of five new fluoroquinolones against anaerobic bacteria. *Journal of Antimicrobial Chemotherapy*, **18**, 693–701.

Gellert, M. (1981). DNA topoisomerases. *Annual Review of Biochemistry*, **50**, 879–910.

Gellert, M., Mizuuchi, K., O'Dea, M. H., Itoh, T. & Tomizawa, J.-I. (1976). DNA gyrase: an enzyme that introduces superhelical turns into DNA. *Proceedings of the National Academy of Sciences, USA*, **73**, 3872–6.

Gellert, M., Mizuuchi, K., O'Dea, M. H., Itoh, T. & Tomizawa, J.-I. (1977). Nalidixic acid resistance: a second genetic character involved in DNA gyrase activity. *Proceedings of the National Academy of Sciences, USA*, **74**, 4772–6.

Gootz, T. D., Brighty, K. E., Anderson, M. E., Haskell, S. L., Sutcliffe, J. A., Castaldi, M. J. & Miller, S. A. (1992). *In vitro* activity and synthesis of CP-99,219, a novel 7-(3-azabicyclo[3,1,0] hexyl)naphthyridone. *Abstract of the 32nd Interscience Conference on Antimicrobial Agents and Chemotherapy*, Anaheim, Abstract 751.

Goueffon, Y., Montay, G., Roquet, F. & Pesson, M. (1981). Sur un nouvel antibacterien de synthese: l'acide ethyl-1 fluoro-6 (methyl-4 piperazinyl-1)-7 oxo-4 dihydro-1, 4 quinoleine-3 carboxylique (1589 R.B.). *Comptes Rendus Academie Sciences Paris*, **292, Ser.III**, 37–40.

Granneman, G. R., Carpentier, P., Morrison, P. J. & Pernet, A. G. (1991).

132   DANIEL T. W. CHU AND LINUS L. SHEN

Pharmacokinetics of temafloxacin in humans after single oral doses. *Antimicrobial Agents and Chemotherapy*, **35**, 436–41.

Hallett, P., Grimshaw, A. J., Wigley, D. B. & Maxwell, A. (1990). Cloning of the DNA gyrase genes under tac promoter control: overproduction of the gyrase A and B proteins. *Gene*, **93**, 129–42.

Hardy, D. J., Swanson, R. N., Hensey, D. M., Ramer, N. R., Bower, R. R., Hanson, C. W., Chu, D. T. W. & Fernandes, P. B. (1988). Comparative antibacterial activities of temafloxacin hydrochloride (A-62254) and two reference fluoroquinolones. *Antimicrobial Agents and Chemotherapy*, **31**, 1768–74.

Høiby, N. (1986). Clinical uses of nalidixic acid analogues: the fluoroquinolones. *European Journal of Clinical Microbiology*, **5**, 138–40.

Hokuriku Pharmaceutical Co. (1986). NY-198. *Drugs of the Future*, **11**, 578–9.

Hooper, D. C. (1986). Potential uses of the newer quinolone antimicrobial agents. In *Microbiology* (Leive, L. ed.). pp. 226–230. American Society of Microbiology, Washington, DC.

Hooper, D. C. & Wolfson, J. S. (1985). The fluoroquinolones: pharmacology, clinical uses, and toxicities in humans. *Antimicrobial Agents and Chemotherapy*, **28**, 716–21.

Horowitz, D. S. & Wang, J. C. (1987). Mapping the active site tyrosine of *Escherichia coli* DNA gyrase. *Journal of Biological Chemistry*, **262**, 5339–44.

Hoshino, K., Sato, K., Kitamura, A., Hayakawa, I., Sato, M. & Osada, Y. (1992). Inhibitory effects of DU-6859, a new fluorinated quinolone, on type II topoisomerases. *Abstracts of the 31st Interscience Conference on Antimicrobial Agents and Chemotherapy*, Chicago, Illinois, Abstract 1506.

Hubschwerlen, C., Pflieger, P., Specklin, J.-L., Gubernator, K., Gmunder, H., Angehrn, P. & Kompis, I. (1992). Pyrimido[1,6-a]benzimidazoles: a new class of DNA gyrase inhibitors. *Journal of Medicinal Chemistry*, **35**, 1385–92.

Hussy, P., Maass, G., Tummler, B., Grosse, F. & Schomburg, U. (1986). Effect of 4-quinolones and novobiocin on calf thymus DNA polymerase alpha primase complex, topoisomerases I and II, and growth of mammalian lymphoblasts. *Antimicrobial Agents and Chemotherapy*, **29**, 1073–8.

Ikeda, S., Yazawa, M. & Nishimura, C. (1987). Antiviral activity and inhibition of topoisomerase by ofloxacin, a new quinolone derivative. *Antiviral Research*, **8**, 103–13.

Janknegt, R. (1986). Fluorinated quinolones. A review of their mode of action, antimicrobial activity, pharmacokinetics and clinical efficacy. *Pharmaceutical Week Bulletin* [Sci], **8**, 1–21.

Karabalut, N. & Drusano, G. L (1993). Pharmacokinetics of the quinolone antimicrobial agents. In *Quinolone Antimicrobial Agents* (Hooper, D. C. & Wolfson, J. S., eds). pp. 195–223. American Society for Microbiology, Washington, D.C.

Kato, J.-I. & Ikeda, H. (1993). Purification and characterization of DNA topoisomerase IV in *Escherichia coli*. In *Molecular Biology of DNA Topoisomerases and its Application to Chemotherapy* (Andoh, T., Ikeda, H. & Oguro, M., eds). pp. 31–37. CRC Press, Boca Raton.

Kato, J.-I., Nichimura, Y., Imamura, R., Niki, H., Hiraga, S. & Suzuki, H. (1990). New topoisomerase essential for chromosome segregation in *E. coli*. *Cell*, **63**, 393–404.

Kato, J.-I., Suzuki, H. & Ikeda, H. (1992). Purification and characterization of DNA topoisomerase IV in *Escherichia coli*. *Journal of Biological Chemistry*, **267**, 25676–84.

Kenny, G. E., Hooton, T. M., Roberts, M. C., Cartwright, F. D. & Hoyt, J. (1989).

Susceptibilities of genital mycoplasmas to the newer quinolones as determined by the agar dilution method. *Antimicrobial Agents and Chemotherapy*, **33**, 103–7.

Kirchhausen, T., Wang, J. C. & Harrison, S. C. (1985). DNA gyrase and its complexes with DNA: direct observation by electron microscopy. *Cell*, **41**, 933–43.

Koga, H., Itoh, A., Murayama, S., Suzue, S. & Irikura, T. (1980). Structure–activity relationships of antibacterial 6, 7-and 7, 8-disubstituted 1-alkyl-1,4-dihydro-4 oxoquinoline-3-carboxylic acids. *Journal of Medicinal Chemistry*, **23**, 1358–63.

Kohlbrenner, W. E., Wideburg, N., Weigl, D., Saldivar, A. & Chu, D. T. W. (1992). Induction of calf-thymus topoisomerase II-mediated DNA breakage by the antibacterial isothiazoloquinolones A65281 and A-65282. *Antimicrobial Agents and Chemotherapy*, **36**, 81–6.

Krausse, R. & Ullmann, U. (1988). Comparative *in vitro* activity of fleroxacin (RD 23-6240) against *Ureaplasma urealyticum* and *Mycoplasma hominis*. *European Journal of Clinical Microbiology and Infectious Diseases*, **7**, 67–9.

Kreuzer, K. N. (1989). DNA topoisomerases as potential targets of antiviral action. *Pharmaceutical Therapy*, **43**, 377–95.

Kreuzer, K. N. & Cozzarelli, N. R. (1979). *Escherichia coli* mutants thermosensitive for deoxyribonucleic acid gyrase subunit A: effect of deoxyribonucleic acid replication, transcription, and bacteriophage growth. *Journal of Bacteriology*, **140**, 424–35.

Leigh, D. A., Walsh, B., Harris, K., Hancock, P. & Travers, G. (1988). Pharmacokinetics of ofloxacin and the effect on the faecal flora of healthy volunteers. *Journal of Antimicrobial Chemotherapy*, **22** (Suppl. C), 115–25.

Lesher, G. Y., Froelich, E. J., Gruett, M. D., Bailey, J. H. & Brundage, R. P. (1962). 1,8-Naphthyridine derivatives. A new class of chemotherapeutic agents. *Journal of Medicinal Chemistry*, **5**, 1063–5.

Maple, P., Brumfitt, W. & Hamilton-Miller, J. M. (1990). A review of the antimicrobial activity of the fluoroquinolones. *Journal of Chemotherapy*, **2**, 280–94.

Maschera, B., Ferrazzi, E., Rassu, M., Toni, M. & Palu', G. (1993). Evaluation of topoisomerase inhibitors as potential antiviral agents. *Antiviral Chemistry and Chemotherapy*, **4**, 85–91.

Matsumoto, J., Miyamoto, T., Minamida, A., Nichimura, Y., Egawa, H. & Nishimura, H. (1984). Pyridonecarboxylic acids as antibacterial agents. 2. Synthesis and structure–activity relationships of 1,6,7,-trisubstituted 1,4,-dihydro-4-oxo-1,8-naphthyridine-3-carboxylic acids, including enoxacin, a new antibacterial agent. *Journal of Medicinal Chemistry*, **28**, 292–301.

Maxwell, A. (1992). The molecular basis of quinolone action. *Journal of Antimicrobial Chemotherapy*, **30**, 409–16.

Mitscher, L.A., Devasthale, P. V. & Zavod, R. M. (1990). Structure-activity relationships of fluoro-4-quinolones. In *The 4-Quinolones: Anti-bacterial Agents In Vitro* (Crumplin, G. C., ed.). pp. 115–146. Springer-Verlag, New York.

Mitscher, L. A., Zavod, R. M., Devasthale, P. V., Chu, D. T. W., Shen, L. L., Sharma, P. N. & Pernet, A. G. (1991). Microbes beware: quinolones. Part 1 & 2. *CHEMTECH*, **21**, 50–6, 249–55.

Mitscher, L. A., Zavod, R. M. & Sharma, P. N. (1989). Structure–activity relationships of the newer quinolone antibacterial agents. In *International Telesymposium on Quinolones* (Fernandes, P. B., ed.). pp. 3–22. J. R. Prous Science Publishers, Barcelona.

Mitscher, L. A., Zavod, R. M., Sharma, P. N., Chu, D. T. W., Shen, L. L. & Pernet, A. G. (1988). Recent advances on quinolone antimicrobial agents. In *Horizons on*

Antibiotic Research (Davis, B. D., Ichikawa, T., Maeda, K. & Mitscher, L. A., eds). pp. 166–193. Japan Antibiotics Research Association, Tokyo.

Miyamoto, T., Matsumoto, J., Chiba, K., Egawa, H., Shibamori, K., Minamida, A., Nishimura, Y., Okada, H., Katasoka, M., Fujita, M., Hirose, T. & Nakano, J. (1990). Pyridonecarboxylic acids as antibacterial agents. 14. Synthesis and structure–activity relationships of 5-substituted 6,8-difluoroquinolones, including sparfloxacin, a new quinolone antibacterial agent with improved potency. *Journal of Medicinal Chemistry*, **33**, 1645–56.

Moellering, R. C. & Neu, H. C. (1989). Introduction: ofloxacin. *American Journal of Medicine*, **87**, 1s.

Morrison, A. & Cozzarelli, N. R. (1979). Site-specific cleavage of DNA by *E. coli* DNA gyrase. *Cell*, **17**, 175–84.

Morrison, A. & Cozzarelli, N. R. (1981). Contacts between DNA gyrase and its binding site on DNA: Features of symmetry and asymmetry revealed by protection from nucleases. *Proceedings of the National Academy of Sciences, USA*, **78**, 1416–20.

Mouton, Y. & Leroy, O. (1991). Review: ofloxacin. *International Journal of Antimicrobial Agents*, **1**, 57–74.

Neu, H. C. (1987). Clinical use of the quinolones. *Lancet*, 1319–22.

Neu, H. C. (1989). Clinical utility of DNA gyrase inhibitors. *Pharmaceutical Therapy*, **41**, 207–21.

Nicolle, L., Urias, B., Brunka, J., Kennedy, J. & Harding, G. (1988). Comparative *in vitro* activity of lomefloxacin and other quinolones for gram negative organisms isolated from complicated urinary infection. Abstract of the 28th Interscience Conference *Antimicrobial Agents and Chemotherapy*, Los Angeles, Abstract 687.

Palu', G., Valisena, S., Ciarrocchi, G., Gatto, B. & Palumbo, M. (1992). Quinolone binding to DNA is mediated by magnesium ions. *Proceedings of the National Academy of Sciences, USA*, **89**, 9671–5.

Paton, J. H. & Reeves, D. S. (1988). Fluoroquinolone antibiotics: microbiology, pharmacokinetics and clinical use. *Drugs*, **36**, 193–228.

Peng, H. & Marians, K. J. (1993a). Decatenation activity of topoisomerase IV during *oriC* and pBR322 DNA replication *in vitro*. *Proceedings of the National Academy of Sciences, USA*, **90**, 8571–5.

Peng, H. & Marians, K. J. (1993b). *Escherichia coli* topoisomerase IV. Purification, characterization, subunit structure, and subunit interactions. *Journal of Biological Chemistry*, **268**, 24481–90.

Pernet, A. G. (1989). A preliminary survey of temafloxacin: microbiological activity, pharmacokinetic profile, clinical efficacy and safety. In *International Telesymposium on Quinolones* (Fernandes, P. B., ed.). pp. 417–425. J. R. Prous Science Publishers, Barcelona.

Pocidalo, J. J., Vachon, F. & Regnier, B. (1985). *Les Nouvelles Quinolones*. Editions Arnette, Paris.

Radl, S. (1990). Structure–activity relationships in DNA gyrase inhibitors. *Pharmaceutical Therapy*, **48**, 1–17.

Reece, R. J. & Maxwell, A. (1991). DNA gyrase: structure and function. *Critical Reviews in Biochemistry and Molecular Biology*, **26**, 335–75.

Riesbeck, K., Bredberg, A. & Forsgren, A. (1990). Ciprofloxacin does not inhibit mitochondrial functions but other antibiotics do. *Antimicrobial Agents and Chemotherapy*, **34**, 167–9.

Rosen, T. (1990). The fluoroquinolone antibacterial agents. In *Progress in Medicinal*

Chemistry (Ellis, G. P. & West, G. B., eds). pp. 235–295. Elsevier Science Publishers, Amsterdam.

Sarma, P. S. (1989). Norfloxacin: a new drug in the treatment of Falciparum malaria. *Annals in Internal Medicine*, **111**, 336–7.

Sato, K., Hoshino, K., Tanaka, M., Hayakawa, I. & Osada, Y. (1992). Antimicrobial activity of DU-6859, a new potent fluoroquinolone, against clinical isolates. *Antimicrobial Agents and Chemotherapy*, **36**, 1491–8.

Sato, K., Matsuura, Y., Inoue, M., Une, T., Osada, Y., Ogawa, H. & Mitsuhashi, S. (1982). *In vitro* and *in vivo* activity of DL-8280, a new oxazine derivative. *Antimicrobial Agents and Chemotherapy*, **22**, 548–53.

Schaad, U. B. (1992). Role of the new quinolones in pediatric practice. *Pediatrics Infectious Disease Journal*, **11**, 1043–6.

Schentag, J. & Domagala, J. (1985). Structure–activity relationships with the quinolone antibiotics. *Research Clinical Forums*, **7**, 9–13.

Shah, P. M. (1987). Quinolones. In *Progress in Drug Research* (Jucker E., ed.). pp. 243–256. Birkhauser Verlag, Basel.

Shah, P. M. (1991). Review: Ciprofloxacin. *International Journal of Antimicrobial Agents*, **1**, 75–96.

Shen, L., Baranowski, J., Chu, D. T. W. & Pernet, A. G. (1989a). DNA unwinding properties of new quinolones – the antagonizing effect of polyamines. Abstract 29. *Abstracts of the ICAAC. Houston Texas*.

Shen, L. L., Baranowski, J. & Pernet, A. G. (1989b). Mechanism of inhibition of DNA gyrase by quinolone antibacterials: Specificity and cooperativity of drug binding to DNA. *Biochemistry*, **28**, 3879–85.

Shen, L. L., Baranowski, J., Wai, T., Chu, D. T. W. & Pernet, A. G. (1989c). The binding of quinolones to DNA: should we worry about it? In *International Telesymposium on Quinolones* (Fernandes, P. B., ed.). pp. 159–170. J. R. Prous Science Publishers, Barcelona, Spain.

Shen, L. L., Kohlbrenner, W. E., Weigl, D. & Baranowski, J. (1989d). Mechanism of quinolone inhibition of DNA gyrase. Appearance of unique norfloxacin binding sites in enzyme–DNA complexes. *Journal of Biological Chemistry*, **264**, 2973–8.

Shen, L. L., Mitscher, L. A., Sharma, P. N. O'Donnell, T. J., Chu, D. W. T., Cooper, C. S., Rosen, T. & Pernet, A. G. (1989e). Mechanism of inhibition of DNA gyrase by quinolone antibacterials. A cooperative drug-DNA binding model. *Biochemistry*, **28**, 2886–94.

Shen, L. L. & Pernet, A. G. (1985). Mechanism of inhibition of DNA gyrase by analogues of nalidixic acid: The target of the drugs is DNA. *Proceedings of the National Academy of Sciences, USA*, **82**, 307–11.

Shiba, K., Saito, A., Shimada, J. & Miyahara, T. (1988). Pharmacokinetics of temafloxacin in humans after single oral doses. *Abstracts of the 28th Interscience Conference Antimicrobial Agents in Chemotherapy, Los Angeles*, Abstract 788.

Siporin, C. (1989). The evolution of fluorinated quinolones: pharmacology, microbiological activity, clinical uses, and toxicities. *Annual Review in Microbiology*, **43**, 601–25.

Siporin, C., Heifetz, C. L. & Domagala, J. M. (1990). *The New Generation of Quinolones*. M. Dekker, New York.

Smith, J. T. (1984). Awakening the slumbering potential of the 4-quinolone antibacterials. *Pharmaceutical Journal*, **233**, 299–305.

Soejima, R. & Shimada, K. (1989). Tosufloxacin tosylate. *Drugs Future*, **14**, 536–8.

Speller, D. & Wise, R. (1988). Preface. *Journal of Antimicrobial Chemotherapy*, **21**, 1.

Sreedharan, S., Oram, M., Jensen, B., Peterson, L. R. & Fisher, L. M. (1990). DNA gyrase *gyrA* mutations in ciprofloxacin-resistant strains of *Staphylococcus aureus*: close similarity with quinolone resistance mutations in *Escherichia coli*. *Journal of Bacteriology*, **172**, 7260–2.

Stahlmann, R., Forster, C. & Van Sickle, D. (1993). Quinolones in children: are concerns over arthropathy justified? *Drug Safety*, **9**, 397–403.

Sugino, A., Peebles, C. L., Cozzarelli, N. R. & Kreuzer, K. N. (1977). Mechanism of action of nalidixic acid: Purification of *E. coli nalA* gene product and its relationship to DNA gyrase and a novel nicking-closing enzyme. *Proceedings of National Academy of Sciences, USA*, **74**, 4767–71.

Sutcliffe, J. A., Gootz, T. D. & Barrett, J. F. (1989). Biochemical characteristics and physiological significance of major DNA topoisomerases. *Antimicrobial Agents and Chemotherapy*, **33**, 2027–33.

Thornsberry, C., Brown, S. D., Bouchillon, S. K., Marler, J., Rich, T. & Yee, Y. C. (1993). *United States Surveillance of Clinical Bacterial Isolates to Ciprofloxacin*. Institute of Microbiology Research, Franklin, Tennessee.

Tjiam, K. H., Wagenvoort, J. H., van Klingeren, B., Piot, P., Stolz, E. & Michel, M. F. (1986). *In vitro* activity of the two new 4-quinolones A56619 and A56620 against *Neisseria gonorrhoeae, Chlamydia trachomatis, Mycoplasma hominis, Ureaplasma urealyticum* and *Gardnerella vaginalis*. *European Journal of Clinical Microbiology*, **5**, 498–501.

Venezia, R. A., Yocum, D. M. & Luzinas, M. J. (1988). *In vitro* activity of lomefloxacin. *Abstracts of the 28th Interscience Conference Antimicrobial Agents Chemotherapy, Los Angeles*, Abstract 686.

Walker, R. C. & Wright, A. J. (1991). The fluoroquinolones. *Mayo Clinical Proceedings*, **66**, 1249–59.

Wang, J. C. (1982). DNA topoisomerases. *Scientific American*, **247**, 94–109.

Wentland, M. P. (1990). Structure–activity relationships of fluoroquinolones. In *The New Generation of Quinolones* (Siporin, C., Heifetz, C. L. & Domagala, J. M., eds). pp. 1–43. Marcel Dekker, New York.

Wigley, D. B., Davies, G. J., Dodson, E. J., Maxwell, A. & Dodson, G. (1991). Crystal structure of an N-terminal fragment of DNA gyrase B protein. *Nature*, **351**, 624–9.

Wise, R., Andrews, J. M. & Edwards, L. J. (1983). *In vitro* activity of Bay 09867, a new quinoline derivative, compared with those of other antimicrobial agents. *Antimicrobial Agents and Chemotherapy*, **23**, 559–64.

Wise, R. & Lockley M. R. (1988). The pharmacokinetics of ofloxacin and a review of its tissue penetration. *Journal of Antimicrobial Chemotherapy*, **22**, 59–64.

Wolfson, J. S. & Hooper, D. C. (1985). The fluoroquinolones: structures, mechanisms of action and resistance, and spectra of activity *in vitro*. *Antimicrobial Agents and Chemotherapy*, **28**, 581–6.

Wolfson, J. S. & Hooper, D. C. (1989a). *Quinolone Antimicrobial Agents*. American Society for Microbiology, Washington, DC.

Wolfson, J. S. & Hooper, D. C. (1989b). Fluoroquinolone antimicrobial agents. *Clinical Microbiology Review*, **2**, 278–424.

Wyler, D. J. (1989). Fluoroquinolones for malaria: the newest kid on the block? *Annals of Internal Medicine*, **111**, 269–271.

Yamagishi, J., Yoshida, H., Yamayoshi, M. & Nakamura, S. (1986). Nalidixic acid-resistant mutations of the *gyrB* gene of *Escherichia coli*. *Molecular and General Genetics*, **204**, 367–73.

Yoshida, H., Bogaki, M., Nakamura, M., Yamanaka, L. M. & Nakamura, S. (1991). Quinolone resistance-determining region in the DNA gyrase *gyrB* gene of *Escherichia coli. Antimicrobial Agents and Chemotherapy*, **35**, 1647–50.

Yoshida, H., Bogaki, M., Nakamura, M. & Nakamura, S. (1990). Quinolone resistance-determining region in the DNA gyrase *gyrA* gene of *Escherichia coli. Antimicrobial Agents and Chemotherapy*, **34**, 1271–2.

Yoshida, H., Nakamura, M., Bogaki, M., Ito, H., Kojima, T., Hattori, H. & Nakamura, S. (1993). Mechanism of action of quinolones against *Escherichia coli* DNA gyrase. *Antimicrobial Agents and Chemotherapy*, **37**, 839–45.

Zechiedrich, E. L., Ullsperger, C., Khodursky, A., Bramhill, D. Y Cozzarelli, N. R. (1994). Roles of topoisomerases in plasmid segregation in *Escherichia coli. FASEB Journal Abstracts of the American Society of Biochemistry and Molecular Biology*, **8**, A1322.

MOLECULAR GENETICS OF ANTIMICROBIALS: A CASE STUDY OF β-LACTAM ANTIBIOTICS

GERALD COHEN and YAIR AHARONOWITZ

Department of Molecular Microbiology and Biotechnology, Faculty of Life Sciences, Tel-Aviv University, Ramat Aviv, 69978, Israel

INTRODUCTION

Molecular genetics of antibiotic production is presently one of the most exciting and challenging areas of research on antimicrobials. Dramatic developments in gene technologies within the last few years has made it possible to clone antibiotic biosynthetic genes, which in turn has led to remarkable insights into their structure, organization, regulation and evolution. This is especially true for the penicillin and cephalosporin class of antibiotics (Ingolia & Queener, 1989; Aharonowitz, Cohen & Martin, 1992; Turner, 1992), the polyketide and macrolide antibiotics (Hopwood & Khosia, 1992; Katz & Donadio, 1993) and the aminoglycoside antibiotics (Distler *et al.*, 1992).

To a large extent, the chemical nature of these clinically valuable compounds, their mode of action, and the biochemical pathways by which they are assembled are well understood. Predictably, the current impetus of research on antimicrobials has switched therefore to elucidating the genetic systems that determine antibiotic synthesis, and with it the promise of answers to some fundamental questions concerning not only how antibiotics are made but also their role in nature and the mechanisms by which they have evolved. Equally important, although from a rather different standpoint, these studies have paved the way for radically new approaches for improving industrial antibiotic-producing strains, for creating by genetic engineering novel and potentially useful metabolites such as the hybrid antibiotics, and for designing innovative screens for drug discovery programmes. In this review it would be unrealistic to attempt to summarize the entire field of molecular genetics of antimicrobials; rather, after discussing some general concepts, we focus on one well-characterized system, that of the β-lactam class of antibiotics.

TOOLS OF MOLECULAR GENETICS

Classical genetic technologies have played a prominent role in studies on antibiotic production by microorganisms (Cohen & Aharonowitz, 1981), most importantly by creating mutants that are either defective in the synthesis of antibiotics or overproduce them. In the former case, they have been invaluable in defining the biosynthetic pathways. In the latter case, they have resulted in impressive increases in antibiotic yields in industrial strain improvement programmes. However, further increases in antibiotic production by these technologies, which depend heavily on random muta-genesis, natural selection and extensive screening protocols, are likely to be severely limited due to the great complexity of most antibiotic-producing systems, and have led to the need for alternative and more rational approaches to strain improvement.

Recently, powerful molecular genetic tools are being applied to the precise manipulation of genes determining antibiotic synthesis, particularly in the streptomycetes and their close relatives among the actinomycetes, and in certain filamentous fungi. Conveniently termed 'enabling technologies' (Hopwood, 1993), they have been instrumental in the cloning, sequence analysis and transfer of antibiotic genes and have resulted in major advances in our knowledge of antibiotic production. We briefly review these key genetic technologies before proceeding to describe their application to the biosynthesis of β-lactam antibiotics.

Cloning of antibiotic genes

Several strategies exist for cloning antibiotic biosynthetic genes. In the 'reverse genetics' technique, mixed oligodeoxynucleotides are synthesized according to the amino-terminal portion, or occasionally an internal region, of the purified protein and used to screen genomic libraries of the producer organism by hybridization. One useful ploy in constructing these oligonuc-leotide probes for *Streptomyces* spp. takes advantage of their high G + C content (70–74%) which necessitates a G or C base in the third position of most codons of open reading frames (Wright & Bibb, 1992). The relative ease with which genomic libraries can now be prepared in plasmid, phage and cosmid vectors makes this approach to gene isolation attractive. Moreover, because the antibiotic biosynthetic genes, and their counterparts for self-resistance, are almost invariably clustered in the chromosome on a continuous DNA segment (see below), identification of one biosynthetic gene rapidly allows by 'chromosomal walking' localization of related genes. Clusters of genes involved in β-lactam biosynthesis were among the first to be mapped in this way (Diez *et al.*, 1990; Smith *et al.*, 1990a; MacCabe *et al.*, 1990).

Once an antibiotic specific gene has been cloned in one producer strain it can serve as a probe for isolating related genes from other strains by hybridization (Malpartida *et al.*, 1987). Thus, a β-lactam biosynthetic gene isolated from a Gram-positive *Streptomyces* sp. was used successfully to clone the homologous genes from other actinomycetes, as well as from more distantly related Gram-negative bacteria and filamentous fungi (Aharonow-itz *et al.*, 1992). In each case, DNA sequencing analysis revealed striking similarities in the organization of nearby β-lactam biosynthetic genes.

A valuable approach for cloning antibiotic biosynthetic genes, when other methods are not practical, involves complementation of a genetic block in the pathway of a non-producer strain to restore production capability. It relies on the availability of biochemically well-defined non-producer mutants and a gene transfer system (see below) that allows adequate expression of the sought-after gene. Complementation cloning of antibiotic genes has been used with great success to isolate individual β-lactam biosynthetic genes and in some cases entire clusters of genes (Chen *et al.*, 1988; Diez *et al.*, 1990).

Disruption of antibiotic genes

The disruption or replacement of an antibiotic specific gene *in vivo* is frequently the most rigorous way for analysing its function in the producer organism. Methodologies are based on phage and plasmid vectors that contain a cloned fragment of the antibiotic gene that is either disrupted or deleted in some essential part (Kieser & Hopwood, 1991). When transferred to the recipient, recombination between the cloned fragment and homologous sequences in the host chromosome generates a non-functional version of the gene. Gene disruption or replacement has been used to 'knock out' individual β-lactam biosynthetic genes as well as gene clusters (Walz & Kuck, 1993; Brakhage *et al.*, 1992).

Transfer of antibiotic genes

A critical aspect of the analysis of antibiotic biosynthetic genes is the ability to transfer individual genes or gene clusters into specific hosts. Gene manipulations are, for instance, conveniently carried out in *Escherichia coli*. Efficient transformation systems are now also available for many *Streptomyces* spp. and for some antibiotic-producing filamentous fungi, including *Aspergillus nidulans*, *Cephalosporium acremonium* (*Acremonium chrysogenum*) and *Penicillium chrysogenum* (Yelton, Hamer & Timberlake, 1984; Sanchez *et al.*, 1987; Skatrud *et al.*, 1987). Typically, they involve the formation of protoplasts for uptake of DNA in the presence of polyethylene glycol and the regeneration of viable cells in medium containing a selective agent that allows only the transformants to grow. As pointed out above, gene transfer procedures enable isolation of antibiotic genes by complementation and their inactivation by disruption; they also provide a route for introducing new genes, or additional copies of existing genes, into the producer strain so as to modify its biosynthetic capacity. More recently a promising new method for gene transfer has become available by the discovery that under certain conditions *E. coli* is able to conjugatively transfer DNA to *Streptomyces* (Mazodier *et al.*, 1989; Bierman *et al.*, 1992).

MOLECULAR GENETICS OF β-LACTAM ANTIBIOTIC SYNTHESIS

We turn to a discussion of the molecular genetics of one of the best characterized antimicrobial systems, the β-lactam antibiotics. Our analysis will, it is hoped, illustrate the application of the key technologies outlined above for furthering our understanding of antibiotic production and for devising rational ways for improving strains to increase yields and to develop new products. Other extensive reviews on the molecular genetics of β-lactams have appeared recently (for example, Aharonowitz *et al.*, 1992; Turner, 1992).

BIOSYNTHESIS OF β-LACTAM ANTIBIOTICS

Beta-lactam antibiotics as a whole comprise the major share of the world-wide clinical market (Cohen & Aharonowitz, 1981). They do in fact consist of several classes of compounds, all of which possess the β-lactam ring but differ in their chemical substituents (Table 1). In the following discussion we confine our attention to the so-called sulphur-containing β-lactams and which historically were the first to be discovered. Penicillins and cephalosporins, important members of this class, were originally thought to be made exclusively by certain filamentous fungi. It came as a surprise therefore when a variety of bacteria were found to produce these antibiotics, as well as the structurally related but more complex cephamycins. Chief among these organisms are the mycelium-forming Gram-positive *Streptomyces* belonging to the actinomycetes, but also some Gram-negative unicellular bacteria (Elander, 1983).

Two crucial findings led to the elucidation of the penicillin and cephalosporin biosynthetic pathway. First, the recognition that a tripeptide δ-(L-α aminoadipyl)-L-cysteinyl-D-valine (ACV) discovered in extracts of *P. chrysogenum* is the earliest precursor in the pathway (Arnstein & Morris, 1960; Fawcett *et al.*, 1976) and second the development of cell-free systems for the assay of biosynthetic enzymes (Konomi *et al.*, 1979; Jensen *et al.*, 1982). Further studies showed that bacteria and fungi make penicillins and cephalosporins in essentially the same way (Fig. 1).

Two early steps in the pathway are common to penicillin and cephalosporin synthesis. Biosynthesis commences with the condensation of three amino acids L-α aminoadipic acid, L-cysteine and L-valine by the enzyme ACV synthetase to form ACV. Isopenicillin N synthase then oxidatively cyclizes the tripeptide to form isopenicillin N which possesses weak antibiotic properties. The fate of isopenicillin N depends on the producer organism. In those fungi which possess an acyltransferase, the aminoadipyl side chain is removed and can be replaced with a variety of acyl derivatives

Table 1. *Distribution of β-lactam antibiotic-producing microorganisms*[a]

Class of β-lactam	Ring structure	Fungi	Bacteria	
			Gram-positive	Gram-negative
Penam		*Aspergillus* *Penicillium* *Epidermophyton* *Trichophyton* *Polypaecilum* *Malbranchea* *Sartorya* *Pleurophomopsis*		
Cephem		*Cephalosporium* *Anixiopsis* *Arachnomyces* *Spiroidium* *Scopulariopsis* *Diheterospora* *Paecilomyces*	*Streptomyces* *Nocardia*	*Flavobacterium* *Xanthomonas* *Lysobacter*
Clavam			*Streptomyces*	
Carbapenem			*Streptomyces*	*Serratia* *Erwinia*
Monobact			*Nocardia*	*Pseudomonas* *Gluconobacter* *Chromobacterium* *Agrobacterium* *Acetobacter*

[a]Only certain species within the microorganisms listed make the indicated β-lactam antibiotic (Elander, 1983)

such as phenylacetyl and phenoxyacetyl to create hydrophobic penicillins (penicillin V and penicillin G, respectively). Other fungi and the bacterial β-lactam producers all possess additional biosynthetic enzymes that expand and modify the ring structure to create cephalosporins and cephamycins. The structure, organization and regulation of these genes is now discussed.

TWO GENES COMMON TO THE SYNTHESIS OF PENICILLINS AND CEPHALOSPORINS

ACV synthetase gene

An unusually large gene of approximately 11.4 kb codes for the first enzyme in the biosynthetic pathway. It is designated *pcb*AB for its involvement in

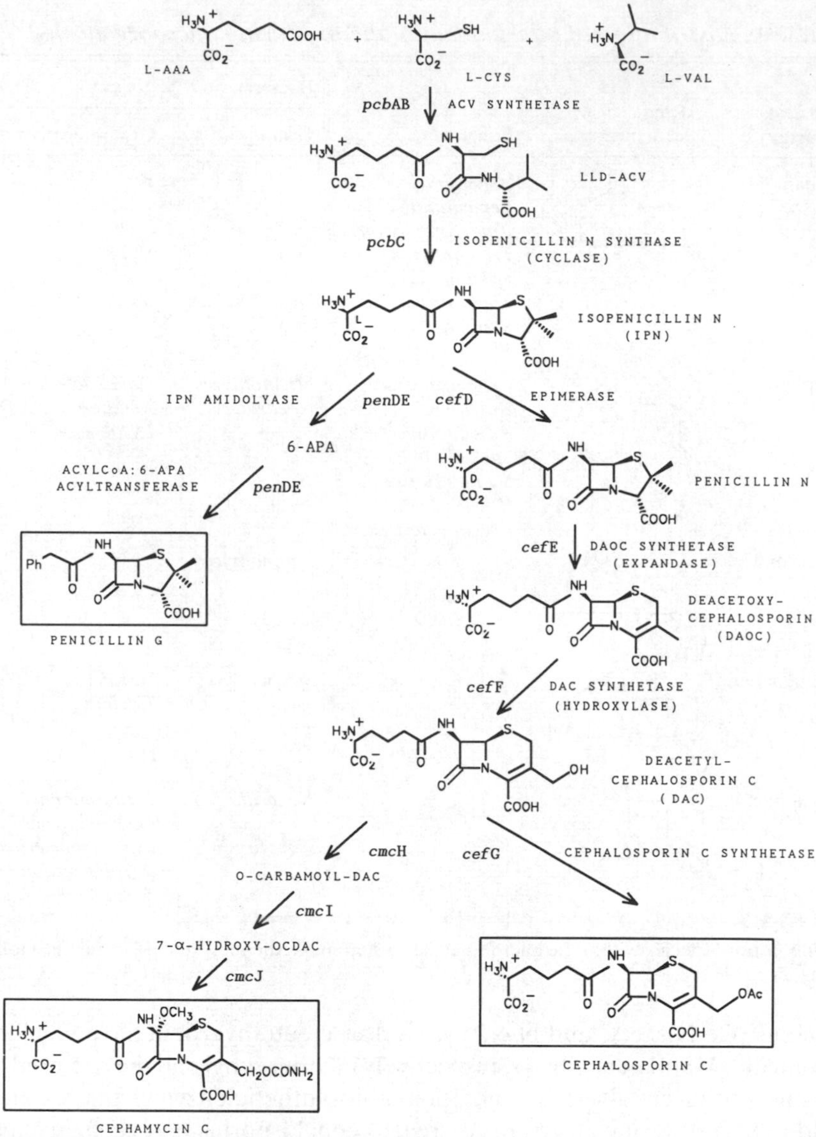

Fig. 1. Biosynthesis of penicillin and cephalosporin antibiotics in bacteria and fungi (Aharonowitz *et al.*, 1992). Abbreviations: *pcb*, penicillin and cephalosporin biosynthesis; *cef*, cephalosporin biosynthesis; *pen*, penicillin biosynthesis; *cmc*, cephamycin biosynthesis; 6-APA, 6-amino penicillanic acid.

penicillin and cephalosporin biosynthesis and because earlier studies suggested that two genetic loci might be required for its function, one for making the α-aminoadipyl-cysteine dipeptide, another for its completion to ACV. More recent genetic and biochemical data indicate that a single gene encodes all the activities needed to synthesize ACV synthetase (ACVS) (Banko, Demain & Wolfe, 1987; Diez *et al.*, 1989). These and other studies *in vitro* show that ACVS is a multivalent enzyme of 405–425 kD size that functionally resembles the antibiotic peptide synthetases (Aharonowitz *et al.*, 1993). It catalyses in a similar manner the ATP-dependent activation of each of the precursor amino acids, binds them as thiol esters, and polymerizes the bound amino acids prior to epimerizing L-valine and releasing ACV. Comparative sequence analysis of ACVS genes with peptide synthetase genes confirms their relatedness (Aharonowitz *et al.*, 1993).

Several strategies were used to isolate ACVS genes. In each case it was assumed that the genes determining β-lactam biosynthesis exist in the genome in a single continuous segment. The view that antibiotic biosynthetic genes exist in gene clusters gradually emerged from studies using actinomycetes. It was demonstrated initially for genes involved in the synthesis of the polyketide antibiotic actinorhodin in *S. coelicolor* (Malpartida & Hopwood, 1984). Subsequently it was shown for important polyketide antibiotics including tetracyclines, rifamycin and avermectin, macrolides like erythromycin, anthracyclines such as adriamycin and other classes of secondary metabolites. The only exception to date to this rule is methylenomycin, an antibiotic whose production is governed by a plasmid (Kirby & Hopwood, 1977)). Furthermore, antibiotic biosynthetic gene clusters typically contain auxiliary genes needed for self-resistance and for regulation of the pathway. Clustering of two β-lactam biosynthetic pathways has also been observed (see below).

We can illustrate how clustering of biosynthetic genes greatly simplifies their isolation by reference to two independent studies of the isolation of the *P. chrysogenum* ACVS gene. In one study, phage and cosmid clones that had previously been shown to contain the IPNS gene (see below) were used to complement *P. chrysogenum* mutants defective in ACVS and blocked for penicillin synthesis. Several clones restored penicillin production and expressed a protein of very high molecular weight that was absent in the mutant host (Diez *et al.*, 1990). DNA sequencing showed that a large open reading frame was located immediately upstream of the IPNS gene (Fig. 2) potentially coding for a polypeptide of about 425 kD. In another study, the *P. chrysogenum* ACVS gene was cloned by heterologous hybridization employing DNA probes derived from a cephalosporin producing *Flavobacterium* sp.. The bacterial and fungal biosynthetic pathways share just two genes, those for ACVS and isopenicillin N synthase (IPNS) (Fig. 1). It was possible after elimination of the IPNS cross-hybridization signals to identify a sequence upstream of the IPNS gene that coded for a large open reading frame with the expected size for a 425 kD protein (Smith *et al.*, 1990*a*; Smith *et al.*, 1990*c*). Some small discrepancies found in the sequence data of the two genes appear to be due to strain differences.

The ACVS genes have been cloned and sequenced from seven bacterial

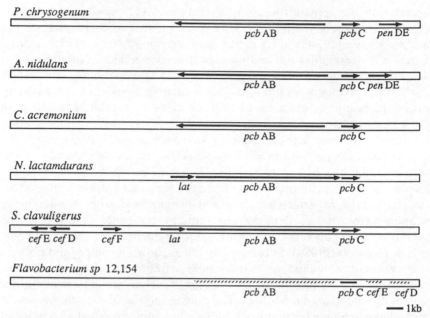

Fig. 2. Physical map of the β-lactam biosynthetic cluster in bacteria and fungi (Aharonowitz *et al.*, 1992). Arrows indicate size of open reading frames of genes, and arrowheads show direction of transcription. Structural genes and products: *pen*DE, acyltransferase; *pcb*C isopenicillin N synthase; *pcb*AB, ACV synthetase; *lat*, lysine ε-aminotransferase; *cef*D, epimerase; *cef*E, expandase; *cef*F, hydroxylase; (see also Fig. 1). In *C. acremonium* a second gene cluster contains *cef*EF (expandase/hydroxylase) and *cef*G (cephalosporin C synthetase) (Gutierrez *et al.*, 1992; Mathison *et al.*, 1993). In *N. lactamdurans*, *cef*D, *cef*E and genes coding for a transmembrane protein and a penicillin binding protein, *cmc*T and *pbp*, respectively, are located immediately upstream of the *lat* gene and oppositely transcribed; four more open-reading frames involved in conversion of deacetoxy cephalosporin to cephalosporin C are located downstream of *pcb*C and transcribed in the same direction; beyond them is a β-lactamase gene, *bla*. A region of approximately 30 kb in *N. lactamdurans* contains 13 genes related to cephamycin biosynthesis and resistance (Coque *et al.*, 1993a).

and fungal sources. All of the genes lack introns. A conspicuous feature of the predicted protein sequences is the presence of three regions, or domains, denoted A, C and V to indicate their respective roles in ACV synthesis, each of about 570 amino acids. Each domain shares extensive sequence similarity with the *Bacillus brevis* peptide synthetases, gramicidin S synthetase I and tyrocidin synthetase I, as well as two other enzymes that carry out ATP-pyrophosphate exchanges, namely firefly luciferase and coumerate CoA ligase. Each of these enzymes has the ability to activate amino acids. A plausible mechanism of action for ACVS is that it functions as a multienzyme thiol-containing template carrying out in sequential fashion peptide bond formation through a pantotheine cofactor (Aharonowitz *et al.*, 1993). ACVS illustrates the principle of colinearly arranged enzyme functions (active sites) integrated into a multifunctional protein. A similar principle operates in the synthesis of polyketide antibiotics (Katz & Donadio, 1993).

Isopenicillin N synthase gene

The first report of the isolation of a β-lactam biosynthetic gene appeared in 1985 (Samson et al., 1985). It described the use of reverse genetics to clone the isopenicillin N synthase (IPNS) gene of the fungus C. acremonium. Shortly afterwards, IPNS genes, designated pcbC, were isolated from other microbial sources by this approach and by heterologous hybridization. At present the IPNS genes from 12 different species have been cloned and sequenced: in each case it was found that the IPNS gene is physically located near to the ACVS gene and other β-lactam biosynthetic genes (Fig. 2).

It is interesting to note that two recent studies describe the use of hybridization techniques to show that IPNS-gene-like sequences appear to be present in some non-producer streptomycetes strains. Thus, S. coelicolor and S. lividans, neither of which are known to be β-lactam producers, respond positively towards IPNS gene probes from producer strains (Shiffman et al., 1988; Garcia-Dominguez, Liras & Martin et al., 1991). These results raise the possibility that β-lactam biosynthetic genes may be more widespread in microorganisms than has been considered until now, and has relevance for the origin of the pathway (see below). There is also evidence that some streptomycetes harbour non-functional copies of the IPNS gene, perhaps due to lack of proper control elements (Shiffman et al., 1988; Garcia-Dominguez et al., 1991). Hybridization techniques could provide a new and sensitive approach for developing screening programmes that detect potential β-lactam producers not on the traditional basis of production but rather on the basis of genetic potential (Aharonowitz, Cohen & Martin, 1992; Hutchinson, 1994).

Sequence analysis of bacterial and fungal IPNS genes and proteins reveal they are closely related (Cohen et al., 1990). For example, the deduced primary amino acid sequences of the two groups of microbial proteins share approximately 60% sequence identity. Very high sequence identity among functionally related procaryotic and eucaryotic proteins is not typical and suggests some important underlying feature. Multiple sequence alignment of nine IPNS proteins shows that the similarity is scattered throughout the molecules (Fig. 3). Unfortunately, this situation precludes the possibility of readily identifying the functionally important domains involved in substrate binding and catalytic activity.

The mechanism of action of IPNS is a subject of great interest for two main reasons. In the first place IPNS is one of a number of non-heme ferrous iron containing dioxygenases that carry out remarkable chemistry (Baldwin & Bradley, 1990; Cooper, 1993). The enzyme catalyses the stereospecific cyclization of ACV to form the β-lactam and thiazolidine rings of isopenicillin N by abstracting four hydrogen atoms from the substrate with formation

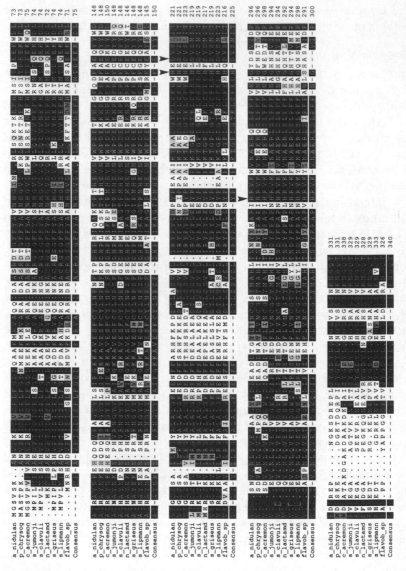

Fig. 3. Multiple sequence alignment of nine bacterial and fungal IPNS proteins. Identical amino acids at each site are shown boxed in black. Arrows indicate amino acid ligands in the IPNS-Fe(II) active site.

of two molecules of water. Other related non-heme dioxygenases are involved in a wide range of secondary metabolic processes, including aliphatic hydroxylations and hetero-ring cyclizations. A common mechanism appears to unite these diverse reactions. In the second place, IPNS possesses a wide latitude of substrate recognition and has been used to create a number of novel types of penicillins employing analogues of ACV (Baldwin & Bradley, 1990; Huffman et al., 1992). Of about 150 chemically synthesized tripeptides that were tested as substrates for IPNS, a little over one half were converted to β-lactams to differing extents and of these some 25% had bioactive properties. These studies demonstrate the potential for creating new antibacterials starting from unnatural substrates. They also throw light on the enzyme–substrate interaction. In this context protein engineering of IPNS may provide a fertile area for developing new IPNS enzymes.

Recent progress in understanding the mechanism of the IPNS reaction has come from studies on the role of the ferrous iron in the active site. Spectroscopic analysis of IPNS has led to a model in which the Fe(II) atom in the active enzyme is coordinated to two histidine and one aspartic acid endogenous ligands and to three exogenous ligands, namely ACV, O_2 and a water molecule (Ming et al., 1991; Randall et al., 1993). In order to identify the endogenous ligands, a site-directed mutagenesis study was carried out in which each of seven conserved histidines and five conserved aspartic acids in the S. jumonjinensis IPNS were replaced by alanines (Fig. 3). The analysis identified His 212, Asp 214 and His 268 as being essential for activity (Cohen et al., 1994). Evidence that these residues are in fact involved in binding to the ferrous atom came from a multiple sequence alignment of 54 Fe(II)-containing dioxygenases, including two other β-lactam biosynthetic enzymes, expandase and hydroxylase (see below). The analysis demonstrated the presence of a recurring motif, His-X-Asp-(53–57 X)-His, in all these enzymes and which in IPNS contains the identical two histidines and one aspartic acid essential for function (Cohen et al., 1994).

A gene specific for synthesis of penicillins

Acyltransferase gene

A single gene present in the penicillin producing fungi, P. chrysogenum and A. nidulans, but notably absent from C. acremonium and bacterial cephalosporin producers, codes for a 40 kD protein containing activities required for removing the α-aminoadipyl side chain of penicillins and replacing it with a hydrophobic group (Barredo et al., 1989b) (Fig. 1). This acyltransferase (AT) gene is designated penDE to indicate its exclusive role in penicillin production. When it was transformed into a P. chrysogenum acyltransferase-deficient mutant penicillin production was restored (Barredo et al., 1989b). A similar transformation of C. acremonium, a cephalosporin producer, led to its acquiring the ability to make benzylpenicillin thus providing one of the first instances of rational modification of a β-lactam biosynthetic pathway (see below).

In contrast to the two early biosynthetic genes for the ACVS and IPNS enzymes, the *P. chrysogenum* and *A. nidulans* AT genes contain three introns located in equivalent positions (Barredo *et al.*, 1989b; Montenegro *et al.*, 1990; Alpin *et al.*, 1993). All three fungal genes are organized and transcribed in the same manner (Fig. 2). That they are sufficient for penicillin production was demonstrated by transferring the *P. chrysogenum* gene cluster into *A. niger* and *Neurospora crassa*, fungi that do not make β-lactam antibiotics, and converting them to penicillin producers (Smith *et al.*, 1990b).

Biochemical analysis of purified acyltransferase shows that it contains one 11 kD (α) and one 29 kD (β) sub-unit per active enzyme. A post-translational processing event cleaves the 40 kD polypeptide between Gly 102 and Cys 103 to yield the heterodimer (Montenegro *et al.*, 1990; Whitman *et al.*, 1990; Alpin *et al.*, 1993).

Genes specific for synthesis of cephalosporins

Epimerase gene

Cephalosporin producers possess a gene that is responsible for the isomerization of the L-α aminoadipyl side chain of isopenicillin N to make penicillin N, which is the substrate for the ring-expansion enzyme (Fig. 1). The epimerase gene is designated *cef*D, for its involvement in cephalosporin synthesis, and codes for a 44 kD protein. It was identified initially by reverse genetics in a cosmid clone of *S. clavuligerus* that contained other β-lactam biosynthetic genes and was used subsequently as a probe to isolate related actinomycete epimerase genes (Kovacevic *et al.*, 1990). The epimerase gene of *S. clavuligerus* is located immediately upstream of the expandase gene (see below) and is expressed from the same transcript (Fig. 2). A slightly different arrangement of the two genes is found in *Nocardia lactamdurans* (*S. lactamdurans*) (Coque *et al.*, 1993b). Profile sequence analysis predicts that the epimerases belong to the family of pyridoxal phosphate enzymes.

Expandase and hydroxylase genes

In the bacterial cephalosporin producers two separate genes, designated *cef*E and *cef*F, code for enzymes that sequentially expand the five-membered thiazolidine ring to the six-membered dihydrothiazine ring and hydroxylate the latter ring respectively (Fig. 1). In contrast, *C. acremonium* possesses a single gene, designated *cef*EF, that specifies an enzyme with both activities. The *S. clavuligerus cef*E and *cef*F genes were isolated from cosmid clones containing other β-lactam biosynthetic genes, using as probes long synthetic oligodeoxynucleotides made according to the N-terminal and internal peptide sequences of the purified enzymes (Kovacevic *et al.*, 1989; Kovacevic & Miller, 1991). The fungal *cef*EF gene was isolated in a similar way, except that it was found to map in a different chromosomal location than the AVCS and IPNS genes (Samson *et al.*, 1987). The former maps on chromosome II, the latter on chromosome VI (see Fig. 2).

Not unexpectedly, the bacterial and fungal genes and their deduced proteins exhibit a high degree of sequence similarity. For example, the fungal expandase/hydroxylase enzyme shares 57% sequence identity with the *S. clavuligerus* expandase. Equally revealing is the high sequence identity (59%) between the bacterial expandase and hydroxylase proteins. Not only are the sequences of the genes alike, but careful studies with highly purified recombinant proteins show that the *S. clavuligerus* hydroxylase contains a weak ring-expansion activity and the expandase has a low hydroxylation activity (Baker *et al.*, 1991). The two enzymes also carry out similar reactions each requiring as cofactors 2-ketoglutarate, ferrous iron and molecular oxygen. The bacterial expandase and hydroxylase may therefore be considered to be bifunctional enzymes resembling the fungal expandase/hydroxylase. These findings are compatible with an ancestral dual function gene, inherited by certain fungi, that underwent gene duplication in some bacteria resulting in the loss of expandase in one copy and hydroxylase in the other.

Cephalosporin C synthetase gene

The biosynthesis of β-lactams in *C. acremonium* terminates with the formation of cephalosporin C; *Streptomyces* and *Flavobacterium* spp. produce a more diverse group of cephamycins and substituted cephalosporins, respectively. All specify a gene, designated *cef*G, that codes for the enzyme cephalosporin C synthetase and that carries out the transfer of the acetyl group from acetyl-CoA to the hydroxyl moiety on the sulphur-containing ring. The *cef*G gene was cloned recently from *C. acremonium* and found to be linked to *cef*EF and transcribed divergently (Gutierrez *et al.*, 1992; Mathison *et al.*, 1993). Indirect evidence points to the *cef*G gene being part of a large cluster of β-lactam genes in cephamycin producers (Chen *et al.*, 1988; Coque *et al.* 1993*a*). An even larger gene cluster occurs in the cephamycin producers, *S. clavuligerus*, *S. jumonjinensis* and *S. katsurahamanus*, that make a second class of β-lactam compounds, the clavulanic acids (Fig. 1), which are potent β-lactamase inhibitors. In these organisms the cephamycin and clavulanic acid biosynthetic genes map next to one another forming a 'super-cluster' of β-lactam genes (Jensen *et al.*, 1993; Ward & Hodgson, 1993).

Genes specific for cephamycin synthesis

Carbamylase, hydroxylase and methylase genes

Further biosynthetic genes are involved in the conversion of cephalosporin C to cephamycin C (Fig. 1). In *S. clavuligerus* these code for enzymes that carry out the stepwise O-carbamylation, 7-α-hydroxylation and O-methylation of the cephalosporin nucleus. Genes responsible for these activities are designated *cmc*H, *cmc*I and *cmc*J, respectively to indicate their role in cephamycin synthesis. At present only the 7-α-hydroxylase gene has been cloned although it seems likely (see above) that all three genes are present in the β-lactam biosynthetic gene cluster in the actinomycetes (Xiao *et al.*, 1993; Coque *et al.*, 1993*a*). Other modifying β-lactam biosynthetic genes must be postulated to account for the production by some streptomycetes of the related cephamycins A and B and by *Flavobacterium* sp. 12,154 of a complex variety of 7-formamido derivations of cephalosporins.

Auxiliary genes of the β-lactam biosynthetic cluster

Analysis of the β-lactam gene cluster in several bacterial cephalosporin producers reveals a more complex organization involving additional genes for the synthesis of α-aminoadipic acid (AAA), the precursor of ACV, as well as genes rendering the organism insensitive to inhibition by its own antibiotics.

Lysine aminotransferase gene

Actinomycete strains that produce cephalosporins and cephamycins possess an enzyme, L-lysine ε-aminotransferase, that mediates the formation of AAA from lysine by removal of the ε-amino group. Fungal β-lactam producers make AAA as an intermediate in the synthesis of ACV and so would not appear to need this enzyme. The gene coding for this activity, designated *lat*, was found to occur exclusively in β-lactam-producing *Streptomyces* and *Nocardia* spp.; it maps upstream of the ACVS and IPNS genes and is transcribed in the same direction (Fig. 2) (Coque *et al.*, 1991; Madduri *et al.*, 1991). Clustering of the *lat* gene with the β-lactam biosynthetic genes represents an effective mechanism to coordinate primary amino acid metabolism with antibiotic production.

Self-resistance genes to β-lactams

Production of an antibiotic poses a potential threat to the viability of the producer strain. A variety of mechanisms have evolved to protect the producer organism by modifying the target of action of the antibiotic, or the antibiotic itself, or by exporting it (Davies, 1994). For example, *Saccharopolyspora erythrae* renders itself insensitive to the anti-ribosomal antibiotic it makes by specific methylation of ribosomal RNA, whereas *S. hygroscopi-*

cus, a producer of the glutamine synthetase inhibitor, bialophos, acetylates an intermediate in the pathway resulting in its conversion to an inactive form.

Very recently, sequence extension of the *N. lactamdurans* β-lactam gene cluster led to identification of three additional genes that code for a typical β-lactamase, a penicillin-binding protein (PBP4) and a transmembrane protein (Coque *et al.*, 1993a). The β-lactamase is similar to the class A β-lactamases of β-lactam-resistant clinical isolates, supporting the view that antibiotic resistance genes in pathogenic bacteria are derived from antibiotic-producing organisms. It is secreted and active against penicillins but not against cephamycin C and is synthesized during the active growth phase prior to formation of the cephamycin biosynthetic enzymes. The PBP4 protein is one of eight such proteins found to accumulate in the cell membrane, none of which binds cephamycin C. It resembles the DD-carboxypeptidases of actinomycetes. A transmembrane protein, encoded by the *cmc*T gene in the cluster, is probably involved in control of β-lactam synthesis and secretion. Based on the above properties of the gene products, it is proposed that they are required for the balanced synthesis and excretion of β-lactams to allow cell survival.

REGULATION OF β-LACTAM BIOSYNTHESIS

After this summary of the spectacular advances in our knowledge of the structure and organization of β-lactam biosynthetic and resistance genes, the reader may be surprised to learn how little we presently know about the molecular mechanisms controlling the pathway. Production of β-lactam antibiotics characteristically takes place during the late stages of the growth cycle in batch culture and appears to reflect the combined effects of low growth rate and nutrient imbalance (Demain, 1991). Numerous studies have shown that the carbon, nitrogen and phosphorus sources, and certain amino acids, can profoundly affect antibiotic yields. Recent efforts to understand the genetic basis for these events have focused on analyzing transcription processes in the biosynthetic gene cluster.

Several research groups have reported that transcriptional regulation is a key feature of the control of β-lactam gene expression. For example, in *C. acremonium* and *S. clavuligerus*, the levels of IPNS mRNA dramatically increased towards the end of the growth phase and correlated well with the onset of IPNS synthesis and antibiotic production (Smith *et al.*, 1990; Petrich *et al.*, 1992). Similar studies in *A. nidulans* showed that induction of IPNS mRNA occurred only after cell growth arrest and when glucose repression was relieved (Penalva *et al.*, 1989; Brakhage *et al.*, 1992). Carbon catabolite repression in *A. nidulans* appears to be controlled by two separate mechanisms, one mediated by a system different from the known carbon regulation system in primary metabolism and another mediated by pH control (Espeso *et al.*, 1993). External alkaline pH, or mutations in the *pac*C gene, which affect pH regulation, act as a physiological signal to trigger penicillin biosynthesis. Certain amino acids have marked effects on the biosynthetic enzymes as well as on expression of β-lactam biosynthetic genes. Lysine was found to inhibit both IPNS and ACVS gene expression in *A. nidulans*, whereas methionine increased levels of transcripts of both genes and expandase/hydroxylase in *C. acremonium* (Brakhage & Turner, 1992; Velasco *et al.*, 1994). Clearly, an enormously complex set of growth, nutrient and other environmental signals can alter the expression of β-lactam biosynthetic genes. Much further work will be necessary to unravel the mechanisms by which these signals operate. However, it may be anticipated that similar studies of other secondary metabolites will reveal some common underlying principles (Chater, 1992).

GENETIC ENGINEERING OF β-LACTAM PATHWAYS

It will be evident from the above discussion that a large number of structural and regulatory genes determine β-lactam production, especially for the more complex cephamycin antibiotics. Current research on β-lactam producer-strain improvement has therefore largely shifted to rationally applying the key molecular technologies, outlined earlier in this review, for the precise manipulation of the genetic components of the β-lactam pathway. One major goal has been to identify and overcome rate-limiting steps in the pathway, either by increasing the copy number of relevant genes or by increasing their expression. A second objective has been to alter existing biosynthetic routes towards desirable metabolites that can otherwise be made only by chemical synthesis (Skatrud, 1992). Examples of each are given below.

Strain improvement

Initial attempts to improve penicillin production in *P. chrysogenum* by introducing extra copies of IPNS genes failed because it later transpired that these high-producer strains contain multiple copies of biosynthetic genes. For instance, one overproducer strain had 8–10 copies of the IPNS gene, another contained a DNA segment with IPNS and AT genes amplified 14-fold (Smith *et al.*, 1989; Barredo *et al.*, 1989a). The first success in β-lactam pathway engineering was achieved through inserting an additional copy of the *C. acremonium* expandase/hydroxylase gene into a high-producer strain (Skatrud *et al.*, 1989). Analysis of that strain indicated that penicillin N accumulated in the fermentation medium suggesting that its conversion to deacetoxycephalosporin C (DAOC) (Fig. 1) might be rate limiting. One transformant, which had a single extra copy of the expandase/hydroxylase gene, integrated into chromosome III, whereas the endogenous gene is located on chromosome II, and produced approximately two-fold more DAOC than the parent strain. In large-scale fermentation the yield of cephalosporin C, the end product, was only 15% above that of the parent strain but still a significant improvement in an industrial producer-strain. Studies with another *C. acremonium* high-producing cephalosporin C strain revealed that additional copies of the cephalosporin C acetyltransferase gene led to increased titres (Mathison *et al.*, 1993). A direct relationship was found between gene copy number and antibiotic levels. Similar gene dosage approaches are finding use in other commercially important β-lactam producers (Veenstra *et al.*, 1991).

Modification of biosynthetic pathways

Modification of existing biosynthetic pathways in industrial strains offers a promising new approach for producing key intermediates for synthesis of valuable cephalosporins and cephamycins. Medically important cephalosporins are presently made by derivatizing the 7-amino group of 7-aminodeacetoxy cephalosporanic acid (7-ADCA). The latter is made by chemical ring expansion of penicillin G followed by enzymatic deacylation to remove the hydrophobic side chain. *P. chrysogenum* is a commercial producer of penicillin G but is unable to make cephalosporins (Fig. 1). It was proposed to carry out the ring expansion step in *P. chrysogenum* by introducing a recombinant version of the *S. clavuligerus* expandase gene. The *C. acremonium* expandase/hydroxylase gene is not suitable since it might lead to hydroxylated products. A first step towards achieving this end was the creation of a hybrid gene in which the flanking control elements of the *P. chrysogenum* IPNS gene were fused to the bacterial expandase gene (Cantwell *et al.*, 1990). Transformants were isolated that possessed levels of expandase activity comparable to that in *C. acremonium*. Providing that the expandase gene can be genetically engineered to accept penicillin G or V as a substrate, then a fermentation route for 7-ADCA production becomes possible.

A similar strategy is being used to enable *P. chrysogenum* to produce DAOC. The *S. clavuligerus* expandase and *S. lipmanii* epimerase genes were joined to the control regions of the *P. chrysogenum* IPNS and AT genes, respectively, and the hybrid constructs were expressed in the host with synthesis of low levels of DAOC (Cantwell *et al.*, 1992). Optimization of this process will require elimination of the endogenous acyl transferase activity. This may prove to be difficult by gene disruption because high-producing *P. chrysogenum* strains contain multiple copies of AT genes. The reverse genetic engineering experiment of cloning the *P. chrysogenum* AT gene into *C. acremonium* has also been carried out, resulting in its conversion to a producer of hydrophobic penicillins (Gutierrez *et al.*, 1991).

A novel application of pathway engineering is the recent modification of

C. acremonium to make 7-aminocephalosporanic acid (7-ACA). Chemical synthesis of 7-ACA is the starting point for production of a number of important cephamycin antibiotics. The conversion of cephalosporin C to 7-ACA was achieved *in vivo* by introducing into *C. acremonium* the D-amino oxidase gene of *Fusarium solani* and the cephalosporin acylase gene of *Pseudomonas diminuta* under the control of the fungal alkaline protease control signals (Isogai *et al.*, 1991). Expression of the hybrid genes led to the stepwise oxidative deamination and hydrolysis of the cephalosporin C side chain with formation of low levels of the desired product, pointing to the feasibility of a fermentation process.

EVOLUTION OF THE β-LACTAM BIOSYNTHETIC PATHWAY

Several features of the β-lactam biosynthetic genes are of interest from the standpoint of understanding the evolution of the pathway. The interpretation of the high degree of sequence similarity of the microbial biosynthetic genes, their organization in gene clusters and the unusual and restricted group of organisms that contain them, has prompted fertile new ideas on the origin of β-lactams.

Sequence divergencies among the bacterial and fungal β-lactam genes indicate a much closer relatedness than that expected for the β-lactam producer organisms. Primary amino acid sequences of IPNS proteins share nearly 60% identity (Cohen *et al.*, 1990). ACVS proteins possess an overall sequence identity of 43% but this increases to approximately 60% if individual domains are compared. Interpretation of the evolutionary significance of this data has been controversial.

Two conflicting hypotheses have been advanced to account for these findings. It was suggested for IPNS that the high sequence identity among the microbial IPNS proteins might reflect functional constraints on the enzyme resulting in a slower rate of molecular evolution. Alternatively, a horizontal gene transfer event, plausibly from the bacteria to the filamentous fungi and occurring after the separation of the procaryotes and eucaryotes, could also explain the sequence data (Weigel *et al.*, 1988; Penalva *et al.*, 1990; Landan *et al.*, 1990). The latter scenario was preferred mainly on the basis of assumptions about rates of changes and the lack of evidence suggesting that the ability to produce β-lactam antibiotics confers a selective advantage for survival of the organism. However, a recent re-examination of a somewhat larger sequence data base, including the distantly related expandase, appears to favour the view that the evolution of IPNS is consistent with a more conventional descent (Smith *et al.*, 1992).

Is there any evidence other than the sequence data that bears on the evolution of the pathway? One curious finding that seems to support horizontal inheritance rather than vertical transmission is the peculiarly limited distribution in nature of β-lactam producing species (Table 1).

Another piece of evidence is the fact that none of the fungal IPNS and ACVS genes contain introns, a situation that is consistent with a bacterial donor for the putative gene-transfer event. Although the acyltransferase genes do contain introns, it is likely that they are fungal in origin and were recruited for penicillin synthesis. A comparison of the nucleotide compositions of fungal and bacterial IPNS genes also hints at a bacterial origin, the fungal genes having a G + C content reminiscent of that of streptomycetes and *Flavobacterium* sp..

Some other comparisons of the biosynthetic genes, in particular the way they are organized and expressed in clusters, is relevant to the evolutionary analysis. The bacterial β-lactam producers contain all the known biosynthetic genes in a single cluster that is chromosomally located. However, their organization is not always the same (Fig. 2). In the fungal producers the ACVS and IPNS genes are linked but in contrast to the bacterial gene clusters they are transcribed in opposing directions. *C. acremonium* has a second gene cluster, containing the expandase/hydroxylase and cephalosporin C synthetase genes, located on a separate chromosome. How these and other features of the microbial β-lactam gene clusters fit into the evolutionary scenarios depicted above remains to be clarified.

CONCLUDING REMARKS

Molecular characterizations of the genes involved in the biosynthesis of the sulphur-containing β-lactam antibiotics is now nearing completion. The genes responsible for penicillin and cephalosporin formation have all been isolated, sequenced and mapped. Other genes necessary for cephamycin production have been identified, as well as the auxiliary genes that connect primary metabolisms to the β-lactam pathway, and that determine resistance to β-lactams and ensure their secretion. Invariably the genes are organized in gene clusters. Much less is known about the mechanisms that control their expression. One of the major tasks of future studies will be to analyse the *cis*- and *trans*-acting regulatory elements that govern expression of the biosynthetic genes. A start in this direction has been made.

The striking sequence similarity of bacterial and fungal biosynthetic genes has provoked stimulating speculation on the origin of the pathway. Horizontal gene transfer between the two groups of microorganisms is an attractive hypothesis to account for the sequence data but it is by no means proved. More sequence data from other β-lactam biosynthetic genes should help to resolve this issue.

Intensive efforts are currently being made to manipulate existing biosynthetic pathways in industrial producer strains so as to increase β-lactam yields and to make valuable intermediates for the semi-synthetic cephalosporins and cephamycins. Rate-limiting steps can be overcome by amplifying the copy number of individual biosynthetic genes or by increasing their expression. Biosynthetic pathways can now be re-routed to produce desirable end products by introducing new genes. Engineering of biosynthetic enzymes to change their substrate specificity and catalytic activity is a crucial aspect of these endeavours. Moreover, these studies have great intrinsic interest in their own right. We anticipate that the next few years will bring fundamental new insights into the mechanism of action of key enzymes in the β-lactam pathway, especially ACVS and IPNS, and will result in the creation of new classes of catalysts for making novel antibiotics.

REFERENCES

Aharonowitz, Y., Bergmeyer, J. & Cantoral, J. M. et al. (1993). δ-(L-α-aminoadipyl)-L-cysteinyl-D-valine synthetase, the multienzyme integrating the four primary reactions in β-lactam biosynthesis, as a model peptide synthetase. Biotechnology, 11, 807–10.

Aharonowitz, Y., Cohen, G. & Martin, J. F. (1992). Penicillin and cephalosporin biosynthetic genes: structure, organization, regulation and evolution. Annual Review in Microbiology, 46, 461–95.

Aplin, R. T., Baldwin, J. E., Cole, S. C., Sutherland, J. D. & Tobin, M. B. (1993). On the production of α, β-heterodimeric acyl-coenzyme A: Isopenicillin N-acetyltransferase of Penicillium chrysogenum. Studies using a recombinant source. FEBS 319, 166–7.

Arnstein, H. R. V. & Morris, D. (1960). The structure of a peptide containing α-aminoadipic acid, cysteine and valine, present in the mycelium of Penicillium chrysogenum. Biochemical Journal, 76, 357–61.

Baker, B. J., Dotzlaf, J. E. & Yeh, W.-K. (1991). Deacetoxycephalosporin C hydroxylase of Streptomyces clavuligerus: purification, characterization, bifunctionality, implication. Journal of Biological Chemistry, 266, 5087–93.

Baldwin, J. E. & Bradley, M. (1990). Isopenicillin N synthase: mechanistic studies. Chemical Review, 90, 1079–88.

Banko, G., Demain, A. L. & Wolfe, S. (1987). ACV synthetase: a multifunctional enzyme with broad specificity for the synthesis of penicillin and cephalosporin precursors. Journal of the American Chemical Society, 109, 2858–60.

Barredo, J. L., Diez, B., Alvarez, E. & Martin, J. F. (1989a). Large amplification of a 35 kb DNA fragment carrying two penicillin biosynthetic genes in high penicillin producing strains of Penicillium chrysogenum. Current Genetics, 16, 453–9.

Barredo, J. L., Van Solingen, P., Diez, B., Alvarez, E., Cantoral, J. M., Kattevilder, A., Smaal, E. B., Groenen, M. A. M., Veenstra, A. E. & Martin, J. F. (1989b). Cloning and characterization of acyl-CoA: 6-APA acyltransferase gene of Penicillium chrysogenum. Gene, 83, 291–300.

Bierman, M., Logan, R., O'Brien, K., Seno, E. T., Rao, R. N. & Schoner, B. E. (1992). Plasmid cloning vectors for the conjugal transfer of DNA from Escherichia coli to Streptomyces spp. Gene, 116, 43–9.

Brakhage, A. A., Browne, P. & Turner, G. (1992). Regulation of Aspergillus nidulans penicillin biosynthesis and penicillin biosynthesis genes acvA and ipnA by glucose. Journal of Bacteriology, 174, 3789–99.

Brakhage, A. A. & Turner, G. (1992). L-lysine repression of penicillin biosynthesis genes acvA and ipnA in Aspergillus nidulans. FEMS Microbiology Letters, 98, 123–8.

Cantwell, C. A., Beckmann, R. J., Dotzlaf, J. E., Fisher, D. L., Skatrud, P. L.,

Yeh, W. K. & Queener, S. W. (1990). Cloning and expression of a hybrid *Streptomyces clavuligerus cef*E gene in *Pencillium chrysogenum*. *Current Genetics*, **17**, 213–21.

Cantwell, C. A., Beckman, R. J., Whiteman, P., Queener, S. W. & Abraham, E. P. (1992). Isolation of deacetoxycephalosporin C from fermentation broths of *Penicillium chrysogenum* transformants: construction of a new fungal biosynthetic pathway. *Proceedings of the Royal Society of London*, **248**, 283–9.

Chater, K. F. (1992). Genetic regulation of secondary metabolic pathways in *Streptomyces*. In *Secondary Metabolites, Their Function and Evolution. Ciba Foundation Symposium, Vol. 171.* (Chadwick, D. J. & Whelan, J., eds.). Wiley, Chichester, UK, pp. 144–62.

Chen, C. W., Lin, H. F., Kuo, C. L., Tsai, H. L., & Tsai, J. F. (1988). Cloning and expression of a DNA sequence conferring cephamycin production. *Bio/ Technology*, **6**, 1222–4.

Cohen, G. & Aharonowitz, Y. (1981). The microbiological production of pharmaceuticals. *Scientific American*, **245**, 141–52.

Cohen, G., Landman, O., Borovok, I., Kreisberg, R. & Aharonowitz, Y. (1994). Genetic and sequence analysis of the ferrous active site of isopenicillin N synthase. Abstract presented at the 7th International Symposium on Genetics of Microorganisms, Montreal.

Cohen, G., Shiffman, D., Mevarech, M. & Aharonowitz, Y. (1990). Microbial isopenicillin N synthase genes: structure, function, diversity and evolution. *Trends in Biotechnology*, **8**, 105–11.

Cooper, R. D. (1993). The enzymes involved in biosynthesis of penicillin and cephalosporin; their structure and function. *Biomedical Chemistry*, **1**, 1–17.

Coque, J. J., Martin, J. F. & Liras, P. (1993*b*). Characterization and expression in *Streptomyces lividans* of cefD and cefE genes from *Nocardia lactamdurans*: the organization of the cephamycin gene cluster differs from that in *Streptomyces clavuligerus*. *Molecular General Genetics*, **236**, 453–8.

Coque, J. J., Liras, P., Laiz, L. & Martin, J. F. (1991). A gene encoding lysine 6-aminotransferase which forms α-aminoadipic acid, a precursor of β-lactam antibiotics, is located in the cluster of cephamycin biosynthetic genes in *Nocardia lactamdurans*. *Journal of Bacteriology*, 173, 6258–64.

Coque, J. J., Liras, P. & Martin, J. F. (1993*a*). Genes for a β-lactamase, a penicillin-binding protein and a transmembrane protein are clustered with the cephamycin biosynthetic genes in *Nocardia lactamdurans*. *EMBO Journal*, **12**, 631–9.

Davies, J. (1994). Inactivation of antibiotics and dissemination of resistance genes. *Science*, **264**, 317–82.

Demain, A. L. (1991). Production of beta-lactam antibiotics and its regulation. *Proceedings of the National Science Council ROC, Taiwan B*, **15**, 251–65.

Diez, B., Barredo, J. L., Alvarez, E., Cantoral, J. M., Van Solingen, P., Groenen, M. A., Veenstra, A. E. & Martin, J. F. (1989). Two genes involved in penicillin biosynthesis are linked in a 5.1 kb *Sal*I fragment in the genome of *Penicillium chrysogenum*. *Molecular and General Genetics*, **218**, 572–6.

Diez, B., Gutierrez, S., Barredo, J. L., Van Solingen, P. M., Van der Voort, L. H. & Martin, J. F. (1990). The cluster of penicillin biosynthetic genes. Identification and characterization of the *pob*AB gene encoding the α-aminoadipyl-cysteinyl-valine synthetase and linkage to the *pcb*C and *pen*DE genes. *Journal of Biological Chemistry*, **265**, 16358–65.

Distler, J., Mansouri, K., Mayer, G., Stockman, M. & Pipersberg, M. (1992). Streptomycin biosynthesis and its regulation in Streptomycetes. *Gene*, **115**, 105–11.

Elander, R. P. (1983). Strain improvement and preservation of β-lactam producing microorganisms. In *Antibiotics Containing the β-Lactam Structure* (Demain, A. L. & Solomon, N. A., eds.). Springer-Verlag, Berlin, **Vol. 1**, pp. 97–146.

Espeso, E. A., Tilburn, J., Arst, H. N. & Penalva, M. A. (1993). pH regulation is a major determinant in expression of a fungal penicillin biosynthetic gene. *EMBO Journal*, **12**, 3947–56.

Fawcett, P. A., Usher, J. J., Huddleston, J. A., Bleaney, R. C., Nisbet, J. J. & Abraham, E. P. (1976). Synthesis of δ-(L-α-aminoadipyl) cysteinylvaline and its role in penicillin biosynthesis. *Biochemical Journal*, **157**, 651–60.

Garcia-Dominguez, M., Liras, P. & Martin, J. F. (1991). Cloning and characterization of the isopenicillin N synthase gene of *Streptomyces griseus* NRRL 3851 and studies of expression and complementation of the cephamycin pathway in *Streptomyces clavuligerus*. *Antimicrobial Agents in Chemotherapy*, **35**, 44–52.

Gutierrez, S., Diez, B., Alvarez, E., Barredo, J. L. & Martin, J. F. (1991). Expression of the *pen*DE gene of *Penicillium chrysogenum* encoding isopenicillin N acyltransferase in *Cephalosporium acremonium*: production of benzyl penicillin by the transformants. *Molecular and General Genetics*, **225**, 56–64.

Gutierrez, S., Velasco, J., Fernandez, F. J. & Martin, J. F. (1992). The *cef*G gene of *Cephalosporium acremonium* is linked to the *cef*EF gene and encodes a deacetyl-cephalosporin C acetyltransferase closely related to homoserine O-acetyltransferase. *Journal of Bacteriology*, **174**, 3056–64.

Hopwood, D. A. (1993). Genetic engineering of *Streptomyces* to create hybrid antibiotics. *Current Opinion in Biotechnology*, **4**, 51–7.

Hopwood, D. A. & Khosia, C. (1992). Genes for polyketide secondary metabolite pathways in microorganisms and plants. In *Secondary Metabolites, Their Function and Evolution. Ciba Foundation Symposium, Vol 171* (Chadwick, D. J. & Wheelan, J., eds.). Wiley, Chichester, UK, pp. 88–112.

Huffman, G. W., Gesellchen, P. D., Turner, J. R., Rothenberger, R. B., Osborne, H. E., Miller, F. D., Chapman, J. L. & Queener, S. W. (1992). Substrate specificity of isopenicillin N synthase. *Journal of Medical Chemistry*, **35**, 897–914.

Hutchinson, R. (1994). Drug synthesis by genetically engineered microorganisms. *Bio/Technology*, **12**, 375–80.

Ingolia, T. D. & Queener, S. W. (1989). Beta-lactam biosynthetic genes. *Medicinal Chemical Review*, **9**, 245–64.

Isogai, T., Fukagawa, M., Aramori, I., Iwami, M., Kojo, H., Ono, T., Ueda, Y., Kohsaka, M. & Imanaka, H. (1991). Construction of a 7-aminocephalosporanic acid (7ACA) biosynthetic operon and direct production of 7ACA in *Acremonium chrysogenum*. *Bio/Technology*, **9**, 188–91.

Jensen, S. E., Alexander, D. C., Paradkar, A. S. & Aidoo, K. A. (1993). Extending the β-lactam biosynthetic gene cluster in *Streptomyces clavuligerus* In *Industrial Microorganisms: Basic and Applied Molecular Genetics* (Baltz, R. H., Hegeman, G. D. & Skatrud, P. L., eds.). A.S.M. Washington, DC, pp. 169–175.

Jensen, S. E., Westlake, D. W., Bowers, R. J. & Wolfe, S. (1982). Cephalosporin formation by cell-free extracts from *Streptomyces clavuligerus*. *Journal of Antibiotics*. (Tokyo), **35**, 1351–60.

Katz, L. & Donadio, S. (1993). Polyketide synthesis: prospects for hybrid antibiotics. *Annual Review in Microbiology*, **47**, 875–912.

Kieser, T. & Hopwood, D. A. (1991). Genetic manipulation of *Streptomyces*: integrating vectors and gene replacement. *Methods in Enzymology*, **204**, 430–58.

Kirby, R. & Hopwood, D. A. (1977). Genetic determination of methylenomycin synthesis by the SCPI plasmid of *Streptomyces coelicolor* A3(2). *Journal of General Microbiology*, **98**, 239–52.

Konomi, T., Herchen, S., Baldwin, J. E., Yoshida, M., Hunt, N. A. & Demain, A. L. (1979). Cell free conversion of delta-(L-alpha aminoadipyl)-L-cysteinyl-D-valine into an antibiotic with the properties of isopenicillin N in *Cephalosporium acremonium*. *Biochemical Journal*, **184**, 427–30.

Kovacevic, S. & Miller, J. R. (1991). Cloning and sequencing of the β-lactam hydroxylase gene (*cef*F) from *Streptomyces clavuligerus*: gene duplication may have led to separate hydroxylase and expandase activities in the actinomycetes. *Journal of Bacteriology*, **173**, 398–400.

Kovacevic, S., Tobin, M. B. & Miller, I. R. (1990). The β-lactam biosynthesis genes for isopenicillin N epimerase and deacetoxycephalosporin C synthetase are expressed from a single transcript in *Streptomyces clavuligerus*. *Journal of Bacteriology*, **172**, 3952–8.

Kovacevic, S., Weigel, H. B., Tobin, M. B., Ingola, T. D. & Miller, J. R. (1989). Cloning, characterization and expression in *Escherichia coli* of the *Streptomyces clavuligerus* gene encoding deacetoxycephalosporin C synthetase. *Journal of Bacteriology*, **171**, 754–60.

Landan, G., Cohen, G., Aharonowitz, Y., Shuali, Y., Grauer, D. & Shiffman, D. (1990). Evolution of isopenicillin N synthase genes may have involved horizontal gene transfer. *Molecular Biology Evolution*, **7**, 399–406.

MacCabe, A. P., Riach, M. B., Unkles, S. Q. & Kinghorn, J. R. (1990). The *Aspergillus nidulans npe*A locus consists of three contiguous genes required for penicillin biosynthesis. *EMBO Journal*, **9**, 279–87.

Madduri, K., Stuttard, C. & Vining, L. (1991). Cloning and locution of a gene governing lysine ε-aminotransferase, an enzyme initiating beta-lactam biosynthesis in *Streptomyces* sp.. *Journal of Bacteriology*, **171**, 299–302.

Malpartida, F., Hallam, S. E., Kieser, H. M., Motamedi, H., Hutchinson, C., Butler, M. J., Sugden, D. A., Warren, M., McKillop, C., Bailey, G. R., Humphreys, G. O. & Hopwood, D. A. (1987). Homology between *Streptomyces* genes coding for synthesis of different polyketides used to clone antibiotic biosynthetic genes. *Nature*, **325**, 818–21.

Malpartida, F. & Hopwood, D. A. (1984). Molecular cloning of the whole biosynthetic pathway of a *Streptomyces* antibiotic and its expression in a heterologous host. *Nature*, **309**, 462–4.

Mathison, L., Soliday, C., Stepan, T., Aldrich, T. & Rambosek, J. (1993). Cloning, characterization and use in strain improvement of the *Cephalosporium acremonium cef*G encoding acetyl transferase. *Current Genetics*, **23**, 33–41.

Mazodier, P., Peter, R. & Thompson, C. (1989). Intergeneric conjugation between *Escherichia coli* and *Streptomyces species*. *Journal of Bacteriology*, **171**, 3583–5.

Ming, L. J., Que, L., Kriaucianas, A., Frolik, C. A. & Chen, V. J. (1991). NMR studies of the active site of isopenicillin N synthase a non-heme iron(II) enzyme. *Journal of Biochemistry*, **30**, 11653–9.

Montenegro, E., Barredo, J. L., Gutierrez, S., Diez, B., Alvarez, E. & Martin, J. F. (1990). Cloning, characterization of the acyl-CoA:6 amino penicillanic acid acyltransferase gene of *Aspergillus nidulans* and linkage to the isopenicillin B synthase gene. *Molecular and General Genetics*, **221**, 322–30.

Penalva, M. A., Moya, A., Dopazo, J. & Ramon, D. (1990). Sequences of isopenicillin N synthetase suggest horizontal gene transfer from prokaryotes to eukaryotes. *Proceedings of the Royal Society of London*, **241B**, 164–9.

Penalva, M. A., Vian, A., Patino, C., Perez-Aranda, A. & Ramon, D. (1989).

Molecular biology of penicillin production in *Aspergillus nidulans*. In *Genetics and Molecular Biology of Industrial Microorganisms* (Hershberger, C. L., Queener, S. W. & Hegeman, G., eds.). *American Society of Microbiology*, pp. 256–61. Washington DC.

Petrich, A. K., Xiaoning, W., Roy, K. L. & Jensen, S. (1992). Transcriptional analysis of the isopenicillin N synthase-encoding gene of *Streptomyces clavuligerus*. *Gene*, 111, 77–84.

Randall, C. R., Zang, Y., True, A. E., Que, L., Charnock, J. M., Garner, C. D., Fujishima, Y., Schofield, C. J. & Baldwin, J. E. (1993). X-ray absorption studies of the ferrous active site of isopenicillin N synthase and related model complexes. *Biochemistry*, 32, 6664–73.

Samson, S. M., Belagaje, R., Blankenship, D. T., Chapman, J. L., Perry, D., Skatrud, P. L., Van Frank, R. M., Abraham, E. P., Baldwin, J. E., Queener, S. W. & Ingolia, T. D. (1985). Isolation, sequence determination and expression in *Escherichia coli* of the isopenicillin N synthetase gene from *Cephalosporin acremonium*. *Nature*, 38, 191–4.

Samson, S. M., Dotzlaf, J. E., Slisz, M., Becker, G. W., Van Frank, R. M., Veal, L. E., Yeh, W. K., Miller, J. R., Queener, S. W. & Ingolia, T. D. (1987). Cloning and expression of the fungal expandase/hydroxylase gene involved in cephalosporin biosynthesis. *Bio/Technology*, 5, 1207–16.

Sanchez, F., Lozano, M., Rubio, V. & Penalva, M. A. (1987). Transformation of *Penicillium chrysogenum*. *Gene*, 51, 97–102.

Shiffman, D., Mevarech, M., Jensen, S. E., Cohen, G. & Aharonowitz, Y. (1988). Cloning and comparative sequence analysis of the gene coding for isopenicillin N synthase in *Streptomyces*. *Molecular and General Genetics*, 214, 562–9.

Skatrud, P. L. (1992). Genetic engineering of β-lactam antibiotic biosynthetic pathways in filamentous fungi. *Trends in Biotechnology*, 10, 324–32.

Skatrud, P. L., Queener, S. W., Carr, L. C., & Fisher, D. L. (1987). Efficient integrative transformation of *C. acremonium*. *Current Genetics*, 12, 337–48.

Skatrud, P. L., Tietz, A. J., Ingolia, T. D., Cantwell, C., Fisher, D. L., Chapman, J. L. & Queener, S. W. (1989). Use of recombinant DNA to improve production of cephalosporin C by *Cephalosporium acremonium*. *Bio/Technology*, 7, 477–86.

Smith, A. W., Ramsden, M., Dodsen, M. J., Harford, S. & Peberdy, J. F. (1990). Regulation of isopenicillin N synthetase (IPNS) gene expression in *Acremonium chrysogenum*. *Bio/Technology*, 8, 237–40.

Smith, D. J., Bull, J. H., Edwards, J. & Turner, G. (1989). Amplification of the isopenicillin N synthetase gene in a strain of *Penicillium chrysogenum* producing high levels of penicillin. *Molecular and General Genetics*, 216, 492–7.

Smith, D. J., Burnham, M. K., Bull, J. H., Hodgson, J. E., Ward, J. M., Browne, P., Barton, B., Earl, A. J. & Turner, G. (1990a). β-Lactam antibiotic biosynthetic genes have been conserved in clusters in prokaryotes and eukaryotes. *EMBO Journal*, 9, 741–7.

Smith, D. J., Burnham, M. K., Edwards, J., Earl, A. J. & Turner, G. (1990b). Cloning and heterologous expression of the penicillin biosynthetic gene cluster from *Penicillium chrysogenum*. *Bio/Technology*, 8, 39–41.

Smith, D. J., Earl, A. J. & Turner, G. (1990c). The multifunctional peptide synthetase performing the first step in penicillin synthesis in *Penicillium chrysogenum* is a 421,073 dalton protein similar to *Bacillus brevis* peptide antibiotic synthetase. *EMBO Journal*, 9, 2743–50.

Smith, M. W., Feng, D. F. & Doolittle, R. F. (1992). Evolution by acquisition: the case for horizontal gene transfer. *TIBS*, 17, 489–93.

Turner, G. (1992). Genes for the biosynthesis of β-lactam compounds in microorganisms. In *Secondary Metabolites: Their Function and Evolution. Ciba Foundation Symposium, Vol. 171* (Chatwick, D. J. & Wheelan, J., eds.). Wiley, Chichester, UK. pp. 113–128.

Veenstra, A. E., van Solingen, P., Bovenberg, R. A. & van der Voort, L. H. (1991). Strain improvement in *Penicillium chrysogenum* by recombinant DNA techniques. *Journal of Biotechnology*, **17**, 81–90.

Velasco, J., Gutierrez, S., Fernandez, F. J., Marcos, A. T., Arenos, C. & Martin, J. F. (1994). Exogenous methionine increases levels of mRNAs transcribed from *pcb*AB, *pcb*C and *cef*EF genes, encoding enzymes of the cephalosporin biosynthetic pathway in *Acremonium chrysogenum*. *Journal of Bacteriology*, **176**, 985–91.

Walz, M. & Kuck, U. (1993). Targeted integration into the *Acremonium chrysogenum* genome: disruption of the *pcb*C gene. *Current Genetics*, **24**, 421–7.

Ward, J. M. & Hodgson, J. E. (1993). The biosynthetic genes for clavulanic acid and cephamycin production occur as a 'super-cluster' in three *Streptomyces*. *FEMS Microbiology Letters*, **110**, 239–42.

Weigel, B. J., Burgett, S. G., Chen, V., Skatrud, P. L., Frolik, C. A., Queener, S. W. & Ingolia, T. D. (1988). Cloning and expression in *Escherichia coli* of isopenicillin N synthetase genes from *Streptomyces lipmanii* and *Aspergillus nidulans*. *Journal of Bacteriology*, **170**, 3817–26.

Whitman, P. A., Abraham, E. P., Baldwin, J. E., Fleming, M. D., Schofield, C. J., Sutherland, J. D. & Willis, A. C. (1990). Acyl coenzyme A: 6 amino-penicillanic acid acyltransferase from *Penicillium chrysogenum* and *Aspergillus nidulans*. *FEBS Letters*, **262**, 342–4.

Wright, F. & Bibb, M. L. (1992). Codon usage in the G + C rich *Streptomyces* genome. *Gene*, **113**, 55–65.

Xiao, X., Hinterman, G., Hausler, A., Barber, P. J., Foor, F., Demain, A. L. & Piret, J. (1993). Cloning of a *Streptomyces clavuligerus* DNA fragment encoding the cephalosporin 7-alpha-hydroxylase and its expression in *Streptomyces lividans*. *Antimicrobial Agents Chemotherapy*, **37**, 84–8.

Yelton, M. M., Hamer, J. E. & Timberlake, W. E. (1984). Transformation of *Aspergillus nidulans* by using a *trp*C plasmid. *Proceedings of the National Academy of Sciences, USA*, **81**, 1470–4.

WHO NEEDS NEW ANTIMICROBIALS?

JOHN F. RYLEY

2 Wych Lane, Adlington, Macclesfield, Cheshire SK10 4NB, UK

ORIGINS OF ANTIMICROBIAL CHEMOTHERAPY

It is an interesting fact that modern antimicrobial chemotherapy has much of its origin in tropical medicine and parasitology. Thus cinchona bark was used by the Indians of Peru for treating malaria, and was introduced into European medicine by the Spaniards in the early seventeenth century. The active principle, quinine, isolated in 1820, remained the only treatment for malaria until well into the twentieth century, and in the current situation of multiple drug resistance is still a valuable agent. Similarly, Ipecacuanha root was used in Brazil, and probably in Asia also, for the treatment of diarrhoea and dysentery. The active constituent emetine, isolated in 1817, was shown in 1891 to have specific therapeutic value in amoebic dysentery, and 100 years later is still in use.

When in the early years of this century a more directed approach to the control of infectious diseases was envisaged, Thomas and Breinl in Liverpool working with atoxyl and Ehrlich in Germany studying a series of synthetic aniline dyes used trypanosomes in mice as their target organisms. It was only later when Ehrlich transferred his attentions to arsenicals and to the idea of molecular modification to achieve a better therapeutic index that he extended his target organisms to include the spirochaetes of syphilis and relapsing fever. It is unfortunate that, in the current developments in antimicrobial chemotherapy, parasitic and tropical diseases are being largely ignored.

CHANGING TARGETS AND PERCEPTIONS

This tendency is well illustrated by the writer's personal experience. Recruited into the pharmaceutical industry in 1952 to work on malaria, he spent 25 years studying the biochemistry, biology and particularly the chemotherapy of the majority of parasitic protozoa causing diseases of medical and veterinary importance. It was like the kiss of death to each project or parasite with which he became involved! The reason the projects were dropped one by one was not however, personal; rather, the company increasingly realized that there was little hope of recouping research costs, let alone making a profit, when working with diseases of the third world. What the author experienced in his company was, by and large, being experienced by colleagues in rival companies. In the climate of battling for economic survival, the pharmaceutical industry has increasingly concentrated its efforts on the most prevalent diseases of the developed world. Only a few companies maintain a limited interest in parasitic diseases – though attitudes are being modified somewhat as a result of AIDS. In the current annual report of the writer's old company it is stated that 'Many diseases have no effective means of treatment or cure. Pharmaceutical companies have an important role to play in researching the products that will treat such diseases in the future.' Although not stated in so many words, it is obvious that such laudable statements are only meant to apply to diseases where the victims – or their health authorities – have the ability to pay, and to pay handsomely. The situation is, however, that the companies *could* play an invaluable role in tackling diseases of the third world without compromising to any great extent their economic targets; the author believes they have a moral responsibility to do so.

GLOBAL CONCERN

It was in the light of this developing situation that the World Health Organization in association with UNDP (United Nations Development Programme) and the World Bank in 1975 established a Special Programme for Research and Training in Tropical Diseases (TDR). Over the years, this programme has been financed by a number of countries, industrial companies, and private and governmental research foundations in addition to its main sponsors. Six target diseases, or rather disease groups, were identified; these include malaria, schistosomiasis, lymphatic filariasis and onchocerciasis, leprosy, African trypanosomiasis, Chagas disease, and leishmaniasis. The idea was to develop new means of disease control by sponsoring research in academic institutions in the western world and by training and research institute support, to develop such capabilities in the endemic countries. Research into each disease has been supervised and funded by separate committees, each with a permanent secretary in Geneva. Research has included, where appropriate, studies on diagnosis, drug treatment, immunology and vaccination, vector control, and social and economic problems. Although appreciable progress has been made in some directions, much remains to be done.

TROPICAL DISEASES OF INTEREST TO TDR

Malaria

Malaria is acknowledged to be by far the most important tropical parasitic disease, causing great suffering and loss of life. Control is becoming increasingly difficult and the epidemiological situation is likely to continue to deteriorate over the next few years. Some 300 million people are believed to be infected with malaria parasites, with 90% of them living in tropical Africa. Of the four species of parasite, *Plasmodium falciparum* accounts for most of the infections in Africa, and for over one-third of infections in the rest of the world. More than two billion people, nearly 40% of the world's population, are at risk. Increasing travel for business or pleasure to the more exotic parts of the world puts increasing numbers of westerners at risk. Many infected people in endemic areas are partially protected from symptomatic disease by concominant immunity acquired through recent previous infection, so that severe illness is found almost exclusively in non-immunes, notably young children, immigrants and visitors, and migratory workers entering highly endemic areas, and in the inhabitants of areas with unstable endemicity. Worldwide it is estimated that there are about 120 million clinical cases of malaria a year with 0.5–1.2 million deaths, mainly among African children below the age of five. Each malaria attack may last for 5–15 days, often incapacitating the patient. The estimated annual direct and indirect cost of malaria in Africa is US$ 1800 million!

Vector control is often not cost-effective in areas where the disease is most severe and where interruption of transmission cannot be sustained. Many years ago, malaria incidence in India was considerably reduced and in Sri Lanka virtually eliminated by the continual use of DDT; immunity in the population decreased in the absence of repeated challenge. When change in political and economic circumstances led to the cessation of massive and continual DDT usage, epidemic malaria quickly returned with horrendous mortality among the now fully susceptible population. Although vector control has its place in the battle against malaria, it is unlikely to be a total solution to the problem. Current TDR research is directed to the use of insecticide-impregnated bed-nets; preliminary results from The Gambia indicate a significant fall in child mortality where such nets are used.

Hopes of the rapid production of a vaccine following Trager and Jensen's cultivation of *P. falciparum in vitro* in 1976 were not realized. TDR has supported work aimed at the development of pre-erythrocytic vaccines, asexual blood stage vaccines and transmission-blocking vaccines. Particular interest is presently centred on the evaluation of the SPf66 Patarroyo vaccine, which has been claimed to give malaria control in 40–60% of a vaccinated population.

In the light of the limitations of vector control and vaccination, there is an urgent need for new antimalarial drugs for both therapeutic and prophylactic use. In view of past experience of the development of drug resistance, this is likely to be a continuing need. TDR has been, or is, involved in the development and clinical evaluation of drugs such as mefloquine, halofantrine, pyronaridine, the primaquine analogue WR 238605, and several derivatives of the Chinese plant extract artemisinin, as well as maintaining a more basic screening programme.

Leprosy

Leprosy has one of the widest distributions of the diseases of interest to TDR and is endemic throughout the tropics. For centuries leprosy has been feared for its crippling mutilations and social stigma. In most infected people the infection remains asymptomatic, but in a minority of people it may lead to clinical disease, with potentially disfiguring deformities. There are estimated to be around 5.5 million cases of leprosy in the world, with about one million new cases each year; no estimates are available of the number of people with subclinical infection. The incidence of deformities is probably around 140 000 new cases per year; some studies suggest that the life expectancy of severely disabled persons is reduced by 50%, mainly due to economic hardship.

Leprosy was first treated with chaulmoogra oil and more recently with dapsone (DDS; diamino diphenyl sulphone), but treatment was long if not for life, and results limited; DDS resistance became a complicating issue. The chemotherapy of leprosy has been transformed over the last decade by the introduction of multidrug therapy (MDT). A variety of drugs, developed and tested in man for more conventional and economically important bacterial infections, have been shown in various permutations and combinations to have marked curative properties in leprosy patients. These drugs include, as well as dapsone, rifampicin, clofazimine, thioamide, and more recently clarithromycin, minocycline and ofloxacin and other fluoroquinolones. TDR is involved in numerous multicentre trials of various multidrug regimens. The advantages of multidrug therapy include more certain cure, higher cure rates and far quicker response to treatment than was obtained with dapsone in the past – and hence better patient compliance. So promising are the results that, in 1991, the World Health Assembly approved the WHO strategy for the global elimination of leprosy as a public health problem; the aim is to bring leprosy prevalence below one case per 10 000 population by the year 2000. The success of the campaign may be limited by the availability of finance (MDT drugs are expensive!) and by difficulties associated with identifying patients with a disease which carries such a social stigma. In addition, freedom from the development of drug resistance and/or the availability of new and more potent antibacterials may affect the ultimate outcome.

It is because of the possibility that multidrug therapy may not be the final solution, and because blocking disease transmission is important, that TDR maintains a research and clinical interest in vaccination against leprosy. Of particular interest are trials with BCG with or without heat-killed *Mycobacterium leprae* produced from armadillos, and trials with vaccines incorporating two Indian atypical cultivable mycobacteria. Longer-term studies are aimed at the production of second-generation sub-unit vaccines.

African trypanosomiasis

Sleeping sickness is endemic in 36 sub-Saharan African countries, occurring in some 200 discrete foci where the disease occasionally resurges. Around 15 000–20 000 new cases of trypanosomiasis are reported each year, but the actual number of new cases is much more likely to be in the range of 200 000–300 000. The disease may be responsible for tens of thousands of deaths annually. Severe epidemics in the early years of the century in Central and East Africa resulted in 750 000 deaths and in the depopulation of several endemic areas. Sleeping sickness is caused by infection with either *Trypanosoma brucei gambiense* or *T. b. rhodesiense*; the two parasites have different geographical distributions and produce different forms of disease – the development and progression of disease with *T. b. gambiense* may take several years in contrast to the more acute *T. b. rhodesiense*. Both forms are ultimately fatal in the absence of treatment. Due to non-specific symptoms and the paucity of facilities for diagnosis in many endemic areas, acute disease is often missed and progression to the neurological, second 'sleeping' stage takes place. Treatment of acute disease with pentamidine (for gambiense) or suramin (for rhodesiense) is fairly simple and effective; treatment of late-stage disease requires the use of arsenicals such as melarsoprol, which is often associated with toxic encephalitis and not insignificant mortality. TDR has been involved in the discovery of trypanocidal activity in the failed anticancer drug eflornithine (DFMO; difluoromethylornithine; Ornidyl) and its subsequent development for clinical use. The compound has been called 'the resurrection drug' because of its ability to rescue comatose patients in the terminal stages of disease. However, the drug has its limitations: it is only effective in the gambiense form of sleeping sickness, and the need to administer intravenous doses of around 20 grammes daily for 14 days is not ideal for an African bush situation – besides being costly!

Economic constraints and political instability in endemic countries are hindering control, and a decline in available resources and a reduction in surveillance and treatment has led to outbreaks of disease. Population movements in certain areas can also be responsible for outbreaks. Populations of the vector tsetse fly are not great, and transmission in communities can be reduced, if not eliminated, in a cost-effective manner by the use of fly traps. The application of deltamethrin to cattle can also reduce tsetse populations and the incidence of animal trypanosomiasis – which is an important disease limiting cattle production and hence affecting the protein nutrition of the population.

Chagas disease

Chagas disease, originally a zoonosis, is endemic in South and Central America, where it is recognized as an important public health problem and is receiving increasing priority for control. The disease is caused by the haemoflagellate *Trypanosoma cruzi*, which is found in the bloodstream, but which multiplies in an aflagellate form in muscular tissues. The disease is mainly transmitted by blood-sucking triatome bugs, which find a favourable habitat in crevices in the walls of poor quality houses in rural areas and in unplanned urban developments; a less frequent mode of transmission is through blood transfusion with unscreened, untreated blood. Acute disease is mild except in very young children, who may develop myocarditis and meningoencephalitis – fatal in 50% of cases. Those who survive the acute disease enter a chronic phase. Symptoms may develop 10–20 years later in a third of those infected, and may result in death from myocarditis, megaoesophagus or megacolon. It is estimated that 16–18 million people are currently infected with *T. cruzi*, 2–3 million of whom are showing symptoms of the chronic disease and a further 3 million of whom will develop – and probably die from – symptoms of the chronic disease in the future. The incidence of new infection is probably close to 1 million cases per year, and the annual mortality around 45 000.

There is no effective chemotherapy for treating infected patients. Nifurtimox and benznidazole have some effect in acute cases, but treatment is prolonged, the drugs are very toxic, and availability is limited. TDR are conducting clinical trials with allopurinol. There is no vaccine for Chagas disease nor any prospect for one in the near future. Traditional vector control methods based on insecticide spraying and newer approaches using fumigant canisters and insecticidal paints are proving to be highly effective, not only in reducing vector densities, but also in reducing or in interrupting domestic transmission and the incidence of infection. Housing improvement schemes have also shown good results, and have the great advantage of not just being a method of control, but of directly improving the living standards of the population. Because of cycling in animal reservoirs it is not feasible to eradicate infection in all endemic areas. It is not uncommon to find a 10–20% seropositivity rate in blood banks in endemic areas. Serological screening and discarding or treating contaminated blood with gentian violet would eliminate this avoidable but significant means of transmission.

Leishmaniasis

Leishmaniasis is endemic in the tropical regions of America, Africa, and the Indian sub-continent, and in the sub-tropics of south-west Asia and the Mediterranean. Leishmaniasis comprises a group of diseases with very different clinical manifestations and public health consequences, ranging from self-healing but disfiguring lesions in a minority of people infected with cutaneous species to severe epidemics with visceral species involving high fatality rates. Cutaneous leishmaniasis, which may be produced by several species of *Leishmania*, is the most prevalent, producing skin ulcers which may take more than a year to heal. Mucocutaneous leishmaniasis initially causes similar lesions, which may heal but reappear to cause hideous tissue destruction, primarily of the nose and mouth. Visceral leishmaniasis is a very severe systemic disease which is nearly always fatal if left untreated. Leishmaniasis is increasingly being seen in AIDS patients.

Of the 350 million people who live in areas where the disease may be caught by the bite of infected sandflies, some 12 million are estimated to be infected, with 3 million displaying clinical symptoms. There are around 1.5 million new cases each year, one third of which are with the visceral disease. Epidemic outbreaks of visceral leishmaniasis have recently occurred in India and the Sudan, and drug resistance is becoming an increasing problem.

Chemotherapy is largely based on antimony preparations such as Pentostam and Glucantime. These have to be given by daily injection for a period of several weeks, and are somewhat toxic; antimony resistance is becoming a problem, particularly in India. Ketoconazole has shown activity in some forms of cutaneous leishmaniasis as has allopurinol, either alone or with antimonials. TDR has been involved in clinical trials with these agents and with topical preparations containing paromomycin (aminosidine). Currently, trials are under way with AmBisome (a liposomal formulation of amphotericin B) and other trials with parenteral aminosidine, alone and with antimony, are being set up under the auspices of TDR.

Vector control by spraying is difficult. Insecticide spraying combined with early diagnosis and treatment of cases could be effective in the control of visceral leishmaniasis in the Indian sub-continent where man-to-man transmission only is involved, but control of infected dogs would have to be included in programmes applicable to Brazil or the Mediterranean region. In places such as Mexico, where the infection is contracted in forests, neither vector not reservoir control would be feasible.

Control of leishmaniasis by vaccination is not yet possible, but research in experimental models and in limited human trials suggests it may one day be feasible. In areas where disfigurement from 'oriental sore' is a distinct possibility, it has often been the custom to inoculate a normally covered area of the body with material from an active sore in another person; the resulting lesion, when it eventually heals, ensures immunity to further infection with that species. TDR-associated trials in humans in Brazil, Venezuela and Iran are utilizing killed organisms of cutaneous species in association with BCG. Safety of doses required for adequate immune response is being assessed; protective efficacy trials have yet to be initiated. Animal models suitable for immunological work are difficult, but some success is being achieved with vervet monkeys in Kenya. Several avenues to produce recombinant live *Leishmania* vaccine, either by selective depletion of leishmanial genes or introducing leishmanial genes in live bacterial vectors, are being pursued.

Other diseases of interest to TDR

Although not strictly microbiological, the TDR programme is also concerned with schistosomiasis and filarial diseases such as onchocerciasis and various forms of lymphatic filariasis. Schistosomiasis is the second most prevalent tropical disease and a leading cause of severe morbidity in several foci in Africa, Asia and South America. Around 200 million people are infected with one or other of the three types of schistosome, which result in liver, intestinal and urinary complications resulting from reactions to schistosome eggs lodged in the tissues. All species are transmitted by water snails. Although there has been limited local success in schistosome control in some areas, overall the disease is spreading, due in part to large-scale irrigation projects which create new habitats for the water snail vectors. Chemotherapy of all three forms of schistosomiasis has been revolutionized by the discovery of the drug praziquantel, so much so that TDR is concentrating the majority of its resources into vaccine development rather than looking for a new drug.

Lymphatic filariasis is one of the most prevalent tropical diseases and one of the most neglected; some 78.6 million people in Asia, Africa and South America are estimated to be infected. The several species of *Wuchereria* and *Brugia* involved are transmitted by various species of mosquito. The most prevalent of the chronic manifestations of disease are hydrocele – a grossly enlarged and hanging scrotum – and lymphoedema of the arms and legs, including the most advanced and feared stage, elephantiasis. Over one million people are deformed with this latter condition.

Onchocerciasis or river blindness affects 18 million people, 95% of whom live in Africa; there are isolated foci also in Latin America and the Yemen. The parasite *Onchocerca volvulus* is transmitted by the bite of blackflies which breed in fast-flowing waters. The last 15 years have seen a decline in the prevalence of onchocerciasis infection and morbidity, largely due to WHO co-ordinated vector control programmes in 11 countries of West Africa supported by outside agencies and based on aerial spraying of larvicides over large areas. The registration of ivermectin (Mectizan) in 1987 as an effective and safe microfilaricide and its provision free of charge by the manufacturer Merck, Sharp and Dohme has enhanced this disease reduction. Annual mass treatment with the drug reduces the microfilaria population in the body, thereby reducing the risk of blindness developing and also reducing the scale of transmission. The disease has been completely eliminated in some areas, but continual epidemiological surveillance is necessary for early detection of possible recrudescences which would require the reintroduction of control measures.

Ivermectin kills microfilariae, and thereby prevents the pathogenesis caused by these microfilariae, and also prevents transmission of the parasite. Ivermectin does not, however, kill the adult worms, which gradually resume microfilaria production following ivermectin treatment. Hence the need for repeated annual or more frequent treatment of populations. It is the adult worms which are responsible for the more serious consequences of lymphatic filariasis. TDR's major concern is thus the discovery of a drug with activity against macrofilariae, a difficult target since the parasite is metabolically inactive apart from its reproductive activity. Diethylcarbamazine has a limited efficacy and some benzimidazoles are under investigation, but the ideal macrofilaricide has yet to be discovered.

INTEGRATED CHEMOTHERAPY (I-CHEM)

Treatment of African sleeping sickness in its early stages relies on pentamidine and suramin, and in its late stages on melarsoprol or DFMO. All these drugs have to be given by injection, the latter ones over a prolonged period of time – an undesirable feature for drugs to be used in the African bush. Moreover, melarsoprol is very toxic and DFMO is only effective against the gambiense form of disease. There is basically no treatment for Chagas disease. The nitro compounds nifurtimox and benznidazole have some utility in acute disease, and although they are orally active, have to be given for prolonged periods of time, which often results in neurotoxicity. Leishmaniasis treatment relies heavily on the use of antimonials. These have to be given by injection daily over several weeks, lead to not insignificant toxicity, and are becoming ever less useful due to drug resistance. In all three areas there is a desperate need for new treatments. Because of the conditions under which treatment usually has to be given, and because of difficulties in ensuring patient compliance, the ideal drugs would be orally active and effective in a single dose, or at least with a short course of treatment.

In 1991 the responsibility for chemotherapy was taken from the three separate TDR committees responsible for African trypanosomiasis, Chagas disease and leishmaniasis, and vested in a new committee, I-Chem (integrated chemotherapy). The priorities of this committee have been to establish integrated screening centres and to identify the most promising targets for new drug discovery. An integrated screen has been set up in London, and two other centres have subsequently been established in Belgium and Switzerland. In each centre a single sample of compound can be tested *in vitro* against the parasites of African trypanosomiasis, Chagas disease, and leishmaniasis, and subsequently evaluated *in vivo* against any of these parasites as appropriate and necessary. Many potential approaches to drug discovery have in the past been pursued, some with TDR support. In many cases the interest of projects has been almost wholly academic, with little more than lip service being paid to drug discovery; many projects have been very speculative and/or basic in nature. In view of the urgent need for treatments for these tropical infections and the very limited resources available to support sponsored research, much attention has been given to target identification; three areas have been selected as being most likely to lead to a new drug for one or more of the diseases in the short to medium term. In each area there is evidence for activity *in vivo*; in no case has a wonder drug yet been identified.

Sterol biosynthesis inhibitors

T. cruzi and the various species of *Leishmania* synthesize ergosterol as their principal sterol. The African trypanosomes by contrast rely on cholesterol, which they obtain from the host. The first two types of parasite should therefore be susceptible to sterol biosynthesis inhibitors in much the same way as pathogenic fungi. This indeed has proved to be the case. Azoles such as ketoconazole have activity against *T. cruzi* and various leishmania *in vitro* and in several animal models; useful clinical activity with ketoconazole is restricted to a few species causing cutaneous leishmaniasis. The triazole ICI 195,739 has remarkable activity against *T. cruzi* in mice, with the potential of producing cure rather than just suppression of infection; it does not appear to have useful activity in leishmaniasis. The enantiomer of ICI 195,739, D0870 is being developed as an antifungal agent, and has recently been made available for evaluation in experimental Chagas disease. The sterol biosynthetic pathway is complex, and quite a number of stages are susceptible to inhibition. Of particular interest are inhibitors of HMGCoA reductase, squalene epoxidase and C-24 methenylation. Although inhibitors of these enzymes have limited antiparasitic activity on their own, they do have the demonstrated potential of synergizing the activity of azoles. Interesting possibilities are also being discovered with mixtures of an azole and other non-sterol-inhibiting antimicrobial agents.

Polyamines/SAM/trypanothione

Trypanothione, a macrocyclic $N^{1,8}$-bisglutathionyl spermidine, is the parasite's equivalent of host glutathione, and is synthesized from polyamines by a parasite-specific pathway. S-adenosyl methionine (SAM) is involved in polyamine synthesis as well as in a host of methylation reactions. The whole area of polyamine/SAM/trypanothione biosynthesis and function would seem to have potential for the development of antiparasitic drugs; apart from the trypanothione component, the area is of considerable interest also for cancer chemotherapy – which is providing much basic information as well as potential inhibitors for the I-Chem programme. The 'resurrection drug' DFMO is an inhibitor of ornithine decarboxylase, and was developed and tested in man with cancer in view.

Protease inhibitors

Inhibition of specific proteases is currently of considerable interest in the search for an HIV antiviral agent, and also in novel approaches to the treatment of hypertension, myocardial infarction, CNS problems, hormonal regulation, arthritis, muscular dystrophy, tissue wasting, bone resorption, cancer, emphysema, and periodontal disease. Over a number of years TDR has supported work on the cysteine proteinases of *T. cruzi* and various species of *Leishmania*; much of the work has been purely academic, but some has had possible relevance for diagnosis, vaccine development, or chemotherapy. Activity against the parasites has been demonstrated *in vitro* with a variety of peptidyldiazomethylketones and peptidylfluoromethylketones, and evidence is now available of limited activity *in vivo* with inhibitors of these types. The enzyme from *T. cruzi* has been crystallized and subjected to X-ray analysis; structural studies on leishmanial proteases are in progress. Based on this structural data, three-dimensional computer models have been established and are being used to scan chemical databases in the hope of identifying new leads, and also to suggest chemical modifications to available inhibitors which might improve selectivity and intrinsic activity.

The current I-Chem programme is providing support in these three areas for basic biochemical work, for synthesis of new potential inhibitors, and for evaluation of inhibitors *in vitro* and *in vivo*. The programme, however, relies heavily on support from outside sources – both industrial and academic – for the provision of compounds which have been synthesized with similar targets in economically more important diseases in view. A surprising number of compounds and leads of current interest have had their origin in programmes directed at cancer chemotherapy.

COMMON LEADS AND TARGETS

Because of the situation described earlier, the TDR programme currently maintains a chemotherapeutic interest in malaria and filariasis as well as the three I-Chem diseases. A recent re-investigation of potential targets in all five disease areas suggested that polyisoprenoids (including sterols), polyamines and proteases were the three targets of prime interest for malaria and filariasis also. Different emphases exist between the five types of parasite, but the overall broad targets are the same. Thus in the area of protease inhibition aspartyl proteinases are probably of even greater significance than cysteine proteinases to the malaria parasite, and a specific aspartic protease inhibitor has been shown to have antimalarial activity *in vitro*. Similarly in the polyamine area inhibitors of polyamine oxidase and putrescine-N'-acetyl transferase, enzymes not as yet of interest with the haemoflagellates, have been shown to be lethal to adult filariae *in vitro*. The logical outcome of this commonality of interest would be the integration of screening programmes; this has in fact taken place at the centre in Belgium. *In vitro* and *in vivo* models for malaria are very similar to those used for haemoflagellates, so that integration of screening is not difficult; suitable models for use in the search for a macrofilaricide are difficult to devise and programme integration is not so readily achieved.

THE NEEDS FOR NEW ANTIMICROBIALS – ECONOMIC VERSUS ETHICAL ISSUES

Is there a need for new antimicrobial agents? Undoubtedly yes! In the field of bacteriology there is a current need, and this is likely to be a continuing need. The need is resistance-driven and is economically sustainable. Whether a company sets its sights on an injectable agent for serious infections in hospitalized patients or on a lower-cost but more widely used oral agent for ambulatory patients is a matter for economic judgement. In either case the need is there, and as resistance to newer agents develops, is likely to remain. With the continual development of new agents and new classes of antibacterials, and particularly in view of the increasing import-ance of drug-resistant tuberculosis, it is likely that new ingredients for multidrug therapy of leprosy will become available, should the components of current multidrug therapy fail.

There is a need for new antiviral drugs for conditions ranging from HIV to the common cold. Research over the past decade suggests that the possi-bility of discovering antiviral agents is not as remote as it seemed some 40–50 years ago. The horror and potential magnitude of the HIV situation – which in some parts of the world at least has an attractive financial component – drives the search for new antivirals; treatments for other viral infections will undoubtedly follow in its wake.

The third world is not immune from bacterial and viral infections which afflict the developed world. AIDS, particularly in Africa, is an even more serious problem to the basic structure of society and family life and to the viability and economy of the region than it is in say the USA or UK, while tuberculosis knows no political or cultural boundaries. To what extent new conventional antimicrobial agents will be able to be used in the third world is a different matter, dependent largely on the economic situations of the victims and the countries in which they live. Pharmaceutical companies, quite rightly, have their expectation of economic return on their investment in research, but they are unable and unwilling to have a differential price structure for their drugs based on the ability of their customers to pay; this makes utilization of many conventional antimicrobials in the third world difficult, if not impossible.

The third world, however, has a number of microbial diseases of tremen-dous importance which are little more than matters of scientific curiosity to those in the temperate regions of the developed world. It is these diseases which have featured largely in the current presentation. The actual *need* for new treatments for these diseases is far greater than that for new treatments for bacterial and viral diseases of the developed world; it is the perceived economic component of that need which is lacking. Were there new and effective drugs available for the treatment of African trypanosomiasis, Chagas disease, the various forms of leishmaniasis, and malaria, there would be problems associated with the ability and willingness of the patients to pay for treatment. There would be little prospect for the discoverer of such remedies to recoup the cost of research leading to such new drugs, let alone opportunity for making profit. Over and above this direct financial problem, many of the victims live in situations where the basic infrastructures necessary to institute diagnosis and treatment are rudimentary or non-existent. Nevertheless, the outstanding need is there.

Only the conscience of the developed world translated into action and aid can remedy this situation. The TDR programme exists to try to help in this situation, both by fostering the discovery of new drugs, vaccines and methods of vector control, and by assisting third world countries to develop the infrastructure necessary to cope with their disease problems. TDR is, however, wholly dependent on financial support from the developed world. But financial input alone is not sufficient. The research programme needs very positive input and co-operation from the scientific community in general and from the pharmaceutical industry in particular. Although it is perfectly understandable that pharmaceutical companies can no longer justify programmes of work directed wholly at tropical diseases (excluding possibly malaria?), yet they could help in such programmes as that being run by TDR by co-operation in sharing ideas and approaches, by supplying compounds for screening and evaluation, particularly recently synthesized ones made with similar targets and pathways in view, and by supplying information regarding the toxicology of their compounds. Although the pharmaceutical companies have their own welfare and that of their share-holders (and pensioners!) very much in view, it is to be hoped that they will consider with compassion – and action – those in the third world so desperately in need of new antimicrobial agents. Directors, workers, share-holders and third-world victims are all human whatever their economic circumstances, and at the end of the day, mortal – with no distinction of class, colour, creed or economic status.

REFERENCE

Much of the factual data in this presentation are taken from contributions written by TDR colleagues in:

Tropical Diseases Research. Progress 1991–1992. (1993). *Eleventh Programme Report of the UNDP/World Bank/WHO Special Programme for Research and Training in Tropical Diseases (TDR)*. World Health Organization, Geneva.

WHY DO WE STILL GET EPIDEMICS?

K. G. KERR and R. W. LACEY

Department of Microbiology, University of Leeds, Leeds LS2 9JT, UK

INTRODUCTION

Human intelligence, culture and technology have left all other plant and animal species out of the competition . . . But we have too many illusions that we can . . . govern . . . the microbes that remain our competitors of last resort for domination of the planet. In . . . natural evolutionary competition, there is no guarantee that we will find ourselves the survivor(s).

(Lederberg, 1988)

Over the last century, great advances in our understanding of the causation, transmission, treatment and prevention of infectious disease have fostered complacency about infection in those societies in which well-nourished and housed individuals have access to vaccines and antimicrobial drugs. But infectious diseases remain the leading cause of mortality worldwide and, even in the prosperous countries of the developed world, infections are responsible for 25% of all physician visits and antimicrobial agents are the second most frequently prescribed class of drug. The financial burdens placed on already overstretched health care budgets of the industrialized nations by infectious disease are colossal – in 1992 in the USA, the costs of nosocomial sepsis alone were \$4.5 billion (Centers for Disease Control, 1992*a,b*).

Continuing socio-cultural change, political and economic upheavals and environmental damage all provide new opportunities for the spread of infectious agents. Furthermore, advances in medical therapeutics have created new groups of patients susceptible to an ever-expanding spectrum of 'opportunistic' pathogens. Moreover, nosocomial and community acquired agents have developed resistance over the last decade (Table 1), primarily because of antibiotic misuse, and outbreaks of infection due to antimicrobial resistant micro-organisms such as penicillin-resistant pneumococci or multiply drug resistant M. tuberculosis are becoming more frequent.

In addition to well-recognized pathogens, 'new' agents capable of causing a wide array of 'new' infections continue to emerge (Table 2). Many of these new pathogens have, on closer examination, been present for decades until uncovered by increasingly sophisticated diagnostic techniques. Although many 'new' organisms are discovered fortuitously, for instance, human parvovirus B19, others are identified following well-publicized outbreaks of illness which appear to be infectious in origin. An example of this is the

Table 1. *Problems of antimicrobial resistance to have arisen during the last decade*

Organism	Compound	Resistance mechanism
Streptococcus pneumoniae	Penicillin	Altered penicillin binding protein
Enterococcus spp.	Glycopeptides	Not known
Klebsiella spp.	Extended spectrum Cephalosporins	β-lactamase
Plasmodium falciparum	Chloroquine	Decreased uptake increased efflux
Human immunodeficiency viruses	Zidovudine	Altered reverse transcriptase
Herpes simplex virus	Acyclovir	Thymidine kinase deficiency

Table 2. *Pathogens of humans to have emerged during the past 30 years*

Viruses
Human immunodeficiency viruses (HIV) 1, 2
Human T-cell lymphotrophic viruses (HTLV) I, II
Human parvovirus B19
Human herpes virus (HHV)6
Hepatitis C virus (HCV)

Bacteria
Borrelia burgdorferi
Legionella pneumophila
Chlamydia pneumoniae
Afipia felis
Helicobacter pylori

Protozoa
Cryptosporidium parvum
Microsporidia
Cyclospora catayensis
Babesia spp.

identification of Legionella pneumophila following an outbreak of atypical pneumonia amongst American Legionnaires at the Belle Vue Stratford Hotel in Philadelphia, USA in 1976. Other discoveries are the product of painstaking, systematic research into long-recognized clinical problems such as the identification of hepatitis C virus as the major cause of post-transfusional hepatitis.

It is also important to note that new manifestations of sepsis caused by 'traditional' pathogens have also been identified and these include staphylo-coccal toxic shock syndrome, the haemolytic uraemic syndrome (following infection with verotoxin-producing *Escherichia coli*), and Brazilian purpuric fever associated with *Haemophilus influenzae* biotype *aegyptius*. Further-

Table 3. *Diseases in which infectious agents have recently been implicated*

Disease	Micro-organism
Guillain–Barre syndrome	*Campylobacter jejuni*
Gastritis, peptic ulcer, gastric carcinoma	*Helicobacter pylori*
Crohn's disease	*Mycobacterium paratuberculosis*
Carcinoma of the uterine cervix	Human papilloma virus
	Herpes simplex virus
Atherosclerosis	*Chlamydia pneumoniae*
Peliosis hepatitis	*Rochalimaea henselae*
Rheumatoid arthritis	Parvovirus B19
Insulin-dependent diabetes mellitus	Coxsackie viruses
Tropical spastic paraparesis	Human T-cell lymphotrophic virus I

more, the number of diseases in which an infectious aetiology is known, or suspected, continues to increase (Table 3).

The predictions made during the 1960s that infections were unlikely to remain serious threats to health (Burnet, 1963) can now, with hindsight, be seen to have been over-optimistic. So, why do we still see epidemics? Epidemics occur when a new infectious agent is introduced into a susceptible population or when existing agents exploit a change in host susceptibility, such as immunosuppression, or the failure of immunization programmes. In both settings, modes of transmission of the pathogen from one host to another must also exist. This is, of course, a gross oversimplification: in practice, the emergence of an epidemic results from complex interactions of many factors. For example, the AIDS pandemic has resulted not only from the introduction of an apparently new viral agent into susceptible populations but also because of a number of human-associated factors such as increases in intercontinental travel and intravenous drug abuse, the changes in sexual behaviour including 'sex tourism', and medical advances such as assisted conception and the therapy of haemophilia.

In many epidemics attention often focuses on the infectious agent – this is particularly so in high profile epidemics when the media reach for well-worn clichés such as 'killer virus' and 'superbug'. But, as several authors (e.g. Morse, 1991) have emphasized, human actions and behaviours are often fundamental in the emergence of epidemics and one commentator has gone so far as to describe some epidemics as 'self-inflicted' (Grist, 1989). Table 4 outlines some of the many factors involved in the emergence of epidemics, a number of which are discussed later in this chapter.

However, the potential devastating effects of prion disease must prove the greatest single infective threat for human and animal populations next century. With hindsight, we can now berate the cannibalistic measures needed to provide an abundance of cheap meat and milk for a dense human population. These issues will be discussed in detail below. The implications

Table 4. *Some factors involved in the emergence of epidemics*

Environmental change	Infection/micro-organism
Changes in land use	Argentine haemorrhagic fever
Deforestation	Oropouche fever
Reforestation	Lyme borreliosis, babesiosis
Climate changes	See text
Demographic/Politico-Economic change	
Rural to urban migration	See text
Urban decay	Diphtheria, tuberculosis
War and civil unrest	See text
Immigration/displaced persons	Tuberculosis, Chagas disease
Increase in child-care/aged person facilities	Viral/bacterial gastroenteritis
Social change	
Sexual behaviour	HIV[a], HSV2, gay bowel syndrome
Drug abuse	HIV[a], HCV[b], HDV[c], *Salmonella* spp.
Travel	*V. cholerae, Plasmodium* spp.
Diet	*Listeria monocytogenes*
Recreational use of water	*Pseudomonas aeruginosa, Legionella* spp.
Mass media 'health scares'	*Bordetella pertussis*
Trade and industry/commerce	
Worldwide transport of goods	*V. cholerae, A. albopictus*
Food handling/packaging/distribution	*E. coli* serotype 0157 H7
Medical	
New treatment modalities	HIV, CMV-related transplant infections
Compromised host defences	*Aspergillus* spp.
Iatrogenic outbreaks	Contaminated IV infusates
Antibiotic misuse/overuse	Methicillin resistant *S. aureus*
Changes in micro-organisms	
'New' micro-organisms	HIV[a]1,2
Changes in antigenicity	Influenza A virus
Changes in pathogenicity	Streptococcal toxic shock syndrome

[a]HIV, human immunodeficiency virus; [b]HCV, hepatitis C virus; [c]HDV, hepatitis D (delta) virus.

for agriculture, and indeed for the entire social fabric of the UK the bovine spongiform encephalopathy (BSE) endemic are grave indeed.

THE ENVIRONMENT

Changes in land use

Since the agrarian and industrial revolutions of the eighteenth century, mankind has inflicted, by accident or design, changes to the atmosphere and to the terrestrial and aquatic environments. Despite pressure from many quarters to halt these changes, there is little doubt that they will continue for the foreseeable future. Changes in land use, particularly in the developing world where there are pressures to feed expanding populations or to maximise income from exports, have led to the emergence of several epidemics of infectious disease. In many instances, the pathogens responsible are vector borne, and the environmental changes have resulted in the destruction, or creation of the natural habitats of these vectors.

An example of the latter is Rift Valley fever (RVF), a mosquito-borne bunyavirus. In 1977, a major outbreak involving approximately 200 000 persons with nearly 600 deaths occurred in Egypt. The epidemic was linked to the construction of the Aswan Dam which was completed in 1970 and led to the flooding of 800 000 hectares of land thus altering the Nile valley water table, in turn creating many more breeding sites for the vector *Aedes linneatropensis* (Meegan & Shope, 1981). An outbreak of RVF, in which 244 individuals died, in Mauritania following construction of the Diama Dam on the Senegal river appears to confirm the association between dam building and RVF (Jouan *et al.*, 1988).

Deforestation and other changes in land use may also be a factor in the emergence of epidemics. Junin virus, an arenavirus, is the causative agent of Argentine haemorrhagic fever. The natural host, the mouse *Calamys musculinus*, is indigenous to the pampas, but when this was cleared to allow the planting of maize the mouse population increased dramatically with the corresponding increase in the number of cases of Argentine haemorrhagic fever (Morse, 1991).

Outbreaks of infection arising from changes in land use are not confined to the so-called developing world, an example being the emergence of Lyme borreliosis in the northeastern USA. In the early eighteenth century, forests were cleared for agricultural purposes; later, however, as agriculture moved westward to the Great Plains, large tracts of land became reforested rapidly and, by 1980, some areas had four times as much woodland than in 1860. Unlike the old forests this new woodland had no predators to regulate deer with a consequent rise in deer populations. Coincident with the return of the deer, humans began to live in forested rural areas and the numbers of visitors to these areas also increased.

The resulting proximity of humans, deer, rodents and ixodal ticks – whose bite transmits the causative agent *Borrelia burgdorferi* – established the conditions necessary for the emergence of Lyme borreliosis in human populations, and Lyme disease is now recognized as the commonest vector-borne disease in the USA (Steere, 1989). The disease can spread explosively as shown in Ipswich Massachusetts, where 35% of all residents who lived adjacent to a nature reserve developed Lyme borreliosis. In this instance it is thought that passerine birds infected with *B. burgdorferi* elsewhere introduced the spirochaete into the area where it spread rapidly through the local tick community (Lastavica *et al.*, 1989).

Climatic change

Although it is difficult to directly implicate climatic changes, such as global warming, in the emergence of new epidemics, the effects on, for example, waterborne organisms such as *Vibrio* spp., and on populations of insect vectors merits careful monitoring (Leaf, 1989). In addition, desertification caused by climatic change may exacerbate migration from rural to urban areas, further aggravating the public health problems associated with many cities in the developing world (see below).

DEMOGRAPHIC AND POLITICO-ECONOMIC CHANGE
War and civil conflict

Since the description by Thucydides of plague in Athens during the Peloponnesean War in 430BC, the association between war and epidemics has been well recognized. Outbreaks of infection may arise in both military and civil populations as a result of complex interactions between such factors as overcrowding, disruption of water supplies and sewage facilities and other breakdowns of public health infrastructure. Such epidemics may be devastating – the influenza A pandemic of 1918–1919 killed at least 20 million people – more than the Great War itself. Movements of combatants into theatres of operation far from home may permit the introduction of new and unfamiliar infectious agents into the susceptible population. An infamous historical example of this is the introduction of smallpox into Mexico by the Spanish army led by Cortez which led to the deaths of an estimated 3.5 million Aztecs. Conversely, susceptible military personnel or refugees, may themselves acquire infections in a particular theatre of operation. Returning combatants may also 'import' infections on their return, recent examples being meliodosis in Vietnam veterans and leishmaniasis in personnel involved in Operation Desert Storm (Centers for Disease Control, 1992*a*).

Population growth and movement

The world's population continues to grow, creating new stresses on already overstretched health-care systems. Economic and other factors mean that this population is becoming increasingly urbanized. By 2000 there will be 425 cities with one million, or more, inhabitants in comparison with 225 such cities in 1985 (World Resources Institute, 1986). Twenty-five cities will have populations of at least 11 million, and indeed, neither London nor Paris will feature in the top 20 cities as defined by population numbers (Last & Wallace, 1992). This relentless growth in the population of many major cities is accompanied by overcrowding, malnutrition, substandard water and sewage services and lack of access to other public health facilities: the potential for epidemics is great.

The emergence of Dengue haemorrhagic fever, serves to highlight many of these problems. Since its appearance in 1956 there have been approximately 30 000 cases of Dengue haemorrhagic fever and Dengue shock syndrome reported annually, but between 1986 and 1990, the number of reported cases increased to 267 000 per annum (Gubler, 1991). This increase is attributable, in part, to densely populated cities in which susceptible individuals are crowded together and poor sanitation which allows the vector, the peridomestic mosquito *Aedes aegypti* which breeds almost exclusively in 'artificial' habitats such as open water storage vessels and discarded tyres, to thrive. These problems are compounded by the lack of effective mosquito control programmes (Hospedales, 1990).

Cities of the so-called developed world, too, have their own problems and inner city decay associated with homelessness, drug abuse and cuts in public health expenditure have been accompanied by a resurgence of tuberculosis in several USA metropolitan areas. Similar circumstances have led to the re-emergence of diphtheria in the former Soviet Union and the outbreak which began in 1990 and is still ongoing is, with over 12 800 cases, the largest in the developed world since the 1960s (Centers for Disease Control, 1993).

Immigration, whether for economic or political reasons, may also be responsible for the introduction of infectious agents into new environments. Changes in immigration patterns (84% of 640 000 immigrants to the USA in 1988 were Latin American or Asian in origin) may further increase the opportunities for agents to spread in this manner (e.g. see Wendel & Gonzaga, 1993).

Other demographic changes

In many industrialized societies over the past three decades a number of factors, including a marked rise in the divorce rate and the number of single-parent families coupled with increased job opportunities for women, have led to a substantial growth in the number of children attending day-care centres. Children in these establishments often have imperfectly developed personal hygiene and often indulge in activities which facilitate the spread of infectious agents. Outbreaks of viral and bacterial gastroenteritis, hepatitis A, giardiasis, cryptosporidiosis, varicella and other infections are well recognized in this setting (Osterholm *et al.*, 1986).

The break-up of the extended family in many societies has led to an increasing tendency for elderly family members to be cared for in nursing homes. Again, sub-optimal personal hygiene together with underlying medical problems, such as achlorhydria, provide opportunities for the spread of infectious agents in these facilities, and outbreaks of infectious gastroenteritis, influenza and tuberculosis have been reported (Harris, Levin & Trenholme, 1984).

Trade and commerce

The volume of international trade continues to grow and is likely to be accelerated as political agreements result in the removal of barriers between trading nations. Unfortunately pathogenic micro-organisms in addition to goods may be exchanged during commercial activities. For example, the current pandemic of cholera is thought to have arrived in South America as a result of ships from the Far East pumping bilge water containing *Vibrio cholerae* into the harbour at Lima. The water rapidly contaminated local fish and shellfish which are often eaten raw ('ceviche'). Following initial sea-food related cases, the organisms were probably spread by faecal contamination of sub-standard water supplies (Gotuzzo *et al.*, 1994).

In addition to pathogenic micro-organisms, the vectors of infectious agents may be transferred by commercial activities. The Asiatic mosquito, *Aedes albopictus*, deposits its eggs in pools of water including those which collect in discarded tyres. In 1985 a consignment of used tyres from Japan was imported into the USA at Houston, Texas, where the mosquito soon established itself (Hawley *et al.*, 1987). Subsequently, importations of *A. albopictus* have been reported in Brazil, South Africa, Nigeria, Albania and Italy. Once in their new environment, the mosquitoes soon establish themselves and spread rapidly. To date, *A. albopictus* is found in at least 23 states of the USA (Centres for Disease Control, 1987). Indigenous *A. aegypti* did not appear to retard the newcomers and, indeed, *A. albopictus* has replaced the endemic *A. aegyptius* populations in several areas. Of particular concern is the ability of this insect to transmit a large number of flaviviruses including dengue 1-4, St Louis encephalitis virus and the alphaviruses which are the agents of western and eastern equine encephalitis. In 1991 the latter was isolated from *A. albopictus* strains involved in an epizootic in Florida – of five human cases, two individuals died (Centers for Disease Control, 1991).

HUMAN BEHAVIOUR – 'SELF-INFLICTED EPIDEMICS'

A number of changes in the way humans behave have contributed to the emergence of new epidemics (see Table 4), and some of these are discussed in greater detail here.

Drug abuse

The use of illegal drugs is a major problem in many communities and these activities provide new opportunities for the spread of infectious agents. The control of outbreaks of drug-related infection is often hampered by the fact that drug abusers may not have access to health care facilities. Even if such access does exist, drug abusers may be reluctant to use it and, even if they do, compliance with treatment regimens may be poor. Intravenous administration of drugs coupled with the sharing amongst individuals of injection paraphernalia has resulted in well-publicized outbreaks of hepatitis B, C, D and the human immunodeficiency viruses. In addition, non-sterile materials may also be associated with outbreaks of infection; for example, outbreaks of systemic candidiasis in drug abusers dissolving heroin in contaminated lemon juice (Podzamczer *et al.*, 1985). An outbreak of sepsis with methicillin-resistant *Staphylococcus aureus* in a USA city was most likely due to excessive prophylactic use of cephalosporins to prevent staphylococcal abscesses at injection sites (Levine, Crane & Zervos, 1986). The economic necessities of a drug habit may also have indirect consequences relevant to the spread of infection such as the marked increase in syphilis (including congenital disease) in many USA cities as a result of the exchange of sexual favours for 'crack' cocaine (Rolfs, Goldberg & Shorra, 1990).

Even apparently 'safe' drugs such as marijuana have been associated with infections, for instance aspergillosis (Hamadeh *et al.*, 1988). In 1982 in the USA, investigation of a multistate outbreak of *Salmonella munchen* enteritis identified contaminated marijuana as the vehicle of infection (Taylor *et al.*, 1982).

Recreational use of water

Over the last decade jacuzzis and whirlpool spas have become a common feature, not only in public facilities, but also in the domestic setting. These have been associated with outbreaks of Pontiac fever caused by *Legionella pneumophila*. Outbreaks of folliculitis and other superficial infections due to *Pseudomonas aeruginosa* are common and may involve large numbers – as many as 250 individuals infected from one source. Aeration, agitation, high temperature and pH of the water coupled with large numbers of bathers serve to shorten the activity of chlorine and permit the survival of particular serotypes of *P. aeruginosa* – such as 011 – in these environments (Gustafson *et al.*, 1983).

Swimming pools, despite modern hygiene measures, continue to play a role in the spread of such infections as cryptosporidiosis.

Concern has also been raised over the use of reclaimed waste water for recreational purposes as is becoming increasingly common in the South Western USA, and indeed outbreaks of shigellosis and hepatitis A have been reported in these settings (Blostein, 1991).

MEDICAL CAUSES OF NEW EPIDEMICS

The ever-expanding array of new investigational and therapeutic modalities is often accompanied by new infections. Many of these are caused by the iatrogenic breaching of host defences such as the transmission of viruses by blood transfusion and organ transplantation. Contaminated intravenous infusates have led to outbreaks of septicaemia with Gram-negative bacteria including *Enterobacter, Citrobacter* and *Serratia* spp. (Maki, 1986). Bypassing of the defences of the upper respiratory tract which occurs during mechanical ventilation may lead to nosocomial pneumonia.

Increased numbers of immunocompromised patients, for example those undergoing cytoreductive therapy for haematological malignancies, have been associated with the emergence of infections associated with opportunistic pathogens. Although many of the latter are autochthonous in origin, others can be acquired from the hospital environment, for instance outbreaks of aspergillosis in haematology units following construction work in or near the hospital (Arnow *et al.*, 1978; Lentino *et al.*, 1982).

Finally, the overuse, or misuse, of antimicrobial agents increases the selection pressure on fungi, viruses, bacteria and protozoa and, in turn, has led to the emergence of nosocomial and community epidemics of infection caused by these resistant organisms (Munoz *et al.*, 1991). International travel facilitates the rapid intercontinental spread of such organisms. This is discussed in greater detail below.

CHANGES IN MICRO-ORGANISMS

The preceding discussion has focused principally on the role of *Homo sapiens* in the development of new epidemics – what of the organisms themselves?

New organisms

> The historiography of epidemic disease is one of the last refuges of the concept of special creationism.
>
> (Lederberg, 1988)

Contrary to the often hysterical views of the popular media who appear ever willing to broadcast sensationalized accounts of the latest 'killer virus' truly new micro-organisms are extremely rare. Many of the so-called new agents identified in the last 20–30 years (Table 2) have, in retrospect, been present for many years until recognized by novel diagnostic techniques, and recent advances such as polymerase chain reaction methodology will permit the examination of archival, and even palaeontological material to elucidate further the evolution of the 'new' pathogens.

Truly new infectious agents such as the human immunodeficiency viruses do occur, but it is perhaps surprising that given the genetic plasticity of micro-organisms – particularly the viruses – this is not a more frequent phenomenon.

An often cited example of microbial evolution as an important factor in the evolution of epidemics is that of antigenic shift in influenza A virus resulting from the reassortment of viral genes in different influenza strains. A possible explanation for this phenomenon relies on the observation that every influenza A haemagglutinin (H) subtype is known to occur in water fowl but these viruses appear incapable of transmission to humans. Pigs, however, can be infected by both avian and human influenza strains and can therefore act as a 'mixing vessel' where reassortment between avian and human strains can occur. These recombinants can then be transmitted to human populations where a lack of neutralizing antibodies to the 'new' H antigens may then give rise to pandemics of influenza. Every recorded pandemic has originated in South East Asia where integrated pig–duck farming has been practised for centuries. Thus it can be argued that even these epidemics are 'man-made' (Kida *et al.*, 1988; Scholtissek & Naylor, 1988).

Changes in pathogenicity and virulence

The media attention which focused on the outbreak of 'flesh eating bugs' in Gloucestershire in 1994, highlighted an apparent change in virulence in group A streptococci (*Streptococcus pyogenes*) which had first been noted in the mid-1980s. Beginning in Salt Lake City, Utah, outbreaks of rheumatic fever were reported in several USA cities. Of particular interest was the fact that these occurred not in poor urban areas, where rheumatic fever was frequently seen, but in affluent, predominantly white, suburban and rural neighbourhoods (Bisno, 1991). In addition, several publications described a severe manifestation of streptococcal sepsis which resembled the staphylo-coccal toxic shock syndrome – streptococcal toxic shock-like syndrome (e.g. see Stevens, 1992). The production of pyrogenic exotoxins by particular group A streptococcal strains was deemed important in the pathogenesis of this syndrome. Nucleotide sequencing of the structural gene for one of these, exotoxin A, from isolates of five genotypically heterogeneous groups of strains gave nearly identical results and it has been suggested that these genes may be spread by bacteriophage-mediated transfer (Musser *et al.*, 1991). More recently it has been suggested that other toxins, which like the pyrogenic endotoxins, can act as super antigens are more important in the pathogenesis of toxic shock-like syndrome (Mollick *et al.*, 1993).

Changes in microbial susceptibility

Micro-organisms are extraordinarily successful in their ability to develop resistance to antimicrobial agents. Although some resistance is inherent, many epidemics are caused by organisms which have acquired resistance to chemotherapeutic agents (see Table 1). The genetic material coding for resistance mechanisms is often easily transferable to other species or genera. Such resistance is important both in community and hospital settings. Although multiple drug resistance is especially common in the latter, for example in *Enterococcus* spp. *Stenotrophomonas maltophilia*, *Burkholderia cepacia*, it is also seen outside the hospital environment such as in multiply-drug-resistant *Mycobacterium tuberculosis*. Even pathogens long regarded as uniformly sensitive to particular agents, for example *Neisseria meningitidis* and penicillin, now show signs of developing resistance (Bowler *et al.*, 1994).

Many factors contribute to the emergence of resistance, for example poor patient compliance in the development of multiply-drug-resistant *M. tuberculosis*, but there can be no doubt that the major determinant is the overuse and abuse of antimicrobials by both the medical and veterinary professions. Once again, the problem seems 'self-inflicted'.

PRION DISEASES: THE INFECTIONS OF THE NEXT CENTURY

Transmissible spongiform encephalopathies are infections mainly acquired by mammals from ingestion of contaminated mammalian products. There is also evidence of vertical transfer. The incubation period varies from less than a year for some laboratory experiments on rodents to several decades in man. Although greatest infectivity is found in the brain at the time of the terminal illness, many organs are involved prior to this. The resultant diseases are sub-acute dementias and paralyses. There is no known treatment, and no practical way of identifying sub-clinical carriers because of the failure of the host to effect a cellular or humoral immune response to the causative agent (Dealler & Lacey, 1990; Kimberlin, 1993). The extreme physical toughness of the presumptive infectious agent or prion, is conducive to its causing a major epidemic in several mammalian species, including man.

These diseases have previously been referred to as slow virus infections, but more recently as transmissible or degenerative spongiform encephalopathies. They afflicted a narrow range of mammals prior to 1986, namely sheep (scrapie), farmed mink (transmissible mink encephalopathy), goats, mule, deer, mountain elk and man (Kimberlin, 1993). The human diseases comprise Kuru in the cannibalistic Fore tribe in New Guinea, very rare familial disorders such as the Gerstmann–Straussler–Scheinker Syndrome, and Creuztfeldt–Jakob Disease (CJD). The latter disease can be due to iatrogenic causes, such as infected growth hormones or grafts and typically has a short incubation period of 4–10 years (Brown *et al.*, 1992). The majority of CJD cases are said to be of uncertain aetiology and are referred to as sporadic. The figures for the UK from 1985 to 1993 are shown in Table 5.

Novel species reported as suffering from spongiform diseases 1986–1994

Since 1986, spongiform encephalopathies have been identified in many novel species of mammal, and even in one bird, the long-living ostrich. The figures for these up to July 1994 are shown in Table 6. The number of

Table 5. *Incidence of Creutzfeldt-Jakob disease (CJD) and Gerstmann–Straussler–Scheinker syndrome (GSS) UK, 1985–1993*

| Year | CJD | | | GSS | Total | Incidence/ million pop* |
	Sporadic	Iatrogenic	Familial			
1985	26	1	1	0	28	0.49
1986	26	0	0	0	26	0.46
1987	23	1	0	1	25	0.42
1988	21	1	1	0	23	0.40
1989	28	1	1	0	30	0.53
1990	26	5	0	0	31	0.54
1991	32	1	3	0	36	0.63
1992	44	2	4	1	51	0.89
1993	34	3	1	1	39	0.70

Source: UK CJD Surveillance Unit, Western General Hospital, Edinburgh.
*Based on UK population of 57.07 million (1988 Census update) and a press release from the Chief Medical Officer (1994).
Note: Because these are the very latest figures, they may be different from those published previously. This is because the Unit is still identifying cases from previous years. The figures for 1993 are incomplete.

Table 6. *Novel species reported as succumbing from spongiform encephalopathies since 1986*

Species	Year of first report	Number
Nyala	1986	1
Cattle	1986	131 000[a]
Gemsbok	1987	1
Arabian oryx	1990	2
Eland	1990	2
Kudu	1990	6
Domestic Cat	1990	47
Ostriches	1991	Many
Puma	1992	1
Cheetah	1992	3
Scimitar horned oryx	1994	1

Data source is the Ministry of Agriculture, Fisheries and Food (MAFF), UK.
[a] These are confirmed as at July 1994. The total number of actual cases probably exceeds 500 000 (Dealler, 1993).

reported diseased animals will obviously be an underestimate, to an uncertain extent, of the exact number of diseased (and in particular, infected) animals. The cause of these infections is thought in the main to be consumption of bovine products, although most of the greater kudu disease was acquired horizontally or vertically (Cunningham *et al.*, 1993). The finding of infected domestic cats may be particularly significant because the causal agent of CJD can, by experiment, be transmitted to cats (Dealler & Lacey, 1990).

The nature of the agent causing bovine spongiform encephalopathy (BSE)
and other transmissible spongiform diseases

The general properties of the agent responsible for BSE resemble those producing similar diseases in other mammalian species. Most recent experimental work on the detailed composition of this group of agents has utilized rodent-adapted scrapie lines because of the need for induction of disease under laboratory conditions. Two main structures have been postulated for the agent: a variant of a virus containing small amounts of nucleic acid (the virino) or a proteinaceous, nucleic-acid-free substance (the prion). The virino proposal is difficult to reconcile with the extreme physical toughness of the agent and the prion theory implies that a protein can somehow persist and replicate itself. Evidence has now accrued to firmly favour the prion hypothesis and this does have relevance to the possible infectivity of BSE for man.

The prion hypothesis proposed by Prusiner (1982) was that the infectious agent (PrP^{sc}) was a modified form of a normal host protein (PrP^c) that was able to catalyse its own synthesis from the normal protein. The hypothesis requires the function of a normal host gene (Prn-P) that is induced to produce PrP^c by PrP^{sc} with a subsequent change in the conformation of PrP^c to yield PrP^{sc}, a crystalline or amyloid product described as scrapie-associated fibrils.

Recent key experiments have been published by Weissman and colleagues from Zurich (Büeler *et al.*, 1993). This group has shown that mice devoid of both sets of normal PrP genes are resistant to scrapie after challenge by the appropriate PrP^{sc} agents. Animals heterozygous for the PrP gene are partially resistant to the disease. After Syrian hamster transgenes had been introduced into mice devoid of their natural PrP genes, the mice became highly susceptible to hamster prions but not to mouse prions.

These experiments not only support Prusiner's prion hypothesis but explain the basis of the 'species barrier' and pave the way for further key experiments. The 'species barrier' refers to the change in property of prion disease as the agent is transferred from one species of mammal to another (Dealler & Lacey, 1990). The changes usually manifest themselves as an unpredictable lengthening in the incubation period, or in the increasing amount of inoculum required to induce disease. However, once a disease is established in a new species by experiment, it can be transferred easily and predictably to further generations. The 'species barrier' is determined by possible variation in the structure of the exogenous prion and in the genome of the host, notably in its PrP gene.

The scale of the BSE epidemic in recent years is thus attributed mainly to bovine-to-bovine rather than ovine-to-bovine spread, with bovine PrP^{sc} inducing the normal bovine PrP gene to synthesize PrP^{c}.

Properties of prions

The two major properties of the prion agent are its small molecular size and its extreme resistance to physico-chemical agents. Experiments studying such resistance rely on the testing for possible residual infectivity into appropriate target animals. This is required because the structure, configuration and amino acid sequence of the prion are not established.

Most disinfectant chemicals, including domestic bleach, do not appear to neutralize the infectivity of the agent (Taylor, 1989), and neither do proteinases found in the animal gut (Prusiner, 1984), DNAse (Pattison, 1988), RNAase (Prusiner, 1984), ultraviolet light (Prusiner, 1989), ionizing irradiation at usable doses (Fraser et al., 1989), or heating to cooking temperatures (Dickinson & Taylor, 1978). Proteinase K has been shown to decrease infectivity, as have specific chemicals that react with protein (Prusiner, 1984). However, the chemicals that react specifically with DNA or RNA (psoralen photoadducts, hydroxylamine) do not have any effect (Prusiner, 1984). Autoclaving at 134°C for 1 h decreases the infectivity of CJD or scrapie material to low levels (Rohwer, 1984) but there is still evidence that full destruction has not taken place. Some infectivity of the scrapie agent can survive baking for 24 h at 160°C in dry heat (Taylor, 1989) or at 360°C for 1 h (Brown et al., 1990). The only ways of ensuring destruction of these agents are with concentrated mineral alkalis.

The American current standard for autoclaving for decontamination of CJD is 132°C for 1 h (Rosenberg, 1986). In Britain several methods are recommended for decontamination: the most common one is 136–138°C for 18 min (Report, 1981). This is aimed to prevent the transfer of the infective agent for CJD between patients by metal electrodes, corneal transplants (Behan, 1982), dental procedures (Adams & Edgar, 1978), or from blood (Manuelidis et al., 1985). Pathologists are often unwilling to undertake post-mortem examinations on patients who are considered as possibly having died of CJD and when they do, gloves, autoclavable gowns and cap, overshoes, and masks are worn along with a disposable plastic apron under the operation gown (Report, 1981).

The BSE epidemic

The number of confirmed cases of BSE are shown in Table 7. The figures for 1992 and 1993 are approximate because the UK Ministry of Agriculture,

Table 7. *Annual number of*
confirmed bovine spongiform
encephalopathy cases in the UK^a

Year	Number
1986	7
1987	413
1988	2185
1989	7136
1990	14180
1991	25025
1992	36000[b]
1993	36000[b]

[a]The total cattle population has dropped
from about 12 million in 1986 to 10 million
in 1993.
[b]Approximate values (see text).

Fisheries and Food (MAFF) has released varying figures on different occasions. The figures for the first part of 1994 seem to be slightly down on 1993 (Report, 1994) but are being affected by reduced compensation paid to farmers from April 1994 and the requirement of two veterinary inspections prior to slaughter of suspect cases.

The cause of the BSE endemic was attributed initially to ovine material – in particular, sheep's brains – entering artificial cattle feed due to changes in the rendering processes used since 1981 (Southwood, 1989). However, no infectivity has been identified from rendered material or from the resulting cattle feed containing these substances known as 'concentrates'. That 97% of all cattle had been fed with these up to 1988 was consistent with both a feed source for BSE or indeed other explanations (Southwood, 1989). However, the description of the disease over such a wide geographical area within the UK, Eire, Isle of Man, Jersey and Guernsey must incriminate the feed in the initiation of the endemic. Because it was believed that the feed was responsible, its use in ruminants was prohibited by MAFF on 18th July 1988 and it was anticipated that this action should cut off the source of the infection, and the disease would die out over the next few years. This expectation was based on the assumption that cattle would be a 'dead-end' host for the disease agent and the total number of afflicted animals was predicted to be 17 000–20 000 with a peak of 350–400 monthly in the early 1990s (Southwood, 1989).

However, in the late 1980s the incidence of sheep scrapie in the UK was not known and it is only since the disease became notifiable in January 1993 that figures have become available. Thus, during 1993 there were just 286 cases confirmed in Great Britain (*Hansard*, 1994a), seemingly far too low an incidence to account for the number of BSE cases in the late 1980s, bearing in mind that such rendered material would be incorporated into the rations of about 10 million cattle, in addition to pigs, poultry and pets.

If sheep scrapie had been thought to have been the cause of BSE, then surely this hypothesis would have been tested by experiment? The following parliamentary question and answer (*Hansard*, 1994*b*) reveals a complete failure to perform such research.

Question Mr. David Hinchliffe (Member of Parliament for Wakefield): 'To ask the Minister of Agriculture, Fisheries and Food, what experiments have been performed in Great Britain over the last 10 years that involve the inoculation or feeding of cattle with infective material from sheep scrapie; and what were the results?'

Answer Mr. Nicholas Soames (for the Minister of Agriculture): 'Since 1984 no experiments involving the inoculating or feeding of cattle with infective material from sheep with scrapie have been undertaken in Great Britain.'

However, such work has been performed and published in the USA (Cutlip *et al.*, 1994). These authors inoculated intracerebrally 18 new-born calves with a pooled suspension made from the brains of 9 sheep infected with scrapie. These included 7 of the Suffolk-type typical of UK breeds. Half of the calves were slaughtered after 1 year and were studied. Those kept longer than 1 year became severely lethargic and weak with appearances comparable to the human motor neurone disease. Moreover, the brain changes were *not* characteristic for BSE (Cutlip *et al.*, 1994).

There is thus no epidemiological nor experimental evidence that BSE – at least in the recent past – was acquired from sheep scrapie. Moreover, the majority of the rendered material re-fed to cattle prior to 1988 was of bovine source (Southwood, 1989). There is also no record of a natural spongiform encephalopathy in pigs, poultry or pets prior to 1990.

Attempts have been made by researchers on behalf of MAFF to transfer BSE from the brain material to eight different species of mammal: cattle, pigs, marmoset monkeys, goats, sheep, mice, mink and golden hamster (Lacey, 1993; Report, 1994). The only species of mammal not to succumb during these experiments was the golden hamster, a species not characterized by longevity (Table 8). Results from the mink experiments included the development of a spongiform encephalopathy after oral feeding but the details have not been published.

In the positive experiments, the interval between challenge with infected brain material and onset of the disease varied from 10 months to 4 years. This is particularly significant for the timing of any prediction of disease in man, since the higher the titre of the challenge agent, in the animal experiments, the more rapid was the onset of the disease. The ages of the animal at challenge were very variable and the routes of transmission included ingestion. As expected for a disease that probably arose in cattle (*vide supra*) the most certain induction of disease was through challenge to

Table 8. *Experimental transfer of infectivity of bovine spongiform encephalopathy to cattle and other species*

Species	Number of animals challenged	Number which developed an SE	Age at challenge (approx.)	Period between challenge and confirmation	Mode of infection[a]
Cattle	16	16	4–5 yrs	17–24 mths	i/c and i/v
Pigs	10	7	2 weeks	17–37 mths	i/c, i/v and i/p
Marmoset	2	2	14 mths	48–49 mths	i/c and i/p
Goats	3	3	4–6 yrs	506–570 days	i/c
Goats	3	1	2–5 yrs	941 days	oral
Sheep	11	4	6–18 mths	440–880 days	i/c
Sheep	12	3	6–18 mths	538–994 days	oral
Mouse	2117	648	4–8 weeks	265–722 days	i/c and i/p
Mouse	460	138	4–8 weeks	265–722 days	i/c and i/p
Mouse	94	5	4–8 weeks	504–596 days	i/p
Mouse	10	10	4–6 weeks	435–540 days	oral

[a] i/c, intracerebral inoculation; i/v, intravenous inoculation; i/p, intraperitoneal inoculation. The material used for all these experiments was brain material from confirmed BSE cases which is the only tissue to date that has shown any infectivity.
Data taken from Lacey (1993).

cattle themselves. The chief conclusion from these experiments is that, as with the spontaneously infected animals, the potential host range of the BSE agent is very wide indeed.

Vertical transfer of BSE

The evidence for vertical transfer of BSE from dam to calf, which has accumulated in the last two years, can be summarized as follows:

1. Vertical transfer of BSE in cattle would be expected by analogy with similar disease in other species, notably scrapie in sheep (Dealler & Lacey, 1990).
2. The number of BSE reported cases in June 1994 at 700 cases per week (MAFF) is far too high, 6 years after the feed ban, to be explained solely by a feed source. There is no reliable documentation that the ban was broken by many feed compounders and farmers.
3. More than 11 000 BSE cattle had been born after the ban by June 24th 1994 (data from MAFF).
4. More than 600 BSE animals have been born of BSE dams (data from MAFF). This is too high to have occurred by chance.
5. Individual case histories support vertical transmission (Dealler & Lacey, 1994).
6. The most frequent age of death of BSE animals being 4 years favours vertical transfer rather than a feed source (Lacey, 1993).
7. No infectivity has been demonstrated in feed.

The most important implication from the occurrence of vertical transfer is that the prion agents must surely be in blood, and hence widely distributed in the animal and therefore in much meat and other beef products consumed.

Thus BSE now comprises a very serious epidemic, aggravated by incorporation of cattle remains into the feed, centralized production of wide peripheral distribution of the relevant products. Subsequently the infection has been perpetuated by vertical and horizontal transfer with the chief mechanism being the ingestion by calves of infected placentas from sub-clinically diseased cows, although this has as yet to be proved by experiment.

BSE – the threat to humans

It has been estimated that the average UK resident has consumed a very considerable amount of bovine prions since 1986, the year BSE was first confirmed (Dealler, 1993). The reasons for this claim are as follows. First, it was only in 1988 that clinically infected cattle were excluded from the food chain. Secondly, while certain cattle organs ('specified offals') were excluded in November 1989 (but not from animals under 6 months of age), many that are potentially infected were not, including liver, kidneys, lungs and, extremely importantly, bones. The bones of old cows provide a major source of the protein gelatine, used in many processed foods and sweets. Thirdly, many potentially infected organs of sub-clinically infected cows are used in burgers, sausages, meat pies, and so on.

On June 30 1994, the Minister of Agriculture announced that the ban on two of the specified offals on entering the human food chain – intestines and thymus – would be extended to calves under 6 months. This decision was based on experimental data and has important implications. First, it indicates that calves are acquiring the infection 6 years after the feed ban, and this surely results from vertical and horizontal transfer on a substantial scale. Secondly, it raises the question why such prohibition was not instituted previously, and thirdly it raises questions about the previous failure to find infectivity in BSE cattle (brain and spinal cord, and now intestines, apart).

Typically, when an animal is spontaneously or artificially infected with a spongiform disease, the presence of infectivity is found in many organs approximately half-way through the incubation period (Dealler & Lacey, 1990). Such infectivity is detected by challenge to further test animals and the outcome of the experiment depends on the amount of infectivity present in the donor diseased animal and the vulnerability of the challenged animal to the infective agent in question.

The latter varies with the host in question, and the difficulty of transfer to some infectivity between one species and another is known as the species barrier. However, regardless of the level of infectivity in the non-brain organs, the final fatal illness is always associated with much greater infectivity in the brain itself. The level of infectivity is assessed by ascertaining the

minimal dilution at which the disease is transferred from the tissue hom-
ogenate. When the target animal is fully vulnerable to the source of
infectivity it is typical for the brain to contain infectivity of 10^9 IU/gm and
other tissues, approximately 10^4 IU/gm (Dealler & Lacey, 1990). Experiments with BSE tissues have employed mice as the assay or target species
and typically titres of only 10^4 and 10^5 IU/gm brain have been found (Fraser
et al., 1992). Thus mice are relatively resistant to the effects of BSE, and it is
inevitable that non-brain tissues will usually not be capable of inducing any
disease in mice, even though such infectivity (as say by inducing disease in
cattle) is likely to be present.

The announcement on June 30th 1994 by MAFF that experiments (begun,
amazingly, as late as January 1992) had shown that the oral feeding of calves
with infected material resulted in the appearance of infectivity in their guts,
should not have come as any surprise. But the reason that infectivity should
be thought to be confined to these organs is quite inexplicable. The cells
from the lymphoid regions of the gut where the infection is thought to reside
circulate freely around the body and are concentrated in many organs
including liver, spleen and bone marrow.

There is still very little known about the distribution of the infective agent,
both temporally and spatially in the BSE cow because of the insensitivity of
the mouse assay used. The prospect does seem very real that as different
organs are sooner or later shown to be infected, then the range of organs
banned will approach the totality of the animal.

However, the consumption of the infectious agent by the human population does not necessarily predict the development of the terminal disease,
CJD. Previously, the main argument used by the UK government to
reassure the public has been that since BSE came from sheep scrapie, and
the human population is not vulnerable to scrapie (Southwood, 1989; Tyrell,
1989), then BSE will not be a danger. However, this argument fails to
appreciate that once such an agent passes from one species to another its
subsequent host range is changed (Dealler & Lacey, 1990). Furthermore,
evidence has been presented here that BSE did not arise from sheep scrapie
(*vide supra*). As far as can be judged, the host range of BSE is very wide
indeed, and there is no evidence that BSE cannot or indeed has not already
caused CJD.

What is the source of sporadic CJD?

The analysis above suggests that the infectious agent for BSE has a very wide
host range, and that the human food chain continues to be contaminated
with the agent. As yet there is no formal identification of the scale of
vulnerability of the human population to BSE. If a source for the sporadic
cases were to be identified that was non-bovine, then some reassurance
might be in order, although not necessarily so.

By definition, sporadic CJD is an infective disease following a long incubation period, that is not acquired by close human contact. Since the only reservoirs of these infections have been mammals, then mammals must provide the likely source of sporadic CJD. Accordingly, the suspicion that sheep scrapie is the cause of sporadic CJD has received closest scrutiny. Indeed, there is very little, if any, association between scrapie and sporadic CJD (Southwood, 1989; Tyrell, 1989). This is not reassuring for the possible relationship between BSE and CJD because the question becomes 'if scrapie is not the cause of sporadic CJD, then what is?' Assuming the likely routes of infection are ingestion or possibly direct contact with the internal organs of mammals (Dealler & Lacey, 1990), domestic pets can be excluded as the source of sporadic CJD. Pigs and poultry are unlikely to provide the source of CJD, since there is no natural spongiform disease in them. Therefore, the suspicion falls on cattle, particularly because the meat from the elderly cow at the end of lactation is fashioned into processed food. This possibility requires that BSE has been prevalent for many decades in many countries prior to 1986. There is some evidence for this in that a disease in the 1940s known as 'stoddy' was described by British farmers (Dealler & Lacey, 1992), and in one of the incidences whereby whole populations of mink were decimated by spongiform disease the disease was acquired from cattle (Dealler & Lacey, 1992). Thus, the worst scenario supposes that BSE and CJD are one and the same, and that the recent increase in CJD numbers (Table 5) may well rise to horrific proportions next century.

We will welcome any scientific objection to this pessimistic hypothesis. So far, the only counter argument received is that 'CJD is a human disease' and that because *Homo sapiens* is not vulnerable to sheep scrapie, he is therefore also not vulnerable to BSE. Meanwhile, should we not initiate an urgent research programme into these issues, including the search for preventative or therapeutic drugs? Unfortunately, the claim that BSE is self-limiting (Southwood, 1989) has prevented such an initiative. There is also a need for the systematic elimination of BSE-infected herds, and their replacement by new BSE-free stock husbanded on new territory.

REFERENCES

Adams, D. H., & Edgar, W. M. (1978). Transmission of agent of Creutzfeldt–Jakob disease. *British Medical Journal*, **1**, 987.

Arnow, P. M., Anderson, R. L., Mainous, P. D. & Smith, E. J. (1978). Pulmonary aspergillosis during hospital renovation. *American Review of Respiratory Diseases*, **118**, 49–53.

Behan, P. O. (1982). Creutzfeldt–Jakob disease. *British Medical Journal*, **284**, 1658–9.

Bisno, A. L. (1991). Group A streptococcal infection and acute rheumatic fever. *New England Journal of Medicine*, **325**, 783–93.

Blostein, J. (1991). Shigellosis from swimming in a park pond in Michigan. *Public Health Report*, **106**, 317–22.

Bowler, L. D., Zhang, Q. Y., Riou, J. Y. & Spratt, B. G. (1994). Interspecies recombination between the *pen A* genes of *Neisseria meningitidis* and commensal *Neisseria* species during the emergence of penicillin resistant *N. meningitidis*: natural events and laboratory simulation. *Journal of Bacteriology*, **176**, 333–7.

Brown, P., Liberski, P. P., Wolff, A. & Gajdusek, D. C. (1990). Resistance of scrapie infectivity to steam autoclaving after formaldehyde fixation and limited survival after ashing at 360°: practical and theoretical implication. *Journal of Infectious Diseases*, **161**, 467–72.

Brown, P., Preece, M. A. & Will, R. G. (1992). 'Friendly five' in medicine: hormones, homografts, and Creutzfeldt–Jakob Disease. *Lancet*, **340**, 24–7.

Büeler, H., Aguzzi, A., Sailer A., Greiner, R.-A., Antenried, P., Agent, M. & Weissmann, C. (1993). Mice devoid of PrP are resistant to scrapie. *Cell*, **73**, 1339–47.

Burnet, M. (1963). In *Natural History of Infectious Diseases*. 3rd edn. Cambridge University Press, London.

Centers for Disease Control (1987). *Aedes albopictus* infestation – United States *Morbidity Mortality Weekly Report*, **36**, 769–73.

Centers for Disease Control (1991). Eastern equine encephalitis virus associated with *Aedes albopictus* – Florida. *Morbidity Mortality Weekly Report*, **41**, 115–21.

Centers for Disease Control (1992a) Viscerotopic Leishmaniasis in persons returning from Operation Desert Storm. *Morbidity Mortality Weekly Report*, **41**, 131–4.

Centers for Disease Control (1992b). Public Health Focus: Surveillance, prevention and control of nosocomial infections. *Morbidity Mortality Weekly Report*, **41**, 783–7.

Centers for Disease Control (1993). Diphtheria outbreak – Russian Federation, 1990–1993. *Morbidity Mortality Weekly Report*, **42**, 840–1.

Cunningham, A. A., Wells, G. A. H., Scott, A. C., Kirkwood, J. K. & Barnett, J. E. F. (1993). Transmissible spongiform encephalopathy in greater kudu (*Tragelophus strepsiceros*). *Veterinary Record*, **132**, 68.

Cutlip, R. C., Miller, J. H., Race, R. B., Jenny, A. L., Katz, J. B., Lehmkuhl, H. D., De Bey, B. M. & Robinson, M. M. (1994). Intracerebral transmission of scrapie to cattle. *Journal of Infectious Diseases*, **169**, 814–20.

Dickinson, A. G. & Taylor, D. M. (1978). Resistance of scrapie agent to decontamination. *New England Journal of Medicine*, **299**, 1413–14.

Dealler, S. F. (1993). Bovine spongiform Encephalopathy (BSE). The potential effect of the epidemic on the human population. *British Food Journal*, **95**, 22–34.

Dealler, S. F. & Lacey, R. W. (1990). Transmissible spongiform encephalopathies: the threat of BSE to man. *Food Microbiology*, **7**, 253–9.

Dealler, S. F. & Lacey, R. W. (1992). Transmissible spongiform encephalopathies. *Encyclopaedia of Microbiology*, **4**, 299–309.

Dealler, S. F. & Lacey, R. W. (1994). Suspected vertical transmission of BSE. *Veterinary Record*, **134**, 151.

Fraser, H., Bruce, M. E., Chree, A., McConnell, I. & Wells, G. A. J. (1992). Transmission of bovine spongiform encephalopathy and scrapie to mice. *Journal of General Virology*, **73**, 1891–7.

Fraser, H., Farquhar, C. F., McConnell, I. & Davies, D. (1989). The scrapie disease process is unaffected by ionising radiation. *Progress in Clinical Biological Research*, **317**, 653–8.

Gottuzzo, E., Cieza, J., Estremadoyro, L., & Seas, C. (1994). Cholera: lessons from the epidemic in Peru. *Infectious Diseases Clinics North America*, **8**, 183–205.

Grist, N. (1989). The first ten years – self-inflicted epidemics. *Journal of Infection*, **19**, 1–3.

Gubler, D. J. (1991). Dengue haemorrhagic fever: a global update. *Virus Information Exchange Newsletter*, **8**, 2–3.

Gustafson, T. L., Band, J. D., Hutcheson, R. J. & Schaffner W. (1983). *Pseudomonas folliculitis*: an outbreak and review. *Reviews of Infectious Diseases*, **5**, 1–8.

Hamadeh, R., Ardehali, A., Locksley, R. M. & York, M. C. (1988). Fatal aspergillosis associated with smoking contaminated marijuana in a marrow transplant recipient. *Chest*, **94**, 432–3.

Hansard PQ (1994*a*). 1468, Priority 27, May 17.

Hansard PQ (1994*b*). 1469, Priority 28, May 17.

Harris, A. A., Levin, S. & Trenholme, G. M. (1984). Selected aspects of nosocomial infections in the 1980s. *American Journal of Medicine*, **77** (supp 1B): 3–10.

Hawley, W. A., Reiter, P., Copeland, T. S., Pumpuni, C. B., and Craig, G. B. (1987). *Aedes albopictus* in North America: probable importation in used tires from Northern Asia. *Science*, **236**, 1113–6.

Hospedales, C. J. (1990). Dengue fever in the Caribbean. *West Indian Medical Journal*, **39**, 131–5.

Jouan, A., Le Guernno, B., Digoutte, J. P., Phillipe, B., Riou, O. & Adam, F. (1988). A rift valley fever epidemic in Southern Mauretania. *Annals of the Institute Pasteur, Virology*, **139**, 307–8.

Kida, H., Shortridge, K. F. & Webster, R. G. (1988). Origin of the hemagglutinin gene of H3N2 influenza viruses from pigs in China. *Virology*, **162**, 160–6.

Kimberlin, R. H. (1993). *Bovine Spongiform Encephalopathy*. Food and Agriculture Organisation (UN) pp. 1–68.

Lacey, R. W. (1993). BSE: the gathering crisis. *British Food Journal*, **95**, 17–21.

Last, J. M. & Wallace, R. B. eds (1992). *Maxcy-Rosenau-Last Public Health and Preventative Medicine* 13th edn. Appleton & Lange. Norwalk, CT.

Lastavica, C. C., Wilson, M. L., Beradi, V. P., Spielman, A. & Deblinger, R. D. (1989). Rapid emergence of a focal epidemic of Lyme disease in coastal Massachusetts. *New England Journal of Medicine*, **320**: 133–7.

Leaf, A. (1989). Potential health effects of global climatic and environmental changes. *New England Journal of Medicine*, **321**, 1577–83.

Lederberg, J. (1988). Medical science, infectious disease, and the unity of humankind. *Journal of the American Medical Association*, **260**, 684–5.

Lentino, J. R., Rosenkranz, M. A., Michaels, J. A., Kurup, U. P., Rose, H. D. & Rytel, M. W. (1982). A retrospective review of airborne disease secondary to road construction and contaminated air conditions. *American Journal of Epidemiology*, **116**, 430–7.

Levine, D. P., Crane, L. R. & Zervos, M. J. (1986). Bacteremia in narcotic addicts at the Detroit Medical Center. II. Infectious endocarditis: a prospective comparative study. *Review Infectious Diseases*, **8**, 374–96.

Maki, D. G. (1986). Infections due to infusion therapy. In *Hospital Infections* (Bennett, J. V. & Brachman, P. S. eds). 2nd edn, Little, Brown & Co. Boston, Mass.

Manuelidis, E. E., Kim, J. H., Mericangas, J. R. & Manuelidis, L. (1985). Transmission to animals of Creutzfeldt-Jakob disease from human blood. *Lancet*, **ii**, 896–7.

Meegan, J. M. & Shope, R. E. (1981). Emerging concepts of Rift Valley Fever. In *Perspectives in Virology* (Pollard, M., Allan, R. eds). Liss Inc. New York, NY.

Mollick, J. A., Miller, G. G., Musser, J. M., Cook, R. G., Grossman, D. & Rich, R. R. (1993). A novel superantigen isolated from pathogenic strains of *Streptococcus* pyogenes with aminoterminal homology to staphylococcal enterotoxins B and C. *Journal of Clinical Investigations*, **92**, 710–19.

Morse, S. S. (1991). Emerging viruses: defining the rules for viral traffic. *Perspectives in Biology and Medicine*, **34**, 387–409.

Munoz, R., Coffey, T. J., Daniels, M., Dowson, C. G., Laible, G., Casal, J., Hakenbeck, R., Jacobs, M., Musser, J. M., Spratt, B. G. & Tomasz, A. (1991). Intercontinental spread of a multiresistant clone of serotype 23F *Streptococcus pneumoniae*. *Journal of Infectious Diseases*, **164**, 302–6.

Musser, J. M., Hauser, A. R., Kid, M. H., Schlievert, P. M., Nelson, K. & Selander, R. K. (1991). *Streptococcus pyogenes* causing toxic shock-like syndrome and other invasive diseases: clonal diversity and pyrogenic exotoxin expression. *Proceedings of the National Academy of Sciences, USA*, **87**, 2668–72.

Osterholm, M. T., Klein, J. O., Aronson, S. S. & Pickering, L. K., eds. (1986) Infectious diseases in child day care: management and prevention. *Review of Infectious Diseases*, **8**, 513–679.

Pattison, I. H. (1988). Fifty years with scrapie: A personal reminiscence. *Veterinary Record*, **123**, 661–6.

Podzamczer, D., Fernandez, Ribera, M., Arruga, J., de Cetis, G. & Gudiol, F. (1985). Systemic candidiasis in drug addicts. *European Journal of Microbiology*, **4**, 509–12.

Press Release, (1994). Chief Medical Officer, June 30.

Prusiner, S. B. (1982). Novel proteinaceous infectious particles cause scrapie. *Science*, **216**, 136–44.

Prusiner, S. B. (1984). Prions. *Scientific American*, **251**, 48–57.

Prusiner, S. B. (1989). Scrapie prions. *Annual Review of Microbiology*, **43**, 345–74.

Report (1981). Report of the Advisory group on the management of patients with spongiform encephalopathy (Creutzfeldt–Jakob disease). Department of Health and Social Security circular DA81, 22.

Report (1994). Bovine spongiform encephalopathy. A progress report. March 1994. Ministry of Agriculture, Fisheries and Food.

Rohwer, R. G. (1984). Scrapie infectious agent is virus like in size and susceptibility to inactivity. *Nature*, **308**, 658–62.

Rolfs, R. T., Goldberg, M. & Shorra, R. G. (1990). Risk factors for syphilis: cocaine use and prostitution. *American Journal of Public Health*, **80**, 853–7.

Rosenberg, R. N. (1986). Precautions in handling tissues, fluids and other contaminated materials from patients with documented or suspected Creutzfeldt–Jakob disease. *Annals of Neurology*, **19**, 75–7.

Scholtissek, C. & Naylor, E. (1988). Fish farming and influenza pandemics. *Nature*, **331**, 215.

Southwood, R. (1989). Report of the working party on bovine spongiform encephalopathy. Ministry of Agriculture, Fisheries and Food.

Steere, A. C. (1989). Lyme disease. *New England Journal of Medicine*, **321**, 586–96.

Stevens, D. L. (1992). Invasive group A streptococcus infections. *Clinical Infectious Disease*, **14**, 2–13.

Taylor, D. M. (1989). Scrapie agent decontamination: implications for bovine spongiform encephalopathy. *Veterinary Record*, **124**, 291–2.

Taylor, D. N., Wachsmuth, K., Shangkuan, Y. H., Schmidt, E. G., Barrett, T., Schrader, J. S., Scherach, C. S., McGee, H. B., Feldman, R. A. & Brenner, D. J. (1982). Salmonellosis associated with marijuana. *New England Journal of Medicine*, **306**, 1249–42.

Tyrell, D. A. J. (1989). Consultative committee on research into spongiform encephalopathies. Interim Report. Ministry of Agriculture, Fisheries and Food and Department of Health.

Wendel, S. & Gonzaga, A. L. (1993). Chagas disease and blood transfusion. *Vox Sanguinis*, **64**, 1–12.

World Resources Institute (1986). *World Resources 1986* International Institute for Environment and Development, Basic Books, New York, NY.

WHY DO MICROORGANISMS PRODUCE ANTIMICROBIALS?

ARNOLD L. DEMAIN

Fermentation Microbiology Laboratory, Department of Biology, Massachusetts Institute of Technology, Cambridge, Massachusetts 02139, USA

INTRODUCTION

Over the years, a number of erroneous views have appeared in the literature such as those which maintain that antimicrobials (antibiotics) are laboratory artefacts, are not produced in nature or lack natural functions. It has always amazed me that the importance of chemical compounds in ecological interactions between plant versus herbivore, insect versus insect, and plant versus plant has been universally accepted (Mann, 1978), but the importance of antimicrobials in microbial interactions has been almost universally denied. However, enough evidence has accumulated (Demain, 1989) to allow us to put aside such old-fashioned views and to use the knowledge gained to further our exploitation of these remarkable natural products to advance medicine, agriculture and the environment. This Chapter summarizes our knowledge of the functions of secondary metabolites with antimicrobial activity.

There is no doubt that antimicrobials are natural products. Over 40% of filamentous fungi and actinomycetes produce antibiotics when they are freshly isolated from nature. Foster, Yasouri & Daoud (1992) reported that 77% of soil myxobacteria had antibiotic activity against *Micrococcus luteus*. Many of these showed antifungal activity and a few were active against Gram-negative bacteria. If one examines soil, straw and agricultural products, they often contain antibacterial or antifungal substances. We may call both of these types 'mycotoxins', but they are nevertheless antimicrobials. Indeed, one of our major public health problems is the natural production of such toxic metabolites in the field and during storage of crops. The natural production of ergot alkaloids by the sclerotial (dormant overwintering) form of *Claviceps* on the seed heads of grasses and cereals has led to widespread and fatal poisoning ever since the Middle Ages (Vining & Taber, 1979; Norstadt & McCalla, 1969). Natural soil and wheat-straw contain patulin, and aflatoxin is known to be produced on corn in the field (Hesseltine, Rogers & Shotwell, 1981). Corn grown in the tropics or semitropics always contains aflatoxin (Hesseltine, 1986). At least five mycotoxins of *Fusarium* have been found to occur naturally in corn: moniliformin, zearalenone, deoxynivalenol, fusarin C and fumonisin (Sydenham *et al.*, 1990). Trichothecin is found in anise fruits, apples, pears and wheat (Ishii *et al.*, 1986). Microbially produced siderophores have been found in soil (Castignetti & Smarrelli, 1986) and microcins (enterobacterial antibiotics) have been isolated from human faecal extracts (Bacquero & Asensio, 1979). The microcins are thought to be important in colonization of the human intestinal tract early in life.

Antibiotics are produced in unsterilized, unsupplemented soil, in unsterilized soil supplemented with clover and wheat straw, in mustard, pea and maize seeds and in unsterilized fruits (for review see Demain, 1980). A further indication of natural antibiotic production is the possession of antibiotic-resistance plasmids by most soil bacteria (Foster, 1983). Nutrient limitation is the usual situation in nature resulting in very low bacterial growth rates, for example, 20 days in deciduous woodland soil (Gray, 1976). Low growth rates favour secondary metabolism.

The widespread production of antimicrobials and the preservation of multigenic biosynthetic pathways in nature indicate that antibiotics serve survival functions in organisms that produce them. There are a multiplicity of such functions, some dependent on antibiotic activity and others independent of such activity. Indeed in the latter case, the molecule may possess antibiotic activity but may be used for an entirely different purpose.

The view that antimicrobials act by improving the survival of the producer in competition with other living species (Demain, 1980, 1989) is supported by Williams *et al.* (1989) who contend that these compounds act via specific receptors (DNA, cell wall synthesizing enzymes, etc.) in competing organisms. Stone and Williams (1992) support the concept that organisms have evolved the ability to produce secondary metabolites because of the selective advantages which the organisms obtain as a result of the functions of these compounds. Their arguments are as follows: (i) only organisms lacking an immune system are prolific producers of these compounds which act as an alternative defence mechanism; (ii) the compounds have sophisticated structures, mechanisms of action, and complex and energetically expensive pathways; (iii) soil isolates produce natural products, most of which have physiological properties; (iv) they are produced in nature and act in competition between microorganisms, plants and animals (Demain, 1980, 1989); (v) clustering of antibiotic biosynthetic genes, which would only be selected for if the product conferred a selective advantage, and the absence of non-functional genes in these clusters; (vi) the presence of resistance and regulatory genes in these clusters; (vii) the clustering of resistance genes in non-producers; and (viii) the temporal relationship between antibiotic formation and sporulation (Katz & Demain, 1977) due to sensitivity of cells during sporulation to competitors and the need for protection when a nutrient runs out. Stone and Williams (1992) call this 'pleiotropic switching', that is a way to express concurrently both components of a two-pronged defence strategy when survival is threatened.

It has been proposed that antibiotics, originally produced by chemical (non-enzymatic) reactions, played important evolutionary roles in effecting and modulating prehistoric reactions, such as primitive transcription and translation, by reacting with receptor sites in primitive macromolecular templates made without enzymes (Davies, 1990). Later on, the small molecules were thought to be replaced by polypeptides but retained their abilities to bind to receptor sites in nucleic acids and proteins. Thus they changed from molecules with a function in synthesis of macromolecules to antagonists of such processes, for example as antibiotics, enzyme inhibitors or receptor antagonists. As evidence, Davies cites examples where antibiotics are known to stimulate gene transfer, transposition, transcription, translation, cell growth, and mutagenesis including examples of antibiotic-dependent mutants.

FUNCTIONS

Competitive weapons

According to Cavalier-Smith (1992), secondary metabolites are most useful to the organisms producing them as agents of chemical warfare, and the selective forces of their production have existed even before the first cell. The antibiotics are more important than macromolecular toxins (for example, colicins and animal venoms) because of their diffusibility into cells and broader modes of action.

Microbial antagonism

One of the first pieces of evidence indicating that one microorganism produces an antibiotic against other microorganisms and that this provides for survival in nature was published by Bruehl, Millar & Cunfer (1969). These workers found that *Cephalosporium gramineum*, the fungal cause of stripe disease in winter wheat, produces a broad-spectrum antifungal antibiotic. Over a three-year period, more than 800 isolates were obtained from diseased plants, each of which was capable of producing the antibiotic in culture. On the other hand, ability to produce the antibiotic was lost during storage on solid medium at 6°C. Thus antibiotic production was selected for in nature but was lost in the test tube. The selection was found to be exerted during the saprophytic stage in soil. These workers further showed that antibiotic production in the straw-soil environment aided survival of the producer culture and markedly reduced competition by other fungi.

Another case involves the parasitism of one fungus on another. The parasitism of *Monocillium nordinii* on the pine-stem rust fungi *Cronartium coleosporioides* and *Endocronartium harkensii* is due to its production of the antifungal antibiotics monorden and monocillins (Ayer *et al.*, 1980).

Competition between bacteria is also effected via antibiotics. Agrocin 84, a plasmid-encoded antibiotic of *Agrobacterium rhizogenes*, is an adenine derivative which attacks strains of plant pathogenic agrobacteria. It is used in the prevention of crown gall, and acts by killing the pathogenic forms (Kerr & Tate, 1984).

An interesting relationship exists between myxobacteria and their bacterial 'diet'. Myxobacteria live on other bacteria and to grow on these bacteria, they require a high myxobacterial cell density. This population effect is primarily due to the need for a high concentration of lytic enzymes and antibiotics in the local environment. Thus, *Myxococcus xanthus* fails to grow on *Escherichia coli* unless more than 10^7 myxobacteria/ml are present (Rosenberg & Varon, 1984). At these high cell concentrations, the parent grows but a mutant which cannot produce its antibiotic fails to grow. This indicates that the antibiotic is involved in the killing and nutritional use of other bacteria.

Antibiotic production was crucial in competition studies carried out in autoclaved sea water by Lemos *et al.* (1991). Four antibiotic-producing marine bacteria and three non-producing marine bacteria were grown in pairs or three-membered cultures. In every case of a non-producer and a producer pair, the non-producer disappeared. In five pairs of producer cultures, one producer survived and the other did not in four of the cases. When non-producers were paired or combined in three-membered cultures, all survived. In three-membered cultures including at least one producer, the producer always survived. This work supports the amensalism concept that antibiotic production aids in survival by killing or inhibiting other strains.

Competition also occurs between strains of a single species. Phenazine production by *Pseudomonas phenazinium* results in smaller colonies and lower maximum cell densities (but not lower growth rates) than those of its non-producing mutants (Messenger & Turner, 1981). Furthermore, the viability of non-producing mutants in various nutrient-limited media is higher than that of the producing parent. Despite these apparent deficiencies, the producing strain wins out in a mixed culture in the above media. The parental strain is able to use its phenazine antibiotic to kill the non-producing cells and due to its resistance to the antibiotic, the parent survives.

Erwinia carotovora subsp. *betavasculorum* is a wound pathogen causing vascular necrosis and root rot of sugarbeet. It produces a broad-spectrum antibiotic which is the principal determinant allowing it to compete successfully in the potato against the antibiotic-sensitive *E. carotovora* subsp. *carotovora* strains. Close correlation was observed between antibiotic production *in vitro* and inhibition of subsp. *carotovora* strains in the plant (Axelrod, Rella & Schroth, 1988).

Bacteria versus *amoebae*

Since protozoa use bacteria as food (Habte & Alexander, 1977) and use these prokaryotes to concentrate nutrients for them, it is not surprising that mechanisms have evolved to protect bacteria against protozoans such as amoebae. Over 50 years ago, Singh (1942) noted that antibiotically active pigments from *Serratia marcescens* and *Chromobacterium violaceum* (prodigiosin and violacein, respectively) protect these species from being eaten by amoebae; in the presence of the pigment, the protozoa either encyst or die. Non-pigmented *S. marcescens* cells are consumed by amoebae but pigmented cells are not. These experiments have been extended to other bacteria such as *Pseudomonas aeruginosa* and to microbial products such as pyocyanine, penicillic acid, phenazines and citrinin (Singh, 1945; Groscop & Brent, 1964; Imshenetskii, 1974). These findings show that antagonism between amoebae and bacteria in nature is crucially affected by the ability of the latter to produce antibiotics.

Microorganisms versus *higher plants*

More than 150 microbial compounds called phytotoxins or phytoaggressins that are active against plants have been reported, and the structure of over 40 are known (Fischer & Bellus, 1993). Many such compounds show typical antibiotic activity against microorganisms (for example, phaseolotoxin, rhizobitoxin, syringotoxin, syringostatin, tropolone and fireblight toxin) and could justifiably be classified either as antibiotics or phytotoxins. These include many phytotoxins of *Pseudomonas* which are crucial in the pathogenicity of these strains against plants (Gasson, 1980). These toxins, which induce chlorosis in plant tissue (Staskawicz & Panopoulos, 1979), include tabtoxinine-β-lactam (a glutamine antagonist produced by *Pseudomonas syringae* pv. '*tabaci*' and *Pseudomonas coronofaciens* which causes wildfire in tobacco and halo blight in oats, respectively) and phaseolotoxin, a tripeptide arginine antimetabolite of *P. syringae* pv. '*phaseolicola*' which causes halo blight in french beans. Phaseolotoxins not only induce chlorosis but are necessary for the systematic spread of *P. syringae* pv. '*phaseolicola*' throughout the plant (Patil, 1974). Other phytotoxic antibiotics include syringomycin from *P. syringae* and the toxic peptides of *Pseudomonas glycinea* and *Pseudomonas tomato* (Strobel, 1977). Syringomycin, a cyclic lipodepsinonapeptide, is involved in bacterial canker of stone fruit trees and holcus spot of maize, and is a broad-spectrum antibiotic against prokaryotes and eukaryotes including *Geotrichum candidum* (Xu & Gross, 1988). Proof of the role of antibiotics as plant toxins has been provided in the case of syringomycin (Mo & Gross, 1991), which disrupts ion transfer across the plasmalemma of plant cells. The antibiotic synthetases are encoded by a series of genes, including *syrB* which appears to encode a subunit of one or both of two proteins, namely SR4 (350 kD) and SR5 (130 kD). Using a *syrB::lacZ* fusion, it was found that the gene is transcriptionally activated by plant metabolites with signal activity, for example, arbutin-β-D-glucopyranoside, salicin, aescalin and helicin, which are all produced by plants susceptible to the pathogen. Activators of genes involved in virulence of *Agrobacterium tumefaciens* (acetosyringone) or nodulation of *Rhizobium* species (flavonoids) were inactive, demonstrating the specificity of the phenomenon.

Production of antibiotics by plant pathogens is beneficial to the producer microbes in their ecological niches (Mitchell, 1991). Tabtoxinine-β-lactam production by strains of *P. syringae* enhances the bacterium's virulence and allows a 10-fold increased population to develop in the plant. Production of coronatine by strains of *P. syringae* (compared to its non-producing mutant) led to larger lesions, longer duration of lesion expansion, higher populations and longer duration of the bacterial population.

Fungi produce a large number of phytotoxins of varied structure such as sesquiterpenoids, sesterterpenoids, diketopiperazines, peptides, spirocyclic lactams, isocoumarins and polyketides (Strobel *et al.*, 1991). The phytotoxins produced by the plant pathogens *Alternaria helianthi* and *Alternaria chrysanthemi* (the pyranopyrones deoxyradicinin and radicinin, respectively) are pathogenic not only to the Japanese chrysanthemum but also to fungi (Robeson & Strobel, 1982). The phytotoxin of *Rhizopus chinensis*, the causative agent of rice seedling blight, is a 16-membered macrolide antifungal antibiotic, rhizoxin (Iwasaki *et al.*, 1984). The fungal pathogen responsible for onion pink root disease, *Pyrenochaeta terrestris*, produces three pyrenocines, A, B and C. Pyrenocine A is the most phytotoxic to the onion and has marked antibacterial and antifungal activity (Sparace, Reeleder & Khanizadeh, 1987).

The plant pathogenic basidiomycete, *Armillarea ostoyae*, which causes a great amount of damage to forests, produces a series of toxic antibiotics when grown in the presence of plant cells (*Picea abies* callus) or with competitive fungi. The antibiotics have been identified as sesquiterpene aryl esters which have antifungal, antibacterial and phytotoxic activities (Peipp & Sonnenbichler, 1992).

Secondary metabolites play a crucial role in the evolution and ecology of plant pathogenic fungi (Scheffer, 1991). Some of the fungi have evolved from opportunistic low-grade pathogens to high-grade virulent host-specialized pathogens, by gaining the genetic potential to produce a toxin. This ability to produce a secondary metabolite has allowed fungi to exploit the monocultures and genetic uniformity of modern agriculture resulting in disastrous epidemics and broad destruction of crops.

With all these weapons directed by microbes against plants, the latter do not take such insults 'lying down'. Plants produce antimicrobials after exposure to plant pathogenic microorganisms in order to protect themselves; these are called phytoalexins (Darvill & Albersheim, 1984). They are low molecular weight, weakly active and indiscriminate, that is they inhibit prokaryotes and eukaryotes including higher plant cells and mammalian cells. There are approximately 100 known phytoalexins. They are not a uniform chemical class and include isoflavonoids, sesquiterpenes, diterpenes, furanoterpenoids, polyacetylenes, dihydrophenanthrenes, stilbenes and other compounds. Their formation is induced following invasion by fungi, bacteria, viruses or nematodes. The compounds which are responsible for the induction are called 'elicitors'. The fungi respond by modifying and breaking down the phytoalexins.

The phytoalexins are just a fraction of the multitude of plant secondary metabolites. Over 10000 of these low molecular weight compounds are known but the actual numbers are probably in the hundreds of thousands. Almost all of the known metabolites which have been tested show some antibiotic activity (Mitscher, 1975). They are thought to function as chemical signals to protect plants against competitors, predators and pathogens, as pollination-insuring agents and as compounds attracting biological dispersal agents (Swain, 1977; Bennett, 1981).

Microorganisms versus insects

Certain fungi have entomopathogenic activity, infecting and killing insects via their production of secondary metabolites. One such compound is bassianolide, a cyclodepsipeptide produced by the fungus, *Beauveria bassiaria*, which elicits atonic symptoms in silkworm larvae (Kanaoka *et al.*, 1979). Another pathogen, *Metarrhizium anisophae*, produces the peptido-lactone toxins known as destruxins (Lee *et al.*, 1975).

Insects fight back against microorganisms by producing antibacterial proteins such as cecropins, attacins, defensins, lysozyme, diptericins, sarcotoxins, apidaecin and abaecin, when infected by bacteria (Kimbrell, 1991). The molecules either cause lysis or are bacteriostatic, and also attack parasites. Social insects appear to protect themselves by producing antibiotics (Dixon, 1992); honey contains antimicrobial substances, and ants produce low molecular weight compounds with broad spectrum antibiotic activity (Trimble *et al.*, 1992).

Microorganisms versus higher animals

Competition may exist between microbes and large animals. Janzen (1977) made a convincing argument that the reason why fruits rot, seeds turn mouldy and meats spoil is that it is 'profitable' for microbes to make seeds, fresh fruit and carcasses as objectionable as possible to large organisms in the shortest amount of time. Among their strategies is the production of antibiotics. In agreement with this concept are the observations that livestock generally refuse to eat mouldy feed and that aflatoxin is much more toxic to animals than to microorganisms. Kendrick (1986) has advanced the same hypothesis. He stated that animals which come upon a mycotoxin-infected food will do one of four things: (i) smell the food and reject it; (ii) taste the food and reject it; (iii) eat the food, become ill and avoid the same fate in the future; or (iv) eat the food and die. In each case, the fungus will be more likely to live than if it produced no mycotoxin.

Corynetoxins are produced by *Corynebacterium rathayi* and cause animal toxicity upon consumption of rye grass by animals. The disease is called 'annual rye grass toxicity'. The relatedness between animal toxins and antibiotics was emphasized by the finding of identity between corynetoxins and tunicamycins, the latter being known antibiotics produced by *Streptomyces* (Edgar *et al.*, 1982).

Metal transport agents

Certain secondary metabolites act as metal transport agents. One group is composed of the siderophores (also known as sideramines) which function in the uptake, transport and solubilization of iron. The other group includes the ionophoric antibiotics which function in the transport of certain alkali-metal ions; these include the macrotetralide antibiotics which enhance potassium permeability of membranes.

Iron-transport factors in many cases are antibiotics. Microorganisms have 'low' and 'high' affinity systems to solubilize and transport ferric iron. The high-affinity systems involve siderophores. The low-affinity systems allow growth in the case of a mutation abolishing siderophore production (Neilands, 1984). The low-affinity system works unless the environment contains an iron chelator, such as citrate, which binds the metal and makes it unavailable to the cell; under such conditions, the siderophore stimulates growth. Over a hundred siderophores have been described. Indeed, all strains of *Streptomyces, Nocardia* and *Micromonospora* examined produce such compounds (Zahner, Drantz & Weber, 1982). Antibiotic activity is due to the ability of these compounds to starve other species of iron when the latter lack the ability to take up the Fe–sideramine complex. Such antibiotics include nocardamin (Keller-Schierlein & Prelog, 1961) and desferritri-acetylfusigen (Anke, 1977).

Most living cells have a high intracellular K^+ concentration and a low Na^+ concentration, whereas extracellular fluids contain high Na^+ and low K^+. To maintain a high K^+/Na^+ ratio inside cells, a mechanism must be available to bring in K^+ against a concentration gradient and keep it inside the cell. Ionophores accomplish this in microorganisms. That production of an ionophore (for instance, a macrotetralide antibiotic) can serve a survival function has been demonstrated. Kanne and Zahner (1976) compared a *Streptomyces griseus* strain which produces a macrotetralide with its non-producing mutant. In low K^+ and Na^+ media, both strains grew and exhibited identical intracellular K^+ concentrations during growth. In the absence of Na^+, both strains took up K^+ from the medium. However, in the presence of Na^+, the mutant could not take up K^+. When the strains were grown in high K^+ concentrations and transferred to a high Na^+, low K^+ resuspension medium, the parent took up K^+ but the mutant took up Na^+ and lost K^+. As a result of these differences, mutant growth was inhibited by a high Na^+, low K^+ environment but the antibiotic-producing parent grew well.

Microbe–plant symbiosis and plant growth stimulants

Almost all plants depend on soil microorganisms for mineral nutrition, especially that of phosphate. The most beneficial microorganisms are those which are symbiotic with plant roots, namely the mycorrhizae, which are highly specialized associations between soil fungi and roots. The ectomycorrhizae, used by 3–5% of plant species, are symbiotic growths of fungi on plant roots in which the fungal symbionts penetrate intracellularly and replace partially the middle lamellae between the cortical cells of the feeder roots. Mycorrhizal fungi, including *Fusarium, Pythium* and *Phytophthera* lead to reduced damage by pathogens such as nematodes. Symbiosis between plants and fungi often involves antibiotics. In the case of ectomycorrhizae, the fungi produce antibiotics which protect the plant against pathogenic bacteria or fungi. One such antibacterial agent was extracted from ectomycorrhizae formed between *Cenococcum graniforme* and white pine, red pine and Norway spruce (Krywolap, Grand & Casida, 1964). Two other antibiotics, diatretyne nitrile and diatretyne 3, were extracted from ectomycorrhizae formed by *Leucopaxillus cerealis* var. *piclina*; they make feeder roots resistant to the plant pathogen, *Phytophthora cinnamomi* (Marx, 1969).

A related type of plant–microbe interaction involves the production of plant growth stimulants by bacteria. Specific strains of the *Pseudomonas fluorescens-putida* group are often used as seed inoculants to promote plant growth and increase yields. They colonize plant roots of potato, sugar beet and radish. Their growth-promoting activity is due in part to antibiotic (siderophore) action by depriving other bacterial and fungal species of iron. For example, they are effective biocontrol agents against *Fusarium* wilt and take-all diseases (caused by *Fusarium oxysporum* F. sp. *lini* and *Gaeumannomyces graminis* var. *tritici*, respectively). Some act by producing the siderophore, ferric pseudobactin, a linear hexapeptide with the structure L-lysine-D-*threo*-β-OH aspartate-L-alanine-D-*allo*-threonine-L-alanine-D-N^6-OH-ornithine (Teintze *et al.*, 1981). The evidence that the ability of fluorescent pseudomonads to suppress plant disease is partly dependent upon production of antibiotically active siderophores (O'Sullivan & O'Gara, 1992) is as follows: (i) the fluorescent siderophore can mimic the disease-suppression ability of the pseudomonad that produces it; (ii) siderophore-negative mutants fail to protect against disease or promote plant growth under field conditions; (iii) antibiotic-negative rhizosphere pseudomonad mutants fail to inhibit plant pathogenic fungi; and (iv) the parent culture produces its antibiotic in the plant rhizosphere.

The role of the antibiotic, phenazine-1-carboxylate, in protecting wheat against take-all disease caused by the fungus *Gaeumannomyces graminis* var *tritici* was demonstrated by Thomashow and Weller (1988) and Hamdan, Weller & Thomashow (1991). The antibiotic is produced by *P. fluorescens* 2-79 isolated from the rhizosphere of wheat, a fluorescent pseudomonad which colonizes the root system. The antibiotic inhibits the fungus *in vitro*. Antibiotic-negative mutants generated by Tn5 insertion did not inhibit the fungus *in vitro* and were less effective *in vivo* (on wheat seedlings). Cloning wild-type DNA into the mutant coordinately restored antibiotic synthesis and action *in vitro* and *in vivo*. The ability of *P. fluorescens* and *P. aureofaciens* to produce phenazine antibiotics also aids in survival of the producing bacteria in soil and in the wheat rhizosphere (Mazzola *et al.*, 1992). Phenazine-negative mutants survive poorly owing to a decreased ability to compete with the resident microflora.

Control of rhizoctonia root rot of pea by inoculated *Streptomyces hygroscopicus* var. *geldanus* is due to production of geldanomycin in the soil (Rothrock & Gottlieb, 1984). The antibiotic was extracted from soil and shown to be active against *Rhizoctonia solani*. Addition of geldanomycin itself to soil controlled disease.

Microbe–nematode symbiosis

Antibiotics play a role in the symbiosis between the bacteria of the genus *Xenorhabdus* and nematodes parasitic to insects. Each nematode species, members of the Heterorhabditidae and Steinernematidae, is associated with a single species of *Xenorhabdus* (Akhurst, 1982). The bacteria live in the gut of the nematode. When the nematode finds an insect host, it enters and when in the insect gut, it releases bacteria which kill the insect, allowing the nematode to complete its life cycle. Without the bacteria, no killing of the insect occurs. The bacteria produce antibiotics to keep the insect from being attacked by putrefying bacteria. Two groups of antibiotics have been isolated. One group is composed of indole derivatives and the other 4-ethyl- and 4-isopropyl-3,5-dihydroxy-*trans*-stilbenes (Paul *et al.*, 1981). The antibiotic produced by *Xenorhabdus luminescens*, the bacterial symbiont of several insect-parasitic nematodes (genus *Heterorhabditis*), has been identified as 3,4-dihydroxy-4-isopropylstilbene (Richardson, Schmidt & Nealson, 1988).

Microbe–insect symbiosis

Symbiosis between intracellular microorganisms and insects involves antibiotics. The brown planthopper, *Nilaparavata lugens*, contains at least two microbial symbionts and lives on the rice plant. One intracellular bacterium is a *Bacillus* sp. which produces polymyxin M (Jigami *et al*. 1986). Another is an *Enterobacter* sp. producing a peptide antibiotic which is selective against *Xanthomonas campestris* var *oryzae*, the white blight pathogen of rice (Fredenhagen *et al*., 1987). The intracellular bacteria increase their survival chances via antibiotic production to protect the insect from invasion by microorganisms and to control competition by bacterial rice pathogens.

Microbe–higher animal symbiosis

Antibiotic production by the commensal bacterium, *Alteromonas* sp., is responsible for protection of embryos of the shrimp, *Palaemon macrodactylus*, from the pathogenic fungus *Lagenidium callinectes* (Gil-Turnes, Hay & Fenical, 1989). The antifungal agent which mediates protection is 2,3-indolinedione (isatin). Similarly, eggs of *Homarus americanus* (the American lobster) are covered with a bacterium which produces tyrosol [2-(*p*-hydroxyphenyl)ethanol] which has antimicrobial activity (Fenical, 1993). The filamentous tropical cyanobacterium, *Microcoleus lyngbyaceus*, contains four specific bacteria on its surface, all of which produce quinone 34, an antibacterial and antifungal compound (Fenical, 1993).

Effectors of differentiation

Development involves two phenomena, growth and differentiation. The latter is the progressive diversification of structure and function of cells in an organism or the acquisition of differences during development (Bennett, 1983). Differentiation encompasses both morphological differentiation (morphogenesis) and chemical differentiation (secondary metabolism). Antimicrobials are not only made during chemical differentiation processes but also function in morphological and chemical differentiation.

Sporulation

Of the various functions postulated for secondary metabolites, the one which has received the most attention is the view that these compounds, especially antibiotics, are important compounds in the transition from vegetative cells to spores. The following observations have made this hypothesis attractive: (i) practically all sporulating microorganisms produce antibiotics; (ii) antibiotics are frequently inhibitory to vegetative growth of their producers at concentrations produced during sporulation; (iii) production of peptide antibiotics usually begins during the late-logarithmic phase of growth and continues during the early stages of the sporulation process in bacilli; (iv) sporulation and antibiotic synthesis are induced by depletion of some essential nutrient; (v) there are genetic links between the synthesis of antibiotics and the formation of spores, and revertants, transductants and transformants of stage-0 asporogenous mutants, restored in their ability to sporulate, also regain the ability to synthesize antibiotic, whilst conditional asporogenous mutants also fail to produce antibiotic at the non-permissive temperature; and (vi) physiological correlations also favour a relationship between the production of an antibiotic and spores. As an example, inhibitors of sporulation inhibit antibiotic synthesis. Furthermore, both processes are repressed by nutrients including glucose. Concentrations of manganese ion of at least two orders of magnitude higher than that required for normal cellular growth are needed for sporulation and antibiotic synthesis by certain species of *Bacillus*.

Much enthusiasm in favour of an obligatory function of antibiotics in sporulation derived from the work of Sarkar and Paulus (1972). They reported that the cessation of exponential vegetative growth of *Bacillus brevis* ATCC 8185 was accompanied by tyrothricin synthesis and a sharp decline of net RNA synthesis. They also reported that both antibiotic components of tyrothricin (tyrocidine and linear gramicidin) were capable of inhibiting purified *B. brevis* RNA polymerase. Sarkar and Paulus advanced the view that antibiotics regulate transcription during the transition from vegetative growth to sporulation by selectively terminating the expression of vegetative genes. Although the inhibition of RNA polymerase by tyrocidine and linear gramicidin was confirmed and further studied by Kleinkauf and co-workers (Schazschneider, Ristow & Kleinkauf, 1974; Ristow, Schazschneider & Kleinkauf, 1975), an obligatory relationship between production of the two antibiotics, inhibition of RNA synthesis, and sporulation has yet to be established. Even in the studies of Ristow *et al.* (1979), in which tyrothricin was found to stimulate sporulation when early log phase cultures were incubated in a glycerol medium lacking nitrogen, this addition stimulated RNA synthesis rather than inhibiting it.

Despite the apparent connections between formation of antibiotics and spores, it has become clear that antibiotic production is not obligatory for spore formation (Demain & Piret, 1979). The most damaging evidence to the antibiotic-spore hypothesis is the existence of mutants which form no antibiotic but still sporulate. Such mutants have been found in the cases of bacitracin (*Bacillus licheniformis*), mycobacillin (*Bacillus subtilis*), linear gramicidin (*B. brevis*), tyrocidine (*B. brevis*), gramicidin S (B. brevis), oxytetracycline (*Streptomyces rimosus*), streptomycin (*S. griseus*), methylenomycin A (*Streptomyces coelicolor*) and patulin (*Penicillium urticae*).

When little to no evidence could be obtained to support the hypothesis that antibiotics are necessary to form spores, Mukherjee and Paulus (1977) next questioned the quality of spores produced without concurrent formation of antibiotics. A mutant was obtained of the tyrothricin-producing *B. brevis* ATCC 8185 which produced normal levels of tyrocidine and spores but did not produce linear gramicidin. The spores were claimed to be less heat-resistant than normal but other workers were unable to confirm these findings (Piret & Demain, 1983). Similarly, mutants producing linear gramicidin but not tyrocidine formed spores of normal quality (Symons & Hodgson, 1982). Studying the *B. brevis* strain which produces gramicidin S, Marahiel *et al.* (1979) reported that non-producing mutants formed heat-sensitive spores but again this was not confirmed (Nandi & Seddon, 1978; Piret & Demain, 1983).

Although antibiotic production is not obligatory for sporulation, it may nevertheless stimulate the sporulation process (Ristow *et al.*, 1982). Transfer of exponential phase populations of *B. brevis* ATCC 8185 (the tyrothricin producer) into a nitrogen-free medium stopped growth and restricted sporulation. Supplementation of the medium with tyrocidine induced sporulation. Neither tyrocidine cleaved by a proteolytic enzyme, nor the component amino acids of tyrocidine nor gramicidin S were active. This indicates that the tyrocidine component of tyrothricin is an inducer of sporulation. Sporulation-associated events of *B. brevis* ATCC 8185 were turned on by linear gramicidin addition when nitrogen limitation was made more severe, for instance, by washing the cells before resuspension in the absence of nitrogen source. In this case, the production of extracellular protease, RNA, dipicolinate and tyrocidine itself were also stimulated. Addition of linear gramicidin also brought about a severe depletion of intracellular ATP. Non-ionophoric analogues of linear gramicidin did not exert the sporulation effect.

Bacilysin, a dipeptide antibiotic, may play a role in the sporulation of *B. subtilis* (Özcengiz & Alaeddinoglu, 1991). A bacilysin-negative strain, NG79, was found to be oligosporogenous, the spores to be sensitive to heat, chloroform and lysozyme, and deficient in dipicolinic acid. When the strain was transduced to bacilysin production, the above characteristics were also returned to wild-type status. Addition of bacilysin to the mutant increased sporulation, resistance and dipicolinic acid content. The concept that bacilysin plays a role in sporulation of *B. subtilis* was supported by the observation that interference in bacilysin production by addition of ammonium ions, Casamino acids, or L-alanine also interfered with sporulation (Basalp, Özcengiz and Alaeddinoglu, 1992). Although glucose interfered with sporulation, and decoyinine induced sporulation with no action on bacilysin production, these observations can be explained as effects on sporulation independent of those on bacilysin synthesis.

Sporulation inducers are present in the actinomycetes. One such compound is the antibiotic pamamycin produced by *Streptomyces alboniger*. Pamamycin inhibits *Staphylococcus aureus* by interfering with uptake of inorganic phosphate and of nucleosides (Chou & Pogell, 1981). In the producing culture, the antibiotic stimulates the formation of aerial mycelia and thus that of conidia (McCann & Pogell, 1979). Pamamycin has been found to be a family of eight homologous compounds varying in size from 593 to 691 daltons (Kondo *et al.*, 1988). The homologue of molecular weight

607 is active as an antibiotic against fungi and bacteria. It induces aerial mycelium formation in a negative-aerial mycelium mutant at 0.1 μg/disc and inhibits vegetative growth of the producing *S. alboniger* at 10 μg/disc. Its structure is that of a macrolide and it has the activity of an anion transfer agent.

Hormaomycin, a peptide lactone antibiotic produced by *S. griseoflavus* induces aerial mycelium and antibiotic formation in other streptomycetes (Andres, Wolf & Zähner, 1990). Carbazomycinal and 6-methylcarbazo-mycinal, inhibitors of aerial mycelium formation, are produced by *Strepto-verticillium* sp. (Kondo, Katayama & Marumo, 1986). An antifungal agent, lunatoic acid, produced by *Cochliobolus lunatus* is an inducer of chlamydo-spore formation (Marumo *et al.*, 1982).

Germination of spores

The close relationship between sporulation and antibiotic formation suggests that certain antimicrobials involved in germination might be produced during sporulation, and that the formation of these compounds and spores could be regulated by a common mechanism or by similar mechanisms.

A number of secondary metabolites are involved in maintaining spore dormancy in fungi. One example of these germination inhibitors is discade-nine [3-(3-amino-3-carboxypropyl)-6-(3-methyl-2-butenylamino)purine] in *Dictyostelium discoideum, Dictyostelium purpureum* and *Dictyostelium mucoroides* (Taya, Yamada & Nishimura, 1980); their function appears to be that of inhibiting germination under densely crowded conditions, and they are extremely potent secondary metabolites. Germination inhibitors have also been found in actinomycetes. The one produced by *Streptomyces viridochromogenes* is an antibiotic which uncouples respiration from ATP production until it is excreted during germination (Eaton & Ensign, 1980). The antibiotic is a specific inhibitor of ATPase and appears to be responsible for maintaining dormancy of the spores. It blocks ATP synthesis in the spores and thus uncouples glucose oxidation from ATP synthesis. Upon addition of the germinating agent Ca^{2+}, the inhibitor is excreted from the spore, the ATPase is activated by the Ca^{2+}, ATP is synthesized as glucose is oxidized and germination ensues. The inhibitor, named germicidin, is 6-(2-butyl)-3-ethyl-4-hydroxy-2-pyrone (Petersen *et al.* 1993).

Considerable evidence has been obtained indicating that gramicidin S (GS) is an inhibitor of the phase of spore germination known as 'outgrowth' in *B. brevis* (Nandi & Seddon, 1978; Marahiel *et al.*, 1979; Lazaridis *et al.*, 1980; Piret & Demain, 1982). The cumulative observations are as follows:

(i) Initiation of germination (observed as the darkening of spores under phase microscopy) is similar in the parent and GS-negative mutants.

(ii) The GS-negative mutant spores outgrow in 1–2 hours whereas parental spores require 6–10 hours. The delay in the parent is dependent on the concentration of spores and hence the concentration of GS.

(iii) Addition of GS to mutant spores delays their outgrowth so that they now behave like parental spores; the extent of the delay is concentration dependent and time dependent.

(iv) Preparation of parental spores on media supporting poor GS production results in spores which outgrow as rapidly as mutant spores.

(v) Removal of GS from parental spores by extraction allows them to outgrow rapidly.

(vi) Addition of the GS-containing extract to mutant spores delays their outgrowth.

(vii) Exogenous GS, which has been hydrolysed by a protease, does not delay outgrowth of mutant spores. Parental spores treated with the protease outgrow rapidly.

(viii) Exponential growth is not inhibited by GS.

(ix) A mixture of parental spores and mutant spores shows parental behaviour, so that the mixture is delayed in outgrowth. This indicates that some of the GS bound externally to parental spores is released into the medium. This release could act as a method of communication by which a spore detects crowded conditions.

(x) Uptake of alanine and uridine into spores and respiration is inhibited by GS (Danders & Marahiel, 1981; Nandi, Lazaridis & Seddon, 1981; Piret & Demain, 1982).

Lobareva, Zharikova & Mesyanzhinov (1977) provided evidence that *B. brevis* excretes over 90% of its GS which is then bound to the outside of the spores. The bound GS can be removed (with difficulty) by water or buffer, the ease of which is dependent on the presence or absence of detergents and pH. They showed that it is not merely a case of insoluble GS in suspension but that soluble GS binds to the cells. Lobareva *et al.* believe that this excretion process is the way *B. brevis* protects itself against GS-mediated uncoupling of electron transport and phosphorylation.

GS antibiotic activity appears to be due to its surface-activity, thus interacting with artificial lipid bilayers, and mitochondrial and bacterial membranes. There is an electrostatic interaction between membrane phospholipids and GS, which causes a phase separation of negatively-charged phospholipids from other lipids, leading to a disturbance of the membrane's osmotic barrier. It is possible that this effect is responsible for the ability of GS to inhibit respiration and the uptake of uridine and alanine during germination of spores of the producing organism and to delay spore outgrowth. GS does not inhibit growth of vegetative cells of the producer *B. brevis* ATCC 9999 or of *E. coli*, but it does inhibit the growth of vegetative cells of *B. subtilis*. However, alanine and uridine uptake into membrane vesicles from all three organisms is inhibited (Danders *et al.*, 1982). It is unclear why vegetative cells of the GS-producer are resistant to GS. Although Danders *et al.* (1982) think that it is due to impermeability, Frangou-Lazaridis and Seddon (1985) point out that exogenous GS added to vegetative cells is incorporated into the resulting spores (Lazaridis *et al.*, 1980) and thus is able to enter vegetative cells. It is of interest that tyrocidine, which has a structure similar to GS, also shows antibiotic action against *B. subtilis*, inhibition of active transport in the three species mentioned above, and delay of spore outgrowth of the GS-producing species, but all to a lesser degree than GS (Danders *et al.*, 1982). Danders and coworkers (1981, 1982) reported that one difference between GS and tyrocidine is that GS does not bind to DNA and inhibit RNA polymerase whereas tyrocidine does. However, there is a serious disagreement between groups on this point. Frangou-Lazaridis and Seddon (1985) found that RNA polymerase from the producer strain is inhibited by GS. They reported that addition of GS to *B. brevis* Nagano or its GS-negative mutant E-1 has no effect on growth or sporulation when added during or after the log phase of growth yet it is permeable. No effect was seen on incorporation of labelled lysine, thymidine or uracil by intact cells or on transcription by permeabilized vegetative cells although the cells were inhibited by rifamycin and actinomycin. However, RNA polymerase was strongly inhibited *in vitro* by GS. Frangou-Lazaridis and Seddon (1985) concluded that transcription was the sensitive step during germination outgrowth. Inhibition was thought to be due to GS complexing with DNA, not with the enzyme. They suggested that DNA and GS are prevented from interacting during growth or that vegetative DNA is in a conformational state that is not vulnerable to GS.

Irrespective of the mode of action of GS in inhibiting germination outgrowth, there are several other questions about this phenomenon which need to be addressed.

First, what would be the value to the producer organism of inhibiting germination in the outgrowth stage during which spores of bacilli were thought to have lost their resistance to factors such as heat? Would not inhibition at this stage make spores more susceptible to attack by other organisms and would not rapid outgrowers (for example, non-producing mutants) be selected? It turns out, however, that phase-dark, initiated spores of GS-producing *B. brevis* are still resistant to heat, starvation, solvents and even to sonication (Dahler, Rosenberg & Demain, 1985). It is also of interest that studies on the survival of *Bacillus thuringiensis* spores in soil have revealed that rapid germination ability confers no survival advantage (Petras & Casida, 1985).

It appears that endogenous GS is the basis of the hydrophobicity of dormant or initiated *B. brevis* spores. After outgrowth ceases, the resulting vegetative cells are hydrophilic (Rosenberg, Brown & Demain, 1985). Since water-insoluble organic matter constitutes the chief source of soil nutrients (Rosenberg & Varon, 1984), it is quite possible that the hydrophobicity of

B. brevis spores and initiated spores aids in their search for nutrients to ensure vegetative growth after germination. If no nutrients are found, it is possible that initiated spores can develop back into normal spores by microcycle sporulation (Vinter & Slepecky, 1965) which may have a role in nature (Maheshwari, 1991).

A second question involves the mechanism by which the outgrowing spores recover from GS inhibition and finally develop into vegetative cells. One possibility is destruction of GS towards the end of the outgrowth stage. *B. brevis* ATCC 9999 produces an intracellular serine protease (Kurotsu *et al.*, 1982; Piret, Millet & Demain, 1983) despite earlier claims that it does not (Slapikoff *et al.*, 1971). This type of enzyme is generally considered to be necessary for sporulation of bacilli. The *B. brevis* enzyme has the ability to cleave GS between valine and ornithine residues. No extracellular proteases are produced by ATCC 9999, a situation very different from most bacilli. Although it would appear that the intracellular enzyme might function to destroy GS and allow vegetative growth from outgrown spores [Marahiel *et al.* (1982) claimed that GS is destroyed at that time], our data indicated that GS is not destroyed as the outgrowing spores develop into vegetative cells (Bentzen *et al.*, 1990). Furthermore, the recovery is not due to selection of spores whose outgrowth is resistant to GS (Piret, 1981). Another possibility is that GS kills outgrowing spores and that the delay in outgrowth is merely the time required by a small population of unkilled spores to germinate and become vegetative cells. Although we have confirmed the finding (Murray, Lazaridis & Seddon, 1985) that GS kills a large proportion of outgrowing spores, the same residual fraction of survivors is seen despite our intentional increasing of the GS concentration (Bentzen & Demain, 1990). This lack of effect of increased concentrations of GS on killing can be contrasted with the increasing delay in outgrowth caused by the increased concentration and makes unlikely a connection between the degree of killing and the length of the outgrowth stage. At this point, it appears that GS (because of its inhibition of oxidative phosphorylation, transport and/or transcription) slows down, but does not totally inhibit, the macromolecular processes of outgrowth until a point is reached where all the outgrown spores have the proper machinery to differentiate into vegetative cells. During this process, GS is excreted into the extracellular medium (Lazaridis *et al.*, 1980; Nandi, Lazaridis & Seddon, 1981).

Thus, it is probable that GS serves the initiated spore as a means of sensing a high population density and preventing vegetative growth until there is a lower density of *B. brevis* spores with which to compete for nutrients. However, proof of such a hypothesis will require experimentation of an ecological nature. Alternative hypotheses might be that GS in and on the dormant and initiated spores protects them from consumption by amoebae or that GS excretion during germination initiation and outgrowth eliminates microbial competitors in the environment and that the delay in outgrowth and death of a part of the outgrowing spore population is merely 'the price the strain must pay' for such protection.

CONCLUSIONS

Secondary metabolites, including antibiotics, are produced in nature and serve survival functions for the organisms producing them. The antibiotics are a heterogeneous group, the functions of some being related to, and others being unrelated to their antimicrobial activities. Antimicrobials serve: (i) as competitive weapons used against other bacteria and fungi, amoebae, plants, insects and large animals; (ii) as metal transporting agents; (iii) as agents of symbiosis between microbes and plants, nematodes, insects or higher animals; and (iv) as differentiation effectors. Although antibiotics do not appear to be obligatory for sporulation, some antimicrobials stimulate spore formation and inhibit or stimulate germination. Formation of antimicrobials and spores may be regulated in a similar way. This could serve to ensure production of antimicrobials during sporulation in order to slow down germination of spores until a less competitive environment and more favourable growth conditions appear, to protect the dormant or initiated spore from consumption by amoebae, or to cleanse the immediate environment of competing microorganisms during germination.

REFERENCES

Akhurst, R. J. (1982). Antibiotic activity of *Xenorhabdus* spp. bacteria symbiotically associated with insect pathogenic nematodes of the families Heterorhabditae and Steinernematidae. *Journal of General Microbiology*, **128**, 3061–5.

Andres N., Wolf, H. & Zähner, H. (1990). Hormaomycin, a new peptide lactone antibiotic effective in inducing cytodifferentiation and antibiotic biosynthesis in some *Streptomyces* species. *Zeitschrift für Naturforschung*, **45c**, 851–5.

Anke, H. (1977). Metabolic products in microorganisms. 163. Desferritriacetylfusigen, an antibiotic from *Aspergillus deflectus*. *Journal of Antibiotics*, **30**, 125–8.

Axelrod, P. E., Rella, M. & Schroth, M. N. (1988). Role of antibiosis in competition of *Erwinia* strains in potato infection courts. *Applied and Environmental Microbiology*, **54**, 1222–9.

Ayer, W. A., Lee, S. P., Tsuneda, A. & Hiratsuka, Y. (1980). The isolation, identification, and bioassay of the antifungal metabolites produced by *Monocillium nordinii*. *Canadian Journal of Microbiology*, **26**, 766–73.

Bacquero, F. & Asensio, C. (1979). Microcins as ecological effectors in human intestinal flora: preliminary findings. In van der Waaij, D. & Verhoef, J., eds, *New Criteria for Antimicrobial Therapy: Maintenance of Digestive Tract Colonization Resistance* pp. 90–4. Exerpta Medica, Amsterdam.

Basalp, A., Özcengiz, G. & Alaeddinoglu, N. G. (1992). Changes in patterns of alkaline serine protease and bacilysin formation caused by common effectors of sporulation in *Bacillus subtilis* 168. *Current Microbiology*, **24**, 129–35.

Bennett, J. W. (1981). Genetic perspective on polyketides, productivity, parasexuality, protoplasts, and plasmids. In *Advances in Biotechnology, Vol. 3. Fermen-

tation products (Vezuna C. & Singh, K., eds), pp. 409–415. Pergamon Press, Toronto.

Bennett, J. W. (1983) Differentiation and secondary metabolism in mycelial fungi. In Bennett, J. W. & Ciegler, A., eds, *Secondary Metabolism and Differentiation in Fungi* pp. 1–32. Marcel Dekker, New York.

Bentzen, G. & Demain, A. L. (1990). Studies on gramicidin S-mediated suicide during germination outgrowth of *Bacillus brevis* spores. *Current Microbiology*, **20**, 165–9.

Bentzen, G., Piret, M. J., Dahler, E. & Demain, A. L. (1990). Transition of outgrowing spores of *Bacillus brevis* into vegetative cells is not dependent on destruction of gramicidin S. *Applied Microbiology and Biotechnology*, **32**, 708–10.

Bruehl, G. W., Millar, R. L. & Cunfer, B. (1969). Significance of antibiotic production by *Cephalosporium gramineum* to its saprophytic survival. *Canadian Journal of Plant Science*, **49**, 235–46.

Castignetti, D. & Smarrelli, Jr. J. (1986). Siderophores, the iron nutrition of plants, and nitrate reductase. *FEBS Letters*, **209**, 147–51.

Cavalier-Smith, T. (1992). Origins of secondary metabolism. In Chadwick, D. J. & Whelan, J., eds, *Secondary Metabolites; Their Function and Evolution* pp. 64–87. Wiley, Chichester.

Chou, W.-G. & Pogell, B. M. (1981). Pamamycin inhibits nucleoside and inorganic phosphate transport in *Staphylococcus aureus*. *Biochemical and Biophysical Research Communications*, **100**, 344–50.

Dahler, E., Rosenberg, E. & Demain, A. L. (1985). Germination-initiated spores of *Bacillus brevis* Nagano retain their resistance properties. *Journal of Bacteriology*, **161**, 47–50.

Danders, W. & Marahiel, M. W. (1981). Control of RNA synthesis by gramicidin S during germination and outgrowth of *Bacillus brevis* spores. *FEMS Microbiology Letters*, **10**, 277–83.

Danders, W., Marahiel, M. A., Krause, M., Kosui, N., Kato, T., Izumiya, T. & Kleinkauf, H. (1982). Antibacterial action of gramicidin S and tyrocidines in relation to active transport, *in vivo* transcription and spore outgrowth. *Antimicrobial Agents and Chemotherapy*, **22**, 785–90.

Darvill, A. G. & Albersheim, P. (1984). Phytoalexins and their elicitors – a defense against microbial infection in plants. *Annual Review of Plant Physiology*, **35**, 243–75.

Davies, J. (1990). What are antibiotics? Archaic functions for modern activities. *Molecular Microbiology*, **4**, 1227–32.

Demain, A. L. (1980). Do antibiotics function in nature? *Search*, **11**, 148–51.

Demain, A. L. (1989). Functions of secondary metabolites. In Hershberger, C. L., Queener, S. W. & Hegeman, G., eds, *Genetics and Molecular Biology of Industrial Microorganisms* pp. 1–11. American Society of Microbiology, Washington DC.

Demain, A. L. & Piret, J. M. (1979). Relationship between antibiotic synthesis and sporulation. In Luckner, M. & Shreiber, K., eds, *Regulation of Secondary Product and Plant Hormone Metabolism* pp. 183–8. Pergamon Press, New York.

Dixon, B. (1992). The buzz about bees. *Bio/Technology*, **10**, 607.

Eaton, D. & Ensign, J. C. (1980). *Streptomyces viridochromogenes* spore germination initiated by calcium ions. *Journal of Bacteriology*, **143**, 377–82.

Edgar, J. A., Frahn, J. L., Cokrum, P. A., Anderton, N., Jago, M. V., Culvenor, C. C. J., Jones, A. J., Murray, K. & Shaw, K. J. (1982). Corynetoxins, causative agents of annual rye grass toxicity: their identification as tunicamycin group antibiotics. *Journal of the Chemical Society, Chemical Communications*, 222–4.

Fenical, W. (1993). Chemical studies of marine bacteria: developing a new resource. *Chemical Reviews*, **93**, 1673–83.

Fischer, H.-P. & Bellus, D. (1993). Phytotoxicants from microorganisms and related compounds. *Pesticide Science*, **14**, 334–46.

Foster, H. A., Yasouri, F. N. & Daoud, N. N. (1992). Antibiotic activity of soil myxobacteria and its ecological implications. *FEMS Microbiology Ecology*, **101**, 27–32.

Foster, T. J. (1983). Plasmid-determined resistance to antimicrobial drugs and toxic metal ions in bacteria. *Microbiological Reviews*, **47**, 361–409.

Frangou-Lazaridis, M. & Seddon, B. (1985). Effect of gramicidin S on the transcription system of the producer *Bacillus brevis* Nagano. *Journal of General Microbiology*, **131**, 437–49.

Fredenhagen, A., Tamura, S. Y., Kenney, P. T. M., Komura, H., Naya, Y., Nakanishi, K., Nishiyama, K., Sugiura, M. & Kita, H. (1987). Andrimid, a new peptide antibiotic produced by an intracellular bacterial symbiont isolated from a brown planthopper. *Journal of the American Chemical Society*, **109**, 4409–11.

Gasson, M. J. (1980). Indicator technique for antimetabolic toxin production by phytopathogenic species of *Pseudomonas*. *Applied and Environmental Microbiology*, **39**, 25–9.

Gil-Turnes, M. S., Hay, M. E. & Fenical, W. (1989). Symbiotic marine bacteria chemically defend crustacean embryos from a pathogenic fungus. *Science*, **246**, 116–18.

Gray, T. R. (1976). Survival of vegetative microbes in soil. *Symposium of the Society of General Microbiology*, **26**, 327–64.

Groscop, J. A. & Brent, M. M. (1964). The effects of selected strains of pigmented microorganisms on small free-living amoebae. *Canadian Journal of Microbiology*, **10**, 579–84.

Habte, M. & Alexander, M. (1977). Further evidence for the regulation of bacterial populations in soil by protozoa. *Archives of Microbiology*, **113**, 181–3.

Hamdan, H., Weller, D. M. & Thomashow, L. S. (1991). Relative importance of fluorescent siderophores and other factors in biological control of *Gaenmannomyces graminis* var. *tritici* by *Pseudomonas fluorescens* 2–79 and M4-80R. *Applied and Environmental Microbiology*, **57**, 3270–7.

Hesseltine, C. W. (1986). Global significance of mycotoxins. In *Mycotoxins and Phycotoxins* (Steyn, P. S. & Vleggaar, R., eds), pp. 1–18. Elsevier Science, Amsterdam.

Hesseltine, C. W., Rogers, R. F. & Shotwell, O. L. (1981). Aflatoxin and mold flora in North Carolina in 1977 corn crop. *Mycologia*, **73**, 216–28.

Imshenetskii, A. A. (1974). The development of general microbiology. *Mikrobiologiya* (English translation), **43**, 185–207.

Ishii, K., Kobayashi, J., Ueno, Y. & Ichinoe, M. (1986). Occurrence of trichothecin in wheat. *Applied and Environmental Microbiology*, **52**, 331–3.

Iwasaki, S., Kobayashi, H., Furukawa, J., Namikoshi, M., Okuda, S., Sato, Z., Matsuda, I. & Noda, T. (1984). Studies on macrocyclic lactone antibiotics. 7. Structure of a phytotoxin 'rhizoxin' produced by *Rhizopus chinensis*. *Journal of Antibiotics*, **37**, 354–62.

Janzen, D. H. (1977). Why fruits rot, seeds mold and meat spoils. *American Naturalist*, **111**, 691–713.

Jigami, Y., Harada, N., Uemura, H., Tanaka, H., Ishikazwa, K., Nakasoto, S., Kita, H. & Sugiura, M. (1986). Identification of a polymyxin produced by a symbiotic microorganism isolated from the brown planthopper, *Nilaparavata lugens*. *Agricultural Biology and Chemistry*, **50**, 1637–9.

Kanaoka, M., Isogai, A. & Suzuki, A. (1979). Synthesis of bassianolide and its two homologs enniatin C and decabassianolide. *Agricultural Biology and Chemistry*, **43**, 1079–83.

Kanne, R. & Zahner, H. (1976). Metabolic products of microorganisms. 151. Comparative studies on intracellular potassium- and sodium concentrations of wild-type and a macrotetralide negative mutant of *Streptomyces griseus. Zeitschrift fur Naturforschung*, **31c**, 115–7.

Katz, E. & Demain, A. L. (1977). The peptide antibiotics of *Bacillus*: chemistry, biogenesis, and possible function. *Bacteriology Reviews*, **41**, 449–74.

Keller-Schierlein, W. & Prelog, V. (1961). Stoffwechsel produkte von actinomyceten 30. Mitt. Uber das Ferrioxamin E, ein Beitrag, zur Konstitution des Nocardamins. *Helvetica Chimica Acta*, **44**, 1981–5.

Kendrick, B. (1986). Biology of toxigenic anamorphs. *Pure and Applied Chemistry*, **58**, 211–8.

Kerr, A. & Tate, M. E. (1984). Agrocins and the biological control of crown gall. *Microbiological Science*, **1**, 1–4.

Kimbrell, D. A. (1991). Insect antibacterial proteins: not just for insects and against bacteria. *BioEssays*, **13**, 657–63.

Kondo, S., Katayama, M. & Marumo, M. (1986). Carbazomycinal and 6-methoxycarbazomycinal as aerial mycelium formation-inhibitory substances of *Streptoverticillium* species. *Journal of Antibiotics*, **39**, 727–30.

Kondo, S., Yasui, K., Natsume, M., Katayama, M. & Marumo, S. (1988). Isolation, physico-chemical properties and biological activity of pamamycin-607, an aerial mycelium inducing substance from *Streptomyces alboniger. Journal of Antibiotics*, **41**, 1196–204.

Krywolap, G. N., Grand, L. F. & Casida, Jr. L. E. (1964). The natural occurrence of an antibiotic in the mycorrhizal fungus *Cenococcum graniforme. Canadian Journal of Microbiology*, **10**, 323–8.

Kurotsu, T., Marahiel, M. A., Muller, K. D. & Kleinkauf, H. (1982). Characterization of an intracellular serine protease from sporulating cells of *Bacillus brevis. Journal of Bacteriology*, **151**, 1466–72.

Lazaridis, I., Frangou-Lazaridis, M., Maccuish, F. C., Nandi, S. & Seddon, B. (1980). Gramicidin S content and germination and outgrowth of *Bacillus brevis* Nagano spores. *FEMS Microbiology Letters*, **7**, 229–32.

Lee, S., Izumiya, N., Suzuki, A. & Tamura, S. (1975). Synthesis of a cyclohexadepsipeptide, protodestruxin. *Tetrahedron Letters*, 883–6.

Lemos, M. L., Dopazo, C. P., Toranzo, A. E. & Barja, J. L. (1991). Competitive dominance of antibiotic-producing marine bacteria in mixed cultures. *Journal of Applied Bacteriology*, **71**, 228–32.

Lobareva, G. S., Zharikova, G. G. & Mesyanzhinov, V. V. (1977). Peculiarities of the binding of gramicidin S to cells of the producer. *Doklady Biological Science* (English translation), **236**, 418–21.

McCann, P. A. & Pogell, B. M. (1979). Pamamycin: a new antibiotic and stimulator of aerial mycelium formation. *Journal of Antibiotics*, **32**, 673–8.

Maheshwari, R. (1991). Microcycle conidiation and its genetic basis in *Neurospora crassa. Journal of General Microbiology*, **137**, 2103–55.

Mann, J. (1978). Secondary metabolism and ecology. In Mann, J. ed., *Secondary Metabolism* pp. 279–304. Clarendon Press, Oxford.

Marahiel, M. A., Danders, W., Kraepelin, G. & Kleinkauf, H. (1982). Studies on the role of gramicidin S in the life cycle of its producer *Bacillus brevis* ATCC 9999. In Kleinkauf, H. & von Dohren, H., eds, *Peptide Antibiotics – Biosynthesis and Functions* pp. 389–97. Walter de Gruyter, Berlin.

Marahiel, M. A., Danders, W., Krause, M. & Kleinkauf, H. (1979). Biological role of gramicidin S in spore functions. Studies on gramicidin S-negative mutants of *Bacillus brevis* ATCC 9999. *European Journal of Biochemistry*, **99**, 49–55.

Marumo, S., Nukina, M., Kondo, S. & Tomiyama, K. (1982). Lunatoic acid a, a morphogenic substance inducing chlamydospore-like cells in some fungi. *Agricultural and Biological Chemistry*, **46**, 2399–401.

Marx, D. H. (1969). The influence of ectotrophic mycorrhizal fungi on the resistance of pine roots to pathogenic infections. II. Production, identification and biological activity of antibiotics produced by *Leucopaxillus cerealis* var. *piclina*. *Phytopathology*, **59**, 411–7.

Mazzola, M., Cook, R. J., Thomashow, L. S., Weller, D. M. & Pierson III, L. S. (1992). Contribution of phenazine antibiotic biosynthesis to the ecological competence of fluorescent pseudomonads in soil habitats. *Applied and Environmental Microbiology*, **58**, 2616–24.

Messenger, A. J. M. & Turner, J. M. (1981). Effect of secondary metabolite production on the growth rate and variability of a pseudomonad. *Society of General Microbiology Quarterly*, **8**, 263–4.

Mitchell, R. E. (1991). Implications of toxins in the ecology and evolution of plant pathogenic microorganisms: bacteria. *Experientia*, **47**, 791–803.

Mitscher, L. A. (1975). Antimicrobial agents from higher plants. *Recent Advances in Phytochemistry*, **9**, 243–82.

Mo, Y.-Y. & Gross, D. C. (1991). Plant signal molecules activate the *syrB* gene, which is required for syringomycin production by *Pseudomonas syringae* pv. *syringae*. *Journal of Bacteriology*, **173**, 5784–92.

Mukherjee, P. K. & Paulus, H. (1977). Biological function of gramicidin: studies on gramicidin-negative mutants. *Proceedings of National Academy Sciences, USA*, **74**, 780–4.

Murray, T., Lazaridis, I. & Seddon, B. (1985). Germination of spores of *Bacillus brevis* and inhibition by gramicidin S: a stratagem for survival. *Letters in Applied Microbiology*, **1**, 63–5.

Nandi, S. & Seddon, B. (1978). Evidence of gramicidin S functioning as a bacterial hormone specifically regulating spore outgrowth in *Bacillus brevis* strain Nagano. *Biochemical Society (UK) Transactions*, **6**, 409–11.

Nandi, S., Lazaridis, I. & Seddon, B. (1981). Gramicidin S and respiratory activity during the developmental cycle of the producer organism *Bacillus brevis* Nagano. *FEMS Microbiology Letters*, **10**, 71–5.

Neilands, J. B. (1984). Siderophores of bacteria and fungi. *Microbiological Science*, **1**, 9–14.

Norstadt, F. A. & McCalla, T. M. (1969). Microbial populations in stubble-mulched soil. *Soil Science*, **107**, 188–93.

O'Sullivan, D. J. & O'Gara, F. (1992). Traits of fluorescent *Pseudomonas* spp. involved in suppression of plant root pathogens. *Microbiological Reviews*, **56**, 662–76.

Özcengiz, G. & Alaeddinoglu, N. G. (1991). Bacilysin production and sporulation in *Bacillus subtilis*. *Current Microbiology*, **23**, 61–4.

Patil, S. S. (1974). Toxins produced by phytopathogenic bacteria. *Annual Review of Phytopathology*, **12**, 259–79.

Paul, V. J., Frautschy, S., Fenical, W. & Nealson, K. H. (1981). Antibiotics in microbial ecology. Isolation and structure assignment of several new antibacterial compounds from the insect-symbiotic bacteria *Xenorhabdus* spp. *Journal of Chemical Ecology*, **7**, 589–97.

Peipp, H. & Sonnenbichler, J. (1992). Secondary fungal metabolites and their biological activities, No 2. Occurrence of antibiotic compounds in cultures of *Armillaria ostoyae* growing in the presence of an antagonistic fungus or host plant cells. *Biological Chemistry Hoppe-Seyler*, **373**, 675–83.

Petersen, F., Zahner, H., Metzger, J. W., Freund, S. & Hummel, R.-P. (1993). Germicidin, an autoregulative germination inhibitor of *Streptomyces viridochromogenes* NRRL B-1551. *Journal of Antibiotics*, **46**, 1126–38.

Petras, S. F. & Casida, Jr. L. E. (1985). Survival of *Bacillus thuringiensis* spores in soil. *Applied and Environmental Microbiology*, **50**, 1496–501.

Piret, J. M. (1981). Functions of the peptide antibiotic, gramicidin S, in its producer, *Bacillus brevis* Nagano. PhD Thesis, Massachusetts Institute of Technology, Cambridge, USA.

Piret, J. M. & Demain, A. L. (1982). Germination initiation and outgrowth of spores of *Bacillus brevis* strain Nagano and its gramicidin S-negative mutant. *Archives of Microbiology*, **133**, 38–43.

Piret, J. M. & Demain, A. L (1983). Sporulation and spore properties of *Bacillus brevis* and its gramicidin S-negative mutant. *Journal of General Microbiology*, **129**, 1309–16.

Piret, J. M., Millet, J. & Demain, A. L. (1983). Production of intracellular serine protease during sporulation of *Bacillus brevis* ATCC 9999. *European Journal of Applied Microbiology and Biotechnology*, **17**, 227–30.

Richardson, W. H., Schmidt, T. M. & Nealson, K. (1988). Identification of an anthraquinone pigment and a hydroxystilbene antibiotic from *Xenorhabdus luminescens*. *Applied and Environmental Microbiology*, **54**, 1602–5.

Ristow, H., Pschorn, W., Hansen, J. & Winkel, U. (1979). Induction of sporulation in *Bacillus brevis* by peptide antibiotics. *Nature*, **280**, 165–6.

Ristow, H., Russo, J., Stochaj, E. & Paulus, H. (1982). Tyrocidine induced sporulation of *Bacillus brevis* in a medium lacking a nitrogen source. In Kleinkauf, H. & von Dohren, H., eds, *Peptide Antibiotics – Biosynthesis and Functions* pp. 381–8. Walter de Gruyter, Berlin.

Ristow, H., Schazschneider, B. & Kleinkauf, H. (1975). Effects of the peptide antibiotics tyrocidine and the linear gramicidin on RNA synthesis and sporulation. *Biochimica et Biophysica Acta*, **63**, 1085–92.

Robeson, D. J. & Strobel, G. A. (1982). Deoxyradicinin, a novel phytotoxin from *Alternaria helianthi*. *Phytochemistry*, **21**, 1821–3.

Rosenberg, E., Brown, D. R. & Demain, A. L. (1985). The influence of gramicidin S on hydrophobicity of germinating *Bacillus brevis* spores. *Archives of Microbiology*, **142**, 51–4.

Rosenberg, E. & Varon, M. (1984). Antibiotics and lytic enzymes. In Rosenberg, E., ed., *Myxobacteria Development and Cell Interactions* pp. 104–25. Springer-Verlag, New York.

Rothrock, C. S. & Gottlieb, D. (1984). Role of antibiosis in antagonism of *Streptomyces hygroscopicus* var. *geldanus* to *Rhizoctonia solani* in soil. *Canadian Journal of Microbiology*, **30**, 1440–7.

Sarkar, N. & Paulus, H. (1972). Function of peptide antibiotics in sporulation. *Nature New Biology*, **239**, 228–30.

Schazschneider, B., Ristow, H. & Kleinkauf, H. (1974). Interaction between the antibiotic tyrocidine and DNA *in vitro*. *Nature*, **294**, 757–9.

Scheffer, R. P. (1991). Role of toxins in evolution and ecology of plant pathogenic fungi. *Experientia*, **47**, 804–11.

Singh, B. N. (1942). Toxic effects of certain bacterial metabolic products on soil protozoa. *Nature*, **149**, 168.

Singh, B. N. (1945). The selection of bacterial food by soil amoebae, and the toxic effects of bacterial pigments and other products on soil protozoa. *British Journal of Experimental Pathology*, **26**, 316–25.

Slapikoff, S., Spitzer, J. L. & Vaccaro, D. (1971). Sporulation in *Bacillus brevis*: studies on protease and protein turnover. *Journal of Bacteriology*, **106**, 739–44.

Sparace, S. A., Reeleder, R. D. & Khanizadeh, S. (1987). Antibiotic activity of the pyrenocines. *Canadian Journal of Microbiology*, **33**, 327–30.

Staskawicz, B. J. & Panopoulos, N. J. (1979). A rapid and sensitive microbiological assay for phaseolotoxin. *Phytopathology*, **69**, 663–6.

Stone, M. J. & Williams, D. H. (1992). On the evolution of functional secondary metabolites (natural products). *Molecular Microbiology*, **6**, 29–34.

Strobel, G. A. (1977). Bacterial phytotoxins. *Annual Review of Microbiology*, **31**, 205–24.

Strobel, G., Kenfield, D., Bunkers, G., Sugawara, F. & Clardy, J. (1991). Phytotoxins as potential herbicides. *Experientia*, **47**, 819–26.

Swain, T. (1977). Secondary compounds as protective agents. *Annual Review of Plant Pathology*, **28**, 479–501.

Sydenham E., Gelderblom, W. C. A., Thiel, P. G. & Marasas, W. F. O. (1990). Evidence for the natural occurrence of fumonisin B_1, a mycotoxin produced by *Fusarium moniliforme* in corn. *Journal of Agricultural and Food Chemistry*, **38**, 285–90.

Symons, D. C. & Hodgson, B. (1982). Isolation and properties of *Bacillus brevis* mutants unable to produce tyrocidine. *Journal of Bacteriology*, **151**, 580–90.

Taya, Y., Yamada, T. & Nishimura, S. (1980). Correlations between acrasins and spore germination inhibitors in cellular slime molds. *Journal of Bacteriology*, **143**, 715–19.

Teintze, M., Hossain, M. B., Barnes, C. L., Leong, J. & van der Helm, D. (1981). Structure of ferric pseudobactin, a siderophore from a plant growth promoting *Pseudomonas*. *Biochemistry*, **20**, 6446–57.

Thomashow, L. S. & Weller, D. M. (1988). Role of a phenazine antibiotic from *Pseudomonas fluorescens* in biological control of *Gaeumannomyces graminis* var. *tritici*. *Journal of Bacteriology*, **170**, 3499–508.

Trimble, J. E., Veal, D. A. & Beattie, A. J. (1992). Antimicrobial properties of secretions from the metapleinal glands of *Myrmecia gulosa* (the Australian bull ant). *Journal of Applied Bacteriology*, **72**, 188–94.

Vining, L. C. & Taber, W. A. (1979). Ergot alkaloids. In Rose, A. H. ed., *Economic Microbiology, vol. 3. Secondary Products of Metabolism* pp. 389–420. Academic Press, London.

Vinter, V. & Slepecky, R. A. (1965). Direct transition of outgrowing bacterial spores to new sporangia without intermediate cell division. *Journal of Bacteriology*, **90**, 803–7.

Williams, D. H., Stone, M. J., Hauck, P. R. & Rahman, S. K. (1989). Why are secondary metabolites (natural products) biosynthesized? *Journal of Natural Products*, **52**, 1189–208.

Xu, G.-W. & Gross, D. C. (1988). Physical and functional analyses of the *syrA* and *syrB* genes involved in syringomycin production by *Pseudomonas syringae* pv. *syringae*. *Journal of Bacteriology*, **170**, 5680–8.

Zahner, H., Drantz, H. & Weber, W. (1982). Novel approaches to metabolite screening. In Bu'Lock, J. D., Nisbet, L. J. & Winstanley, D. J., eds, *Bioactive Microbial Products: Search and Discovery* pp. 51–70. Academic Press, London.

ANTIMALARIALS: FROM QUININE TO ATOVAQUONE

MARY PUDNEY

Biology Division, The Wellcome Research Laboratories, Langley Court, Beckenham, Kent BR3 3BS, UK

INTRODUCTION

Malaria is one of the world's oldest, most severe, widespread and complex infectious diseases, affecting over 200 million people and killing somewhere between 1 and 2.5 million people, mostly children, annually. In the 1950s, it was thought that malaria, like smallpox, could be totally eradicated using a combination of antimalarial drugs and pesticides to eliminate the insect vector. Despite a co-ordinated international effort, this was not achieved (Bruce-Chwatt, 1979) and today, control is the more modest aim. In fact, both endemic and subsequently imported malaria has increased during the past two decades due to the greater mobility of human populations, and to a large extent the appearance and rapid spread of drug resistance (Kain, 1993). For such a devastating disease there are remarkably few effective drugs for treatment, and *Plasmodium falciparum* has developed resistance to all those that are generally available. Chloroquine resistance is now widespread in most of the malarious tropics (Wernsdorfer, 1994), and the first reports of chloroquine resistance in *Plasmodium vivax* have been confirmed from Oceania (Schuurkamp *et al.*, 1992). Fortunately, the highly multidrug-resistant *P. falciparum* is confined to the western and eastern borders of Thailand, and mefloquine, halofantrine and quinine can still be relied on in most places. The Chinese compounds from the plant *Artemesia annua* (artemisinin or qinghaosu, artemether and artesunate) are proving to be rapidly effective (White & Nosten, 1993) and the new atovaquone/ proguanil combination is remarkably effective in clinical trials (Hutchinson *et al.*, personal communication).

Considerable advances have been made recently both in understanding mechanisms of action and the mechanisms of resistance to the available drugs. This has led, in turn, to the investigation of resistance reversal and to the identification of potential novel targets for antimalarial attack. These could eventually lead to new drugs, although the prospects for the immediate future are poor.

MALARIA PARASITE

The disease is caused by species-specific plasmodia. Four species are able to infect humans, *Plasmodium falciparum, P. vivax, P. ovale* and *P. malariae*.

Table 1. *Site of action of antimalarial drugs*

Drug	Site of action			Clinical utility
	Blood schizonts	Tissue schizonts	Hypnozoites	
Quinine	+	−	−	T
Chloroquine	+	−	−	T,P
Primaquine	−	+	+	T[*1]
Pyrimethamine	+	+	−	P
Proguanil	+	+	−	P
Mefloquine	+	−	−	T,P
Halofantrine	+	−	−	T
Artemisinin	+	−	−	T
Atovaquone/Proguanil	+	+[*]	−	T[a],P[b]

T = treatment P = prophylaxis
[*] Experimental only; [a] Clinical trials; [*1] Plus blood schizontocide [b] Not examined clinically.

P. falciparum is the most widespread, severe and often fatal form, and together with *P. vivax* (the commonest form in the Americas and Asia) account for most of the malaria worldwide. Infection is brought about by a mosquito bite during which sporozoites are introduced into the bloodstream. A few minutes later the parasites sequester themselves in liver parenchymal cells where they transform into the pre-erythrocytic form. The parasites then develop by asexual schizogony (tissue schizonts) into merozoites. This form invades erythrocytes, causing the clinical symptoms of malaria (cyclic fevers) owing to the repeated rupture of infected erythrocytes during blood schizogony. Some merozoites further transform into gametocytes. When taken up during a blood meal the gametocytes continue a sexual reproductive cycle in the mosquito gut, resulting in sporozoites which can infect another human when the mosquito bites again. *P. vivax* and *P. ovale* cause relapsing malaria owing to a further stage, the hypnozoite which remains dormant in liver cells. These become 'reactivated' and initiate further cycles of erythrocytic schizogony and thus fevers, in a disease relapse long after the original parasites have been removed from the bloodstream.

The parasite life cycle can be interrupted at various points and the stage at which a particular drug works determines how it is used clinically (see Table 1). The most widely used drugs are those which work against the blood schizonts, thus preventing disease symptoms. Such agents can be used for both treatment and suppressive prophylaxis of malaria by killing parasites as they enter the bloodstream from the liver. In the cases of vivax and ovale malaria such therapy is insufficient, and in order to avoid relapses over a period of some years, it is necessary to eliminate the long-lived hypnozoites using primaquine.

HISTORY OF SYNTHETIC ANTIMALARIALS

The roots, leaves and flowers of many plants have been used for the treatment of malaria since Hippocrates first described the symptoms in 400 BC. In China, Qing hao (*Artemesia annua*) and yingzhaosu have been used successfully for at least 2000 years. In the West, the first potent remedy (an infusion of the bark of the 'fever tree', Cinchona) dates from the seventeenth century. The crude bark was used for 200 years, until 1820, when two French chemists isolated the active principle, quinine. Quinine was widely used and it was not until war threatened the stocks of quinine in Germany that considerable efforts were directed at developing synthetic alternatives. Ehrlich's observation of the effect of methylene blue on malaria was the starting point which led in 1925 to the first synthetic antimalarial of the 8-aminoquinoline series pamaquine (Plasmochin). However, limited activity against human malaria led to the quinoline ring being replaced by acridine, a yellow dye, giving Atebrin (mepacrin). Mepacrin was marketed in 1932 and routinely used throughout the Second World War, playing a considerable role in the course of modern history.

In 1934, Andersag discovered the 4-aminoquinoline chloroquine, whilst working for Bayer in Germany, but it was considered too toxic and was abandoned. Samples of the 4-aminoquinoline series were used in a huge co-ordinated effort in the USA (the allies were also cut off from their quinine supplies) in which over 17 000 compounds were tested (see Coatney, 1963). Chloroquine resulted from this effort, and by 1946 had become the drug of choice for all malaria the world over. During the course of this intensive research programme the 8-aminoquinoline, primaquine, was also found and this was effective against relapsing vivax malaria. In Britain, synthesis of pyrimidine derivatives led to the production of proguanil in 1945, which proved to be an outstanding causal prophylactic in falciparum malaria, and a satisfactory suppressive of vivax malaria. Because of its slow activity, the availability of mepacrine, and the fact that it seemed to induce resistance in some strains of plasmodia, proguanil was not used for treatment, but became firmly established as a causal prophylactic for the prevention of malaria for those working in the tropics. Further work on the 2,2-diaminopyrimidines as folic acid synthesis inhibitors, led to the identification of pyrimethamine. Despite its wide safety margin, it was soon found that resistance appeared relatively rapidly both experimentally and in the field, and that there was cross-resistance with proguanil. Thus, although widely used in combination, pyrimethamine is not used as a single entity.

In spite of some drawbacks of the new compounds, it seemed in the 1950s that the arsenal of antimalarial drugs was almost complete and research in anti-malarials declined markedly. However, the achievements of the eradication programme in Europe, North America and Northern Asia could not be reproduced in the tropics, and the growing conflict in South East Asia revived interest in antimalarial drugs in the following decade. The appearance of resistance to the 4-aminoquinolines and other available synthetic compounds revealed the relative poverty of the arsenal of drugs and the narrow margin of safety in the treatment of disease caused by drug-resistant parasites. The most comprehensive scientific effort was launched in 1963 by the US army who screened over a quarter of a million potential antimalarial compounds between 1963 and 1976. This led to the discovery of four new chemical groups as potentially valuable new antimalarials; 4-quinolinemethanols; 9-phenanthrenemethanols; 2,4-diaminoquinazolines; and 2,4-diaminotriazines. The two analogues of quinine, WR142,490 (Walter Reed Army Institute of Research), a 4-quinolinemethanol (mefloquine) and WR171,669, a phenanthrenemethanol (halofantrine) were subsequently developed and marketed by Roche and Smith Kline Beecham, respectively. Both compounds are blood schizonticides effective against multi-drug-resistant falciparum malaria (Sweeney, 1981; Horton, Parr & Bokor, 1990).

During the 1960s, the sulfones and sulfonamides also re-entered the history of malaria chemotherapy, with the discovery of the synergistic action of sulfonamides or sulfones given together with pyrimethamine or proguanil. These drugs act synergistically owing to their action on different enzymes in the parasite's folic acid cycle; pyrimethamine inhibiting dihydrofolate reductase while the sulphonamides/sulphones effect dihydropteroate synthetase (Schmidt et al., 1977; Ferone, 1984; Scholer, Leimer & Richle, 1984). The most frequently used combinations are pyrimethamine-sulfadoxine (Fansidar) and pyrimethamine-dapsone (Maloprim). The major classes of compounds currently used as antimalarials are listed in Table 2 and the structures are shown in Fig. 1.

DRUG RESISTANCE

Antimalarial resistance generally refers to *P. falciparum*, since the failure or success of medication may determine the patients survival. This is rarely so with other species. The phenomenon of drug resistance in *P. falciparum* has been known since 1910 when some patients in Brazil failed to respond to quinine. Fortunately, resistance to quinine has not yet developed into a widespread problem despite 350 years of use, and it remains an important and effective antimalarial, especially when combined with tetracycline. The exception is Thailand, where resistance to quinine was already substantial in 1982 when the routine use of quinine combined with tetracycline was introduced. Quinine resistance has continued to increase.

Resistance to pyrimethamine was encountered soon after it was used for prophylaxis and now extends to all four species of human plasmodia (Wernsdorfer, 1991, 1994). The use of the combination of pyrimethamine with a sulphonamide or a sulphone for treatment has proved useful in

Table 2. *Antimalarial compounds used against* P. falciparum

Cinchona alkaloids	Quinine
	Quinidine
4-Aminoquinolines	Chloroquine
	Amodiaquine
4-Quinoline methanols	Mefloquine
Qinghaosu and derivatives	Artemisinin
	Artesunate
	Arteether
	Artemether
Other antibacterial drugs	Tetracycline
	Sulfa drugs
	Clindamycin
Acridine derivatives	Quinacrine
	Pyronaridine
9-Phenanthrenemethanols	Halofantrine
DHFR inhibitors	Pyrimethamine
	Proguanil and analogues
	Cycloguanil
	Trimethoprim

delaying the spread of pyrimethamine-resistant parasites (Peters, 1990), but resistance to the combination is now widespread in S.E. Asia, Western Oceania and S. America, and is increasing in West Africa (Wernsdorfer, 1994). Because the parasites remained susceptible to chloroquine, the early observations caused no great concern. However, in the 1960s the first of many reports started appearing on the phenomenon of chloroquine-resistant falciparum infections both in South America and South East Asia, and these have now spread throughout most of the malaria endemic world. Chloroquine had become the standard antimalarial for treatment and prophylaxis of all the human malarias and, as a consequence, one of the most widely used drugs in the world. At first, resistance was low grade and local, but during ensuing years treatment failures increased in numbers and degree. By the beginning of the 1980s, chloroquine was no longer useful in many parts of South East Asia and South America, and the ominous first reports of resistance to chloroquine were emerging from the East coast of Africa. Over the past 10 years, resistance has moved from the east to the western coasts of Africa (Bjorkman & Phillips-Howard, 1990). Few countries in the tropics are now unaffected (Wernsdorfer, 1994). Despite this, chloroquine continued to be used for falciparum malaria in much of sub-Saharan Africa, although recent evidence indicates it can no longer be considered effective therapy in these areas (Boland, Werc & Campbell, 1993).

Until recently, the other human plasmodial species have remained largely susceptible to chloroquine. *P. vivax* is the second in importance as the cause

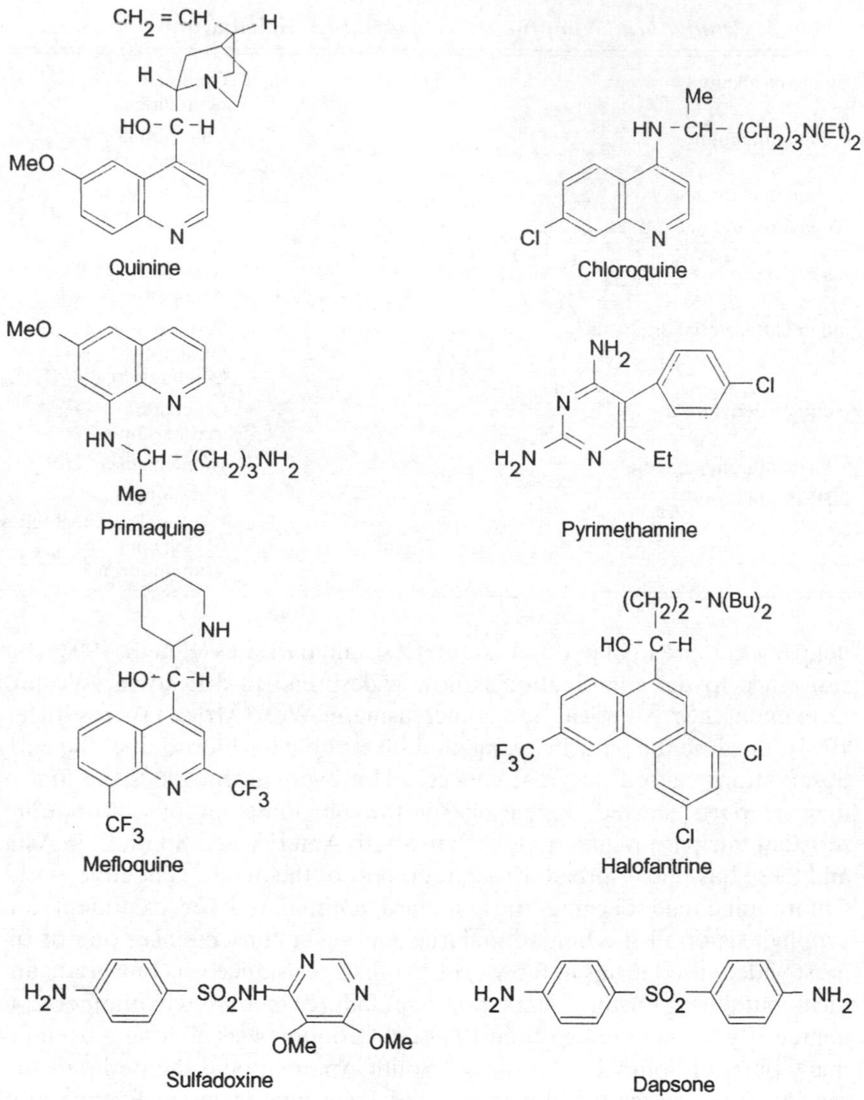

Fig. 1. Structures of drugs in common use.

of malaria worldwide, and is a frequent cause of imported malaria into non-endemic areas (Freedman, 1992). For over 40 years chloroquine has been the treatment of choice for vivax malaria, but since 1989, cases failing standard treatment have been reported from Papua New Guinea, Irian Jaya, Indonesia, Colombia and recently from Brazil (White, 1992). Few studies have yet been carried out to determine alternative drugs. Relapse of *P. vivax* after standard courses of primaquine is also commonly reported from Papua New Guinea, Thailand and other parts of South East Asia, and less commonly from India, Africa and Columbia, and are being treated with up to twice the standard dose (Luzzi *et al.*, 1992; Bunnag *et al.*, 1994). There is still no evidence of drug resistance in *P. ovale* and *P. malariae*.

The molecular action of chloroquine has been the subject of debate for over 40 years. Chloroquine and other quinoline antimalarials are concentrated in parasitized erythrocytes, particularly in the acid food vacuole. Resistant parasites do not concentrate chloroquine because they pump the drug out of the cell 40–50 times faster than the sensitive parasites. It has now been shown that chloroquine and quinine specifically block the parasite enzyme heme polymerase (Slater & Cerami, 1992). Free heme is toxic, damaging biological membranes and inhibiting a variety of enzyme systems. The drug sensitive enzyme catalyses the detoxifying polymerization of heme to the malaria pigment hemozoin. The process that pumps the quinoline antimalarial drugs out of resistant parasites has many similarities to the multiple drug resistance (MDR) mechanism in tumour cells (for review see Upcroft, 1994). Two *P. falciparum* genes with homologies to mammalian MDR genes have been identified, although these were not linked to chloroquine resistance in a genetic cross that localized the chloroquine resistance mechanism to a segment of *P. falciparum* chromosome. Thus multiple drug resistance may be one of several discrete processes leading to chloroquine resistance (for review see Wellems, 1991; Bray & Ward, 1993).

The molecular basis for resistance to the dihydrofolate reductase (DHFR) inhibitors such as pyrimethamine, has in contrast been well characterized. It has always been thought that there was cross-resistance between the DHFR inhibitors, but this does not appear to be true in all cases. All pyrimethamine-resistant *P. falciparum* isolates have a Ser108–Asn108 mutation, but interestingly the mutation does not confer resistance to cycloguanil. In contrast, parasites with the paired Ser108–Thr108 and Ala16–Val16 mutation are cycloguanil-resistant but not pyrimethamine-resistant (Foote & Cowman, 1994). Other less common paired mutations confer cross-resistance to both drugs.

In order to delay the emergence of resistance to the new drug mefloquine, it was recommended that mefloquine be combined with pyrimethamine and sulfadoxine. Extensive experience with this drug cocktail in SE Asia has, however, provided clear evidence for the failure of this strategy. Over a 6-year period *P. falciparum* resistance to mefloquine developed rapidly, despite its use only in the triple combination. Mefloquine is well tolerated in prophylactic dose; however, in treatment doses, neuropsychiatric reactions are 10–60 times more frequent than for chloroquine, and early vomiting can be a cause of treatment failures. Mefloquine has been in use since clinical trials in 1975, and at present treatment failure is not a significant problem except in areas of the Thai-Myanmar/Cambodia borders, where cure rates have fallen from 98% in 1986 to 71% in 1990 (Nosten *et al.*, 1991). Presently, mefloquine is an effective and well-tolerated drug for malaria chemoprophylaxis in all endemic areas except the Thai–Cambodian borders, and is effective against multidrug-resistant malaria, although the reports of neuropsychiatric adverse events have limited enthusiasm for its use in therapeutic doses. An additional concern is the development of cross-resistance, with recent evidence showing that mefloquine resistance may drive halofantrine and possibly quinine resistance (White & Nosten, 1993; Peel et al., 1993; Ketrangsee et al., 1992).

Halofantrine is more active in vitro than mefloquine, and is the most recent antimalarial for the treatment of multidrug-resistance malaria (clinical studies began in 1984). It is available as an oral formulation but has variable bioavailability which is increased with food. Mechanisms of action and resistance are not known. Resistance to halofantrine is already present in Congo and Thailand, where a high dose regimen has proved more effective (Brasseur et al., 1993). Unfortunately, following the deaths of two patients, halofantrine has been shown to put predisposed individuals at risk for cardiotoxicity. The drug should not now be given to those who might have a long Q–T interval, and should not be taken with food or by patients deficient in thiamine. The development of halofantrine resistance has paralleled mefloquine resistance in SE Thailand (Wongsrichanalai et al., 1992).

RESISTANCE-REVERSING AGENTS

Chloroquine is an excellent antimalarial drug for sensitive infections. It is cheap, rapidly effective, well tolerated, and requires relatively few doses. In 1987 it was observed that treatment of cultures of chloroquine-resistant P. falciparum with the calcium channel blocker verapamil (see Fig. 2) rendered the parasites fully susceptible to chloroquine (Martin, Oduola & Milhous, 1987). The rationale behind this was the similarity between multidrug resistance patterns seen in cancer and malaria patients (Karcz & Cowman, 1991). With the discovery of the efflux mechanism that pumps chloroquine out of resistant parasites came the observation that a variety of drugs already in clinical use (calcium channel antagonists, tricyclic antidepressants and phenothiazines) could inhibit the efflux mechanism. A plethora of structurally and functionally unrelated compounds can be shown to reverse drug resistance in MDR cancer cells and a similar situation is emerging from the work with drug-resistant plasmodia. A combination of desipramine and chloroquine has been used to treat monkeys infected with the highly drug-resistant Vietnam Smith strain of P. falciparum. Chloroquine alone had little effect on this infection but the drug combination rapidly inhibited parasite growth although recrudescences later occurred (Bitonti et al., 1988). Antihistamines, such as cyproheptadine, ketotifen and pizotyline, with structural similarities to desipramine were also found to be effective

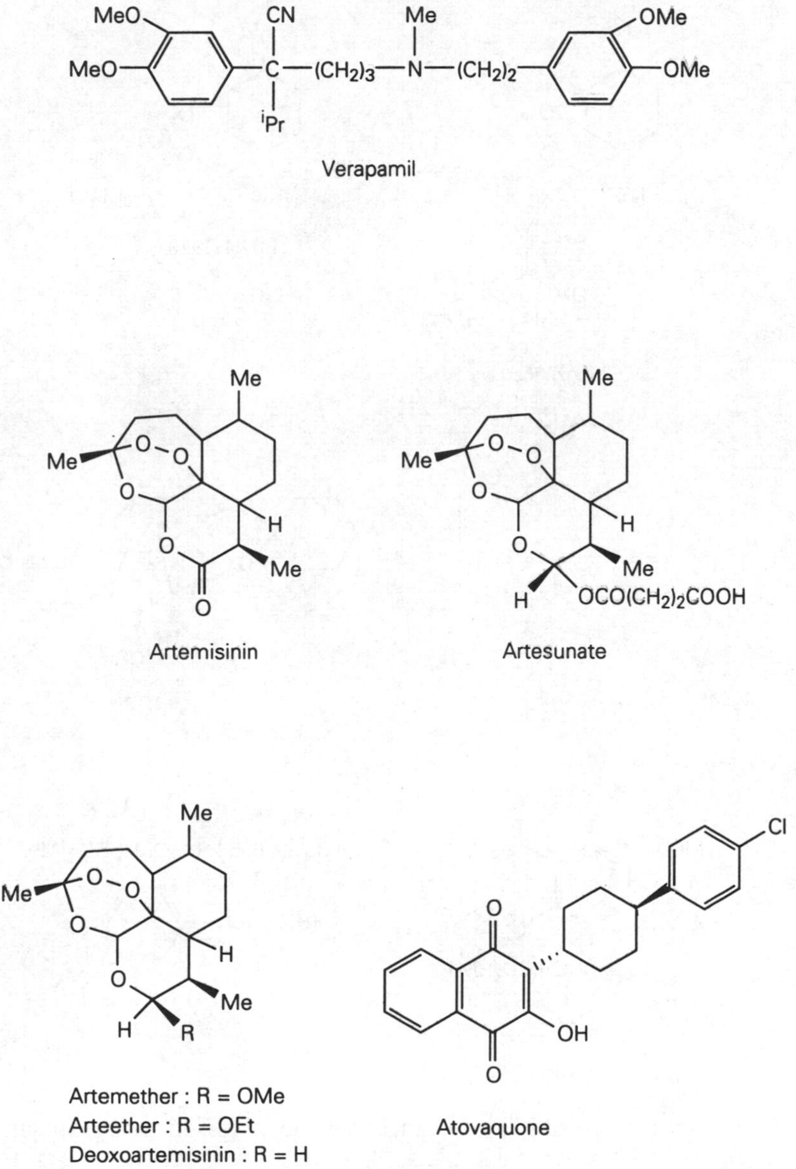

Verapamil

Artemisinin

Artesunate

Artemether : R = OMe
Arteether : R = OEt
Deoxoartemisinin : R = H

Atovaquone

Fig. 2. Structures of newer antimalarials and resistance reversing agents.

reversing agents against chloroquine-resistant plasmodia *in vitro* (Peters *et al.*, 1990). Unlike desipramine and verapamil, these compounds were intrinsically active against malaria parasites. Several other compounds have been described in the patent literature (Hudson, 1992), but as yet there is no

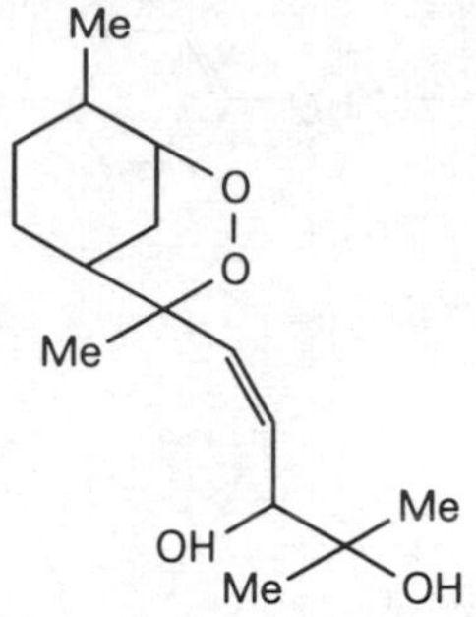

Yingzhaosu

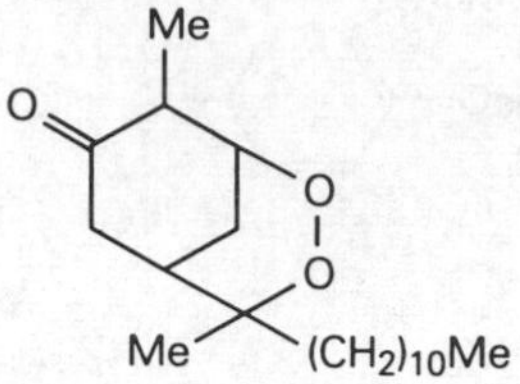

Ro 41-3823

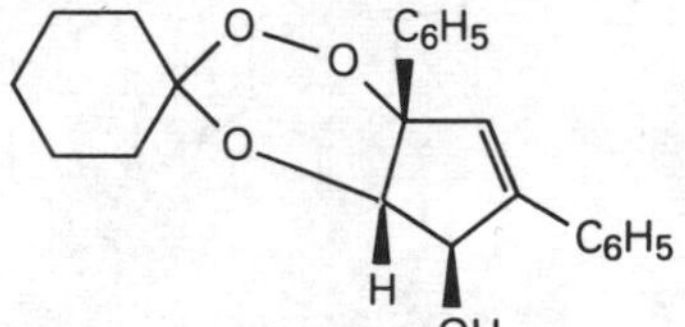

7

1, 2, 4-trioxane

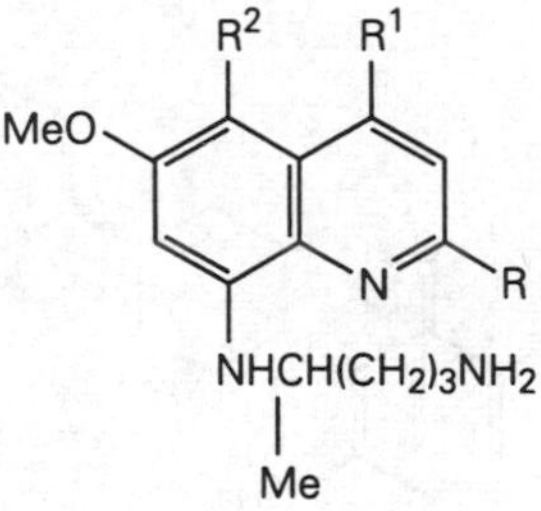

Primaquine : R = R^1 - R^2 = H

WR 238, 605 : R = OMe, R^1 = Me

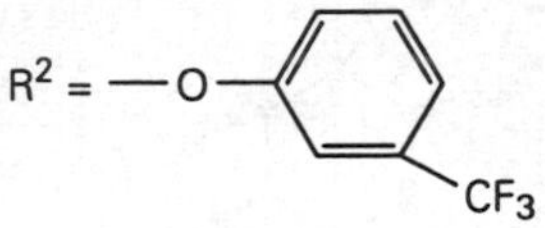

R^2 = —O—

Fig. 2. *Continued*.

evidence that chloroquine resistance can be reversed safely in clinical practice.

DRUGS UNDER DEVELOPMENT

Artemesinin and derivatives

An essential pre-requisite of any new antimalarial is the ability to work against chloroquine-resistant parasites. To maximise the chances of achieving this, it is desirable that candidate compounds be unrelated functionally or structurally to chloroquine. Artemisinin is a naturally occurring sesqui-terpene peroxide, structurally unrelated to any known antimalarial. Extracts of *Artemisia annua* (wormwood) have been used for centuries but it was not until 1972 that the active principle was isolated. In China the compound has been used to treat thousands of patients with vivax and falciparum malaria (see Brossi *et al.*, 1988; Butler & Wu, 1992). It was well tolerated, fast acting and effective against chloroquine-resistant disease but not against the liver stages of malaria. Because artemisinin is poorly absorbed and recrudescence rates are high, numerous derivatives have been investigated as alternatives. Qinghaosu is available as the parent compound artemisinin and as three semi-synthetic derivatives: a water soluble hemisuc-cinate salt (artesunate); and two oil soluble compounds, artemether (methyl ether) and arteether (ethyl ether). Arteether is being developed by the World Health Organisation with WRAIR and Beyeer BV for its lipophilic properties, a potential advantage in cerebral malaria, and its presumed less toxic metabolites, for development as an i.m. sesame oil solution for emergency treatment of severe malaria. Artemisinin and its derivatives lead to faster parasite and fever clearance times than any other antimalarials, and shorter coma-resolution times have been reported with artemether com-pared to quinine in Malawian children with cerebral malaria. However, high recrudescent rates are observed when used as monotherapy, and recent studies have examined combinations of qinghaosu derivatives and meflo-quine, which display synergy *in vitro*, and combination therapy has resulted in 100% cure rates in Thailand (White & Nosten, 1993). Combinations of artesunate and mefloquine appear to be the most active drug regimen against multidrug-resistant falciparum malaria. However, present pre-clinical and toxicity data are insufficient to meet current drug registration requirements necessary for these drugs to be licensed and distributed and the possibility of sub-clinical neurotoxicity (Brewer *et al.*, 1994) is of particular concern and will need further clinical and neuropathological investigation. Currently, Rhône Poulenc Rorer are seeking registration for Artemether, whilst WHO will seek registration for arteether.

Artemesinin's activity appears to be mediated, at least in part, by activated oxygen radicals. These toxic oxygen by-products may react with high concentrations of heme present within the parasite resulting in selective toxicity. The unique structure of artemisinin coupled with its potent antima-larial activity has attracted much chemical attention, and a wide variety of derivatives and related compounds have been synthesized (see Zaman & Sharma, 1991; Butler & Wu, 1992; White, 1994). Biological studies confirm the contention that a peroxide moiety is essential for activity and a wide range of structurally simple cyclic peroxides have been produced (Jefford *et al.*, 1988), including a number of bicyclic peroxides related to yingzhaosu such as Roche's Ro 41-3823. Recently, simpler tricyclic 1,2,4-trioxanes related structurally to artemisinin have been synthesized (for example,

WR279137 and WR279138) and shown to have good activity *in vitro* and *in vivo* against *P. falciparum* (Posner *et al.*, 1992, 1994). Clinical evaluation is anticipated in the near future. Artemisinin derivatives are the most rapidly acting of all the antimalarial drugs, they have the broadest stage specificity of action, and produce a more rapid clinical and parasitological response than other antimalarial drugs.

Atovaquone

Atovaquone (566C80, Wellcome) is a member of a family of hydroxynaphthoquinones whose antimalarial potential was first recognized over 50 years ago (Hudson, 1984). Previous attempts to develop these compounds have been thwarted by problems arising from poor oral absorption and rapid metabolism in man. Because of this, the compounds were first evaluated for activity against *P. falciparum in vitro* and then for metabolic stability using human liver microsome preparations (Hudson, 1988; Hudson *et al.*, 1991). One compound, atovaquone, was found to have both outstanding antimalarial potency and resistance to metabolism (Hudson, 1993). The mode of action of hydroxynaphthoquinones involves selective inhibition of malarial electron transport systems (Fry & Pudney, 1992). This causes a blockade of *de novo* pyrimidine biosynthesis at the locus of the membrane-bound enzyme complex dihydroorotate dehydrogenase (Hammond, Burchell & Pudney, 1985). The parasites cannot replicate because they are unable to take up pre-formed pyrimidines. This unique site of action for an antimalarial agent coupled with atovaquone's lack of structural similarity to any of the established drugs make it likely that cross-resistance to chloroquine, mefloquine, etc. would not be a problem. This contention was supported by a wide variety of *in vitro* and *in vivo* data, including results against chloroquine-resistant *P. falciparum* in Aotus (Hudson *et al.*, 1991) and recently against a large number of clinical isolates in Zambia (Basco, Ramiliariso & Le Bras, 1994). In addition to the blood schizonticidal activity seen in these assays, atovaquone also has potent activity against the liver stages of rodent malaria (Davies *et al.*, 1989, 1993) indicating potential as a causal prophylactic agent. Activity is also seen against the early gametocyte stage (Fleck *et al.*, personal communication) and against mosquito stages (Fowler *et al.*, 1994), factors which could have important implications for transmission.

Clinical evaluation of atovaquone in healthy volunteers has shown the compound to be very well tolerated and to achieve readily blood levels considered appropriate for antimalarial activity. It has an elimination half-life of some 70 hours and does not appear to be metabolized to any significant extent. In Thailand, where drug resistance is endemic, the compound has shown cure rates of about 70% against acute uncomplicated falciparum malaria comparable to the most effective single agents. This was not considered adequate for a new compound, particularly as some of the isolates from recrudescent patients exhibited decreased sensitivity to atovaquone. In order to overcome this, and to make atovaquone unavailable as a single entity for malaria, the use of another compound in a fixed combination was examined.

Proguanil was selected from a wide range of compounds tested in combination with atovaquone *in vitro*, because of the consistent potentiation observed in different strains of malaria and its excellent safety profile (Luzzi & Peto, 1993). In clinical trials a fixed combination of atovaquone/proguanil has achieved consistently high cure rates. Optimization of the dosing regimen has led to the use of 1000 mg atovaquone and 400 mg proguanil once daily for 3 days. This resulted in rapid resolution of clinical symptoms and parasitaemia, with a cure rate of 100% in 24 patients. An excellent profile of tolerance and safety to this combination has been established and a phase III programme in key endemic areas is under way. On-going studies in Thailand and Zambia have confirmed the earlier studies and a cure rate of 100% in approximately 200 patients achieved to date, in contrast to 9 mefloquine failures in Thailand, and 1 Fansidar failure in Zambia in comparable numbers of patients (Looareesuwan, Hutchinson & Canfield, personal communication). The potential for the prophylactic use of this combination where both compounds have tissue schizontocidal activity and sporontocidal activity (experimental only in the case of atovaquone) has yet to be evaluated clinically, but with the excellent safety profile of both compounds (atovaquone has been used at high levels in AIDS patients for the treatment of *Pneumocyctis carinii* pneumonia with very few adverse events, and is now marketed (trade name MEPRON) for this indication (Artymowicz & James, 1993)) this looks most promising.

Pyronaridine

Pyronaridine, another schizonticidal agent developed in China during the mid-1970s, originated from the mepacrine nucleus, and has already undergone extensive trials in man against *P. falciparum* and *P. vivax* in China. Pyronaridine is highly effective against chloroquine-resistant *P. falciparum*, and appears to show no cross-resistance with 4-aminoquinoline and quinoline methanol antimalarials. The mode of action is unknown. Although there is no cross-resistance with chloroquine, parasites rapidly develop resistance to pyronaridine, consequently a triple combination of pyronaridine, sulfadoxine and pyrimethamine has been tried as a method of prolonging the use of the new drug (Fu & Xiao, 1991).

Experimental antimalarials

Daphnetine a Chinese coumarin compound derived from the bark of ash trees, has been used in man for other indications, and has been shown experimentally to have antimalarial activity (Yang *et al.*, 1992).

A novel series of pyridones related to the anticoccidial agent metichlor-pindol have shown potent activity against *P. falciparum in vitro* and *P. yoelii* in mice. The most active compounds, had IC_{50}s of 2×10^{-8} M against the human parasite *in vitro* and showed potentiation with atovaquone (Pudney *et al.*, 1992). The mode of action appears to be the same as atovaquone. Although both act at complex III of the electron transport system, parasites resistant to atovaquone were still sensitive to the 4-pyridones. A combination of a 4-pyridone with atovaquone could be extremely valuable in preventing the emergence of drug-resistant parasites.

The 8-aminoquinoline primaquine is an extremely important drug as it is the only commonly available treatment which can be used to eliminate the long-lived hypnozoite stage of vivax and ovale malaria. However, the compound is not ideal as it lacks blood schizonticidal activity and at high doses haemopoietic toxicity can be problematic. Several analogues of primaquine have been identified as possible successors with WR 238,605 receiving the most detailed pre-clinical evaluation (Milhous *et al.*, 1988). In animal models this compound is more active as a tissue schizonticide than primaquine but less toxic, (Peters, Robinson & Milhous, 1993). It also has a longer half-life, is well absorbed orally and is sufficiently promising to warrant clinical evaluation. Other primaquine analogues are also emerging with interesting activity.

Several other approaches are under investigation but will not be discussed in detail in this review. They include improvements in the chemistry to find a more efficient iron chelator as it is known that chelation of free iron with desferrioxamine inhibits *in vitro* parasite growth and has been shown to clear parasites in adults with asymptomatic malaria. In the clinic, antibacterials such as doxycycline and clindamycin are being investigated mostly for prophylaxis or in combination with standard antimalarials because of the slow activity when used alone. Azithromycin, a macrolide antibiotic, has recently been shown to have some efficacy as a causal prophylactic (Kuschner *et al.*, 1994). Protease inhibitors are also being studied, and a number of compounds unrelated to any of the established antimalarials continue to appear in the patent literature (Gutteridge, 1989; Hudson, 1992), but it will take some time before the initial activity is confirmed in man.

CONCLUSIONS

The spread of drug and insecticide resistance in the many parts of the world where malaria is endemic has created a situation where there are less effective means of checking the disease now than 20 years ago (Oaks *et al.*, 1991). More detailed information on the molecular biology of the parasite, mode of action of antimalarials and drug-resistance is providing a sound basis for novel approaches for future chemotherapeutic attack. However, the time delay between the initial discovery, and proof of efficacy in the clinic, particularly with the declining contribution by many of the major pharmaceutical companies, in contrast to the inevitable increase in resistance to the newer drugs currently available, means that the prospects for the immediate future are poor. Judicious use of the newer compounds, such as artemisinin and analogues, and atovaquone/proguanil, the speedy development of the trioxanes and new 8-aminoquinolines and the use of mixtures of the more established drugs, will buy a limited amount of time. However, effective control of malaria will be achieved only when it is possible to use an integrated strategy of vaccination, vector control and drug treatment.

REFERENCES

Artymowicz, R. J. & James, V. E. (1993). Atovaquone. A new antipneumocystis agent. *Clinical Pharmacy*, **12**, 563–70.
Basco, L. K., Ramiliariso, O. & Le Bras, J. (1994). *In vitro* activity of atovaquone (566C80) against the African clones and isolates of *Plasmodium falciparum*. (Submitted).
Bitonti, A. J., Sjoerdsma, A., McCann, P. P., Kyle, D. E., Oduola, A. M. J., Rossan, R. N., Milhous, W. K. & Davidson, D. E. (1988). Reversal of chloroquine resistance in malaria parasite *Plasmodium falciparum* by desipramine. *Science*, **242**, 1301–3.
Bjorkman, A. & Phillips-Howard, P. A. (1990). The epidemiology of drug-resistant malaria. *Transactions of the Royal Society of Tropical Medicine and Hygiene*, **84**, 177–80.
Boland, P., Were, J. & Campbell, C. (1993). Beyond chloroquine: Implications of drug resistance for evaluating malaria therapy efficacy and treatment policy in Africa. *Journal of Infectious Diseases*, **167**, 932–7.
Brasseur, P., Bittsindou, P., Moyou, R., Eggelte, T., Samba, G., Penchenier, L. & Druile, P. (1993). Fast emergence of *Plasmodium falciparum* resistance to halofantrine. *Lancet*, **341**, 901–2.
Bray, P. G. & Ward, S. A. (1993). Malaria chemotherapy: resistance to quinoline containing drugs in *Plasmodium falciparum*. *FEMS Microbiology Letters*, **113**, 1–8.
Brewer, T. G., Peggins, J. O., Grate, S. J., Petras, J. M., Levine, B. S., Weina, P. J., Swearengen, J., Heiffer, M. H. & Schuster, B. G. (1994). Neurotoxicity in animals due to arteether and artemether. *Transactions of the Royal Society of Tropical Medicine and Hygiene*, **88**, Suppl. 1, 33–6.
Brossi, A., Venugopalan, B., Gerpe, L. D., Yeh, H. J. C., Flippen-Anderson, J. L., Buchs, P., Luo, X. D., Milhous, W. & Peters, W. (1988). Arteether, a new antimalarial drug: synthesis and antimalarial properties. *Journal of Medicinal Chemistry*, **1**, 645–50.
Bruce-Chwatt, L. J. (1979). Man against malaria: conquest or defeat. *Transactions of the Royal Society of Tropical Medicine and Hygiene*, **73**, 605–17.
Bunnag, D., Karbwang, J., Thanavibul, A., Chittamus, S., Ratanapongse, Y., Chalermrut, K., Bangchang, K. N. & Harinasuta, T. (1994). High dose of primaquine in primaquine-resistant vivax malaria. *Transactions of the Royal Society of Tropical Medicine and Hygiene*, **88**, 218–19.
Butler, A. R. & Wu, Y. L (1992). Artemisinin (Qinghaosu): a new type of antimalarial drug. *Chemical Society Reviews*, **21**, 85–90.

Coatney, G. R. (1963). Pitfalls in a discovery: the chronicle of chloroquine. *American Journal of Tropical Medicine and Hygiene*, **12**, 121–8.

Cowman, A. F. & Foote, S. J. (1990). Chemotherapy and drug resistance in malaria. *International Journal for Parasitology*, **20**, 503–13.

Davies, C. S., Pudney, M., Matthews, P. J. & Sinden, R. E. (1989). The causal prophylactic activity of the novel hydroxynaphthoquinone 566C80 against *Plasmodium berghei* infections in rats. *Acta Leidensa*, **58**, 115–28.

Davies, C. S., Pudney, M., Nicholas, J. C., Sinden, R. E. (1993). The novel hydroxynaphthoquinone 566C80 inhibits the development of liver stages of *Plasmodium berghei* cultured *in vitro*. *Parasitology*, **106**, 1–6.

Ferone, R. (1984). Dihydrofolate reductase inhibitors In: Peters, W., Richards W. H. G. eds. *Handbook of Experimental Pharmacology Vol. 68/II Antimalarial Drugs II* Springer-Verlag, Berlin, Heidelberg, New York, Tokyo, pp. 207–221.

Fleck, S. L., Pudney, M. & Sinden, R. E. The effect of atovaquone (566C80) on the maturation and viability of *Plasmodium falciparum* gametocytes *in vitro*. (Submitted).

Foote, S. J. & Cowman, A. F. (1994). The mode of action and the mechanism of resistance to antimalarial drugs. *Acta Tropica*, **56**, 157–71.

Fowler, R. E., Billingsley, P. R., Pudney, M. & Sinden, R. E. (1994). Inhibitory action of the antimalarial compound atovaquone (566C80) against *Plasmodium berghei* ANKA in the mosquito *Anopheles stephensi*. *Parasitology*, **108**, 383–8.

Freedman, D. (1992). Imported malaria – here to stay. *American Journal of Medicine*, **93**, 239–42.

Fry, M. & Pudney, M. (1992). Site of action of the antimalarial hydroxynaphthoquinone, 2-[trans-4-(4'-chlorophenyl)cyclohexyl]-3-hydroxy-1,4-naphthoquinone (566C80). *Biochemical Pharmacology*, **43**, 1545–53.

Fu, S. and Xiao, S. H. (1991). Pyronaridine: a new antimalarial drug. *Parasitology Today*, **7**, 310–13.

Ginsburg, H. (1991). Enhancement of the antimalarial effect of chloroquine on drug-resistant parasite strains – a critical examination of the reversal of multi-drug resistance. *Experimental Parasitology*, **73**, 227–32.

Gutteridge, W. E. (1989). Antimalarial drugs currently in development. *Journal of the Royal Society of Medicine*, **82**, Suppl. 17, 63–8.

Hammond, D. J., Burchell, J. R. & Pudney, M. (1985). Inhibition of pyrimidine biosynthesis *de novo* in *Plasmodium falciparum* by 2-(4'-t-butyl)cyclohexyl-3-hydroxy-1,4-naphthoquinone *in vitro*. *Molecular and Biochemical Parasitology*, **14**, 97–109.

Horton, R. J., Parr, S. N. & Bokor, L. C. (1990). Clinical experience with halofantrine in the treatment of malaria. *Drugs under Experimental and Clinical Research*, **16**, 497–503.

Hudson, A. T. (1984). Menoctone, hydroxyquinolinequinones and similar structures. In *Handbook of Experimental Pharmacology Vol. 68/II Antimalarial Drugs II* (Peters, W., Richards W. H. G. eds). Springer-Verlag, Berlin, Heidelberg, New York, Tokyo, pp. 343–361.

Hudson, A. T. (1988). Antimalarial hydroxynaphthoquinones. In *Topics in Medicinal Chemistry* (Leeming, P. R., ed.). Special Publication No. 65. Proceedings of the 4th SCI-RSC Medicinal Chemistry Symposium, Royal Society of Chemistry London pp. 266–283.

Hudson, A. T. (1992). Malaria chemotherapy–current treatment and recent advances. *Current Opinion in Therapeutic patents*, **2**, 227–41.

Hudson, A. T. (1993). Atovaquone – a novel broad-spectrum anti-infective drug. *Parasitology Today*, **9**, 66–8.

Hudson, A. T., Dickins, M., Ginger, C. D., Gutteridge, W. E., Holdich, T., Hutchinson, D. B. A., Pudney, M., Randall, A. W. & Latter, V. S. (1991) 566C80: a potent broad spectrum anti-infective agent with activity against malaria and opportunistic infections in AIDS patients. *Drugs under Experimental and Clinical Research*, **17**, 427–35.

Hutchinson, D. B. A., Looareesuwan, S. & Farquhar, J. (1992). Evaluation of the hydroxynaphthoquinone, 566C80, in the treatment of uncomplicated *P. falciparum* malaria. *Lecture given at the British Society for Parasitology, 4th Malaria Meeting, London*.

Jefford, C. W., McGoran, E. C., Boukouvalas, J., Richardson, G., Robinson, B. L. & Peters, W. (1988). Synthesis of new 1,2,4-trioxanes and their antimalarial activity. *Helvetica Chimica Acta*, **71**, 1805–12.

Kain, K. C. (1993). Antimalarial chemotherapy in the age of drug resistance. *Current Opinion in Infectious Diseases*, **6**, 803–11.

Karcz, S. & Cowman, A. F. (1991). Similarities and differences between the multidrug resistance phenotype of mammalian tumour cells and chloroquine resistance in *Plasmodium falciparum*. *Experimental Parasitology*, **73**, 233–40.

Ketrangsee, G. S., Vihaykadga, S., Yamokgul, P., Jatapadma, S., Thimasarn, K. & Rooney, W. (1992). Comparative trial on the response of *Plasmodium falciparum* to halofantrine and mefloquine in Trat Province, eastern Thailand. *South East Asian Journal of Tropical Medicine and Public Health*, **23**, 55–8.

Kuschner, R. A., Heppner, D. G., Andersen, S. L., Wellde, B. T., Hall, T., Schneider, I., Ballou, W. R., Foulds, G., Sadoff, J. C., Schuster, B. & Taylor, D. N. (1994). Azithromycin prophylaxis against a chloroquine-resistant strain of *Plasmodium falciparum*. *Lancet*, **343**, 1396–7.

Luzzi, G., Warrell, D., Barner, A. & Dunbar, E. (1992). Treatment of primaquine resistant *Plasmodium vivax* malaria. *Lancet*, **340**, 310.

Luzzi, G. & Peto, T. E. (1993). *Drug Safety*, **8**, 295–311.

Martin, S. K., Oduola, A. M. J. & Milhous, W. K. (1987). Reversal of chloroquine resistance in *Plasmodium falciparum* by verapamil. *Science*, **235**, 899–901.

Milhous, W. K., Theoharides, A. D. Schuster, B. G., Puri, S. K., Dutta, G. P., Heisey, G. B., Kyle, D. E., Oduola, A. M. J., Dhar, M. M., Heiffer, M. H., Reid, W. A. & Davidson, Jr D. E. (1988). New alternatives to primaquine. *XIIth International Congress for Tropical Medicine and Malaria, Amsterdam*: Abstract FrS-12-4.

Nosten, F., Ter Kuile, F., Chongsuphajaisiddhi, T., Luxemburger, C., Webster, H. K., Edstein, M., Phaipun, L., Kyaw, Lay Thew & White, N. J. (1991). Mefloquine-resistant falciparum malaria on the Thai–Burmese border. *Lancet*, **337**, 1140–4.

Oaks, Jr S. C., Mitchell, V. S., Peason, G. W. & Carpenter, C. C. J. (1991). Eds. *Malaria Obstacles and Opportunities*. National Academy Press, Washington DC, pp. 1–22.

Peel, S. I., Merritt, S., Handy, J. & Baric, R. (1993). Derivation of highly mefloquine-resistant lines from *Plasmodium falciparum in vitro*. *American Journal of Tropical Medicine and Hygiene*, **48**, 385–97.

Peters, W., (1990). The prevention of antimalarial drug resistance. *Pharmacological Therapeutics*, **47**, 499–508.

Peters, W., Ekong, R., Robinson, B. L., Warhurst, D. C. & Pan, X. Q. (1990). The chemotherapy of rodent malaria XLV. Reversal of chloroquine resistance in rodent and human plasmodium by anti-histaminic agents. *Annals of Tropical Medicine and Parasitology*, **87**, 547–52.

Peters, W., Robinson, B. L. & Milhous, W. K. (1993). The chemotherapy of rodent malaria. L.I. studies on a new 8-aminoquinoline, WR 238605. *Annals of Tropical Medicine and Parasitology*, **87**, 547–52.

Posner, G. H., Oh, C. H., Gerana, L. & Milhous, W. K. (1992). Extraordinary potent antimalarial compounds: new structurally simple, easily synthesized, tricyclic 1,2,4-trioxanes. *Journal of Medicinal Chemistry*, **35**, 2459–67.

Posner, G. H., Chang, H., Oh, H., Webster, K., Ager, A. L. & Rossan, R. N. (1994). New, tricyclic 1,2,4-trioxanes: evaluations in mice and monkeys. *American Journal of Tropical Medicine and Hygiene*, **50**, 522–6.

Pudney, M., Yeates, C., Pearce, J., Jones, L. & Fry, M. (1992). New 4-pyridone antimalarials which potentiate the activity of atovaquone (566C80). In *VIIIth International Congress for Tropical Medicine and Malaria, Jomtien, Pattaya, Thailand, November 29-December 4*, Vol 2 p 149. Abstract.

Schmidt, L. H., Harrison, J., Rossan, R. N., Vaughan, D. & Crosby, R. (1977). Quantitative aspects of pyrimethamine-sulfonamide synergism. *American Journal of Tropical Medicine and Hygiene*, **26**, 837–49.

Scholer, H. J., Leimer, R. & Richle, R. (1984). Sulphonamides and sulphones. In *Handbook of Experimental Pharmacology Vol. 68/II Antimalarial Drugs II* (Peters, W. & Richards, W. H. G. eds). Springer-Verlag, Berlin, Heidelberg, New York, Tokyo, pp. 123–206.

Schuurkamp, G. J., Spicer, P. E., Kereu, R. K., Bulungol, P. K. & Rieckmann, K. H. (1992). Chloroquine-resistant *Plasmodium vivax* in Papua New Guinea. *Transactions of the Royal Society of Tropical Medicine and Hygiene*, **86**, 121–2.

Shao, B. R., Huang, Z. S., Shi, S. H. & Meng, F. (1991). A 5-year surveillance of sensitivity *in vivo* of *Plasmodium falciparum* to pyronaridine/sulphadoxine/pyrimethamine in Diaoluo area, Hainan Province. *South East Asian Journal of Tropical Medicine and Public Health*, **22**, 65–67.

Slater, A. F. G. & Cerami, A. (1992). Inhibition by chloroquine of a novel haem polymerase enzyme activity in malaria trophozoites. *Nature*, **355**, 167–9.

Sweeney, T. R. (1981). The present status of malaria chemotherapy: mefloquine, a novel antimalarial. *Medical Research Review*, **1**, 281–301.

Upcroft, P. (1994). Multiple drug resistance in pathogenic protozoa. *Acta Tropica*, **56**, 195–212.

Webster, H. K., Thaithong, S., Pavanand, K., Yongvanitchit, K., Pinswasdi, C. & Boudreau, E. F. (1985). Cloning and characterization of mefloquine-resistant *Plasmodium falciparum* from Thailand. *American Journal of Tropical Medicine and Hygiene*, **34**, 1022–7.

Wellems, T. E. (1991). Molecular genetics of drug resistance in *Plasmodium falciparum* malaria. *Parasitology Today*, **7**, 110–12.

Wellems, T. E. (1992). How chloroquine works. *Nature*, **355**, 108–9.

Wernsdorfer, W. H. (1991). The development and spread of drug-resistant malaria. *Parasitology Today*, **7**, 297–303.

Wernsdorfer, W. H. (1994). Epidemiology of drug resistance in malaria. *Acta Tropica*, **56**, 143–56.

White, N. J. (1992). Antimalarial drug resistance: the pace quickens. *Journal of Antimicrobial Chemotherapy*, **30**, 571–85.

White, N. J. (1994). Artemisinin: current status. In Artemesinin. *Transactions of the Royal Society of Tropical Medicine and Hygiene*, **88**, Suppl, 1–65.

White, N. J. & Nosten, F. (1993). Advances in chemotherapy and prophylaxis of malaria. *Current Opinions in Infectious Diseases*, **6**, 323–30.

Wongsrichanalai, C., Webster, K. H., Wimonwattrawatee, T., Sookto, P., Chuanak, N., Wernsdorfer, W. & Thimasarn, K. (1992). Emergence of multidrug resistance *Plasmodium falciparum* in Thailand: *in vitro* tracking. *American Journal of Tropical Medicine and Hygiene*, **47**, 112–6.

Yang, Y. Z., Ranz, A., Pan, H. Z., Zhang, Z. N., Lin, X. B. & Meshnick, S. R. (1992). Daphnetin, a novel antimalarial agent with *in vitro* and *in vivo* activity. *American Journal of Tropical Medicine and Hygiene*, **46**, 15–20.

Zaman, S. S. & Sharma, R. P. (1991). Some aspects of the chemistry and biological activity of artemisinin and related antimalarials. *Heterocycles*, **32**, 1593–638.

GENETIC ENGINEERING OF MICROBES: VIRUS INSECTICIDES–A CASE STUDY

DAVID H. L. BISHOP, MARK L. HIRST,
ROBERT D. POSSEE and JENNIFER S. CORY

*Institute of Virology and Environmental Microbiology,
Mansfield Road, Oxford OX1 3SR, UK*

INTRODUCTION

Baculoviruses are large, arthropod-specific viruses with a genome consisting of a circular, double-stranded DNA. In the current classification scheme, two genera of baculoviruses are recognized, the nucleopolyhedroviruses and the granuloviruses. Both forms make occluded and non-occluded virus particles. In addition to these, there are other similarly organized viruses which only make non-occluded particles. They are described simply as non-occluded baculoviruses since their taxonomic status is unresolved. Little information is available on the molecular or genetic characteristics of these viruses.

For the nucleopolyhedroviruses and granuloviruses, the non-occluded viruses are primarily responsible for spreading an infection within a host and between cells and tissues, including those maintained in culture (Blissard & Rohrmann, 1990). The function of the occlusion bodies, otherwise termed polyhedral inclusion bodies (PIBs or polyhedra), is to provide protection of the virus between hosts. For the nucleopolyhedroviruses, the occlusion body contains many virions and has a polyhedral shape that may vary in diameter from 0.15 to 1.5 μm. For the granuloviruses the occlusion bodies are generally ovicylindrical (*ca.* $0.3 \times 0.5\,\mu$m diameter) and contain one, or rarely two, virions. Both forms of occlusion body have an outer calyx containing carbohydrate which aids in the persistence of viruses in the environment and outside the host species. The main protein component of polyhedra is called polyhedrin protein (*ca.* 28 kD). It exists in a crystalline form in the occlusion body. For the granuloviruses, the main protein component is termed granulin.

Virions within occlusion bodies consist of one or more rod-shaped nucleocapsids with a distinct structural polarity and enclosed within an envelope. The nucleocapsids average 30–35 nm in diameter and are 250–350 nm in length, depending on the virus. The occlusion bodies of the nucleo-polyhedroviruses are formed within the nucleus of infected cells and those of granuloviruses appear to be formed after the nucleus of the infected cell degenerates. The envelopes of viruses within the occlusion bodies lack peplomers. For the non-occluded forms of the virus, the envelope has surface peplomers consisting of a single glycoprotein species. These virus particles are formed at the surface of infected cells and are otherwise termed budded viruses.

Most baculovirus isolates have come from insects. Many have been reported from members of the Lepidoptera, several others have come from members of the Hymenoptera. Although more than 400 baculovirus isolates are described in the scientific literature, the taxonomic status of individual isolates is largely unknown. The reason is that many, but not all, baculoviruses are cross-infective between host species of the same order of the class Insecta (e.g. between members of the Lepidoptera, including members representing different families of Lepidoptera, see Table 1). While individual baculoviruses are not cross-infective between insect species representing different orders, little information is available on the actual number of different baculovirus species that exist, or on the number of hosts that a particular baculovirus species may infect (see Table 1).

A virus species is defined as 'a polythetic class of viruses that constitutes a replicating lineage and occupies a particular ecological niche' (Van Regenmortel, 1989). The current system of nomenclature of baculovirus species only describes them on the basis of their host of origin. This is comparable to naming rabies viruses as species on the basis of the different hosts from which the virus has been recovered (foxes, dogs, skunks, bats, humans, etc.). The present system of nomenclature of baculoviruses is a similar nonsense that is only perpetuated owing to the lack of information on the species status of individual isolates. Due to the cross-infectivities, many of the so-called viruses are probably the same virus species recovered from different hosts. Since only a handful of baculoviruses have been characterized at the genetic level, the taxonomic status of other isolates is not known. Nevertheless, based on the descriptions in the literature, it is clear that many insect hosts, in particular members of the Lepidoptera and Hymenoptera, are susceptible to baculovirus infections, although the actual number of these virus species is not known.

The infection course of baculoviruses is indicated in schematic form in Fig. 1. For the larval stages of a susceptible insect the occluded virus is acquired during feeding from contaminated substrates. Trans-ovarial transmission and infection of neonates from contaminated surfaces of ova may also occur. In the high alkaline pH of the insect midgut, ingested polyhedra degrade and the released virus particles infect a subset of midgut cells. The receptors that mediate the entry of viruses to these cells are unknown. Such cells also provide enzymes and other materials that are required for the virus to replicate. From these cells, and depending on the virus, the infection may spread to other cells and tissues within the larval host via budded progeny viruses. Tracheae and fat body tissues are particularly susceptible to infection within a host. Eventually, occluded viruses are formed which are

Table 1. *Host-range of parent AcNPV and genetically modified AcST-3*[a]

	AcNPV	AcST-3
Permissive species (>50% deaths at doses of 10^3 PIBs)	AcNPV	AcST-3
Arctiidae		
Estigmene acrea (Salt Marsh Caterpillar)	98%	94%
Noctuidae		
Autographa gamma (Silver Y)	100%	71%
Spodoptera exigua (Small Mottled Willow)	100%	42%
Trichoplusia ni (Cabbage Looper)	100%	100%
Yponomeutidae		
Plutella xylostella (Diamondback Moth)	100%	100%
Semi-permissive species (doses of >10^3; e.g. % deaths at doses of 10^5 PIBs)	AcNPV	AcST-3
Noctuidae		
Agrotis segetum (Turnip Moth)	32%	24%
Agrotis puta (Shuttle-shaped Dart)	88%	92%
Amphipyra tragopogonis (Muse Moth)		71%
Apamea epomidion (Clouded Brindle)	33%	58%
Aporophyla nigra (Black Rustic)	90%	67%
Autographa jota (Plain Golden Y)	26%	10%
Caradrina morpheus (Mottled Rustic)	63%	13%
Ceramica pisi (Broom Moth)	22%	18%
Colocasia coryli (Nut Tree Tussock)	10%	60%
Heliothis armigera (American Bollworm)	67%	28%
Heliothis zea (Cotton Bollworm)	77%	25%
Lacanobia w-latinum (Light Brocade)	69%	37%
Mamestra brassicae (Cabbage Moth)	63%	2%
Mythimna separata (Rice Army Worm)	23%	14%
Noctua pronuba (Large Yellow Underwing)	30%	33%
Noctua janthina (Lsr Broad-Bord'd Yellow U'Wing)	36%	7%
Panolis flammea (Pine Beauty Moth)	42%	14%
Orthosia cruda (Small Quaker)	93%	
Orthosia stabilis (Common Quaker)	45%	30%
Rusina ferruginea (Brown Rustic)	33%	10%
Spodoptera frugiperda (Fall Armyworm)	88%	66%
Spodoptera littoralis (Mediterranean Brocade)	12%	NP
Xylocampa areola (Early Grey)	74%	
Nymphalidae		
Melitaea cinxia (Glanville Fritillary)	80%	80%
Polygonia c-alba (Comma)	15%	5%
Sphingidae		
Laothoe populi (Poplar Hawk Moth)	NP	12%
Mimas tiliae (Lime Hawk Moth)	88%	92%
Non-permissive species (<5% deaths at doses of 10^5 PIBs)	AcNPV	AcST-3
Geometridae		
Idaea aversata (Riband Wave)	NP	NP
Ourapteryx sambucaria (Swallow-tailed Moth)	NP	NP
Pelurca comitata (Dark Spinach)	NP	NP
Lymantriidae		
Euproctis similis (Yellow-tail)	NP	NP
Noctuidae		
Acronicta megacephala	NP	NP
Acronicta rumicis (Knot Grass)	NP	NP

Continued

252   DAVID H. L. BISHOP *et al.*

Table 1. *Continued*

Diarsia mendica (Ingrailed Clay)	NP	NP
Dicestra triflolii (The Nutmeg)	NP	NP
Hada nana (The Shears)	NP	NP
Hoplodrina ambigua (Vine's Rustic)	NP	NP
Lacanobia oleracea (Bright-line Brown-eye)	NP	NP
Ochropleura plecta (Flame Shoulder)	NP	NP
Orthosia gothica (Hebrew Character)	NP	NP
Polia nebulosa (Grey Arches)	NP	NP
Xestia c-nigrum (Setaceous Hebrew Character)	NP	NP
Notodontidae		
Fercula fercula (Sallow Kitten)	NP	NP
Nymphalidae		
Aglais urticae (Small Tortoiseshell)	NP	NP
Argynnis paphia (Silver Washed Fritillary)	8%	NP
Pieridae		
Pieris brassicae (Large White)	NP	NP
Sphingidae		
Smerinthus ocellata (Eyed Hawk Moth)	NP	NP

[a]Only data for species tested simultaneously with both viruses are included, specifically focusing on those species that earlier tests had indicated were likely to be infected at some dose of AcNPV (plus some other non-permissive species as controls, and previously indicated as not infectible by AcNPV). Not shown are the uniformly negative results of tests involving non-lepidopteran insects (bees, ants, beetles, lacewings, ladybirds, sawflies, etc). Tests with other modified or wild-type AcNPV have indicated that (1) the following species fall into the 'Permissive' category (% deaths @ 10^3 PIBs): *Sp. albula* (100%), *Apamea sordida* (68%); and (2) the following fall into the 'Semi-permissive' category (unless otherwise indicated, % deaths @ 10^5 PIBs): *Ag. clavis* (62%), *Au. pulchrina* (40%), *Charanyca trigrammica* (12%), *Eurodryas aurinia* (100%), *Inachis io* (18%), *Manduca sexta* (100%), *My. pudorina* (20%), *Noctua comes* (12% @ 10^3 PIBs), *O. incerta* (85%), *Panaxia dominula* (32% @ 10^3 PIBs), *Phlogophora meticulosa* (51%), *X. xanthographa* (22% @ 10^3 PIBs). Other lepidopteran and other non-lepidopteran insects that previous studies indicated were not infectible by AcNPV were 'Non-permissive', or not tested. In these tests, second instar larvae were fed virus polyhedra (PIBs) in a 24 h period and then placed onto virus-free diet. The larvae were observed until death or pupation. The % mortalities at the indicated doses for virus-confirmed infections are given. NP indicates that significant numbers of insects ($\leq 4\%$) were not susceptible to virus infection at the highest dose tested.

While subjective, experience indicates that the likelihood of any 'Semi-permissive' species acquiring high enough doses under field conditions to initiate an epizootic is unlikely, depending on the species and their behaviour. Semi-permissive species, like the permissive species, may contribute to virus maintenance in the environment.

liberated on the death of the infected host to spread the infection to other susceptible species. From an infected fifth instar caterpillar up to 10^9 PIBs of a nucleopolyhedrovirus may be recovered, depending on the virus, with each polyhedron including many infectious virus particles. Younger larvae produce less virus.

There are various means by which baculoviruses are spread in the environment, including both biotic and abiotic intermediates. The former includes insectivores such as birds, rodents as well as predatory and

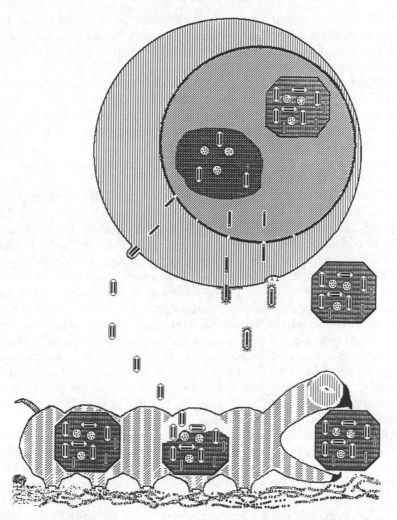

Fig. 1. Schematic of a nucleopolyhedrovirus infection of a lepidopteran larva. Virus in the form of polyhedra is acquired from contaminated food, the polyhedrin degraded in the alkaline midgut of the caterpillar and viruses liberated to infect selected midgut cells. Progeny viruses are released by budding from the surface membranes of such cells to spread the infection to other cells and tissues (tracheae, fat bodies) via the haemolymph. Late in infection, polyhedra are formed in the nuclei of infected cells and released from the larvae upon the death of the organisms, to contaminate and spread the infection to other susceptible lepidopteran larvae.

scavenging insects. These vectors are not infected by the viruses but may pass them in an undegraded form through their excreta. Abiotic distribution includes elution from the cadavers of infected caterpillars both on to leaves and into soil, as well as other forms of passive transfer. While virus infectivity is lost on exposure to the elements (in particular to ultraviolet irradiation in exposed places such as on plant leaves), the rate of loss of infectivity depends on where the virus is located. In soil, virus may remain in an infectious form for some time, but may be generally unavailable to leaf-eating caterpillars.

The main features of the virus infection process have been studied in detail using *Autographa californica* nuclear polyhedrosis virus (AcNPV) as a model baculovirus in view of the ease with which it may be cultured *in vitro* using *Spodoptera frugiperda* cell lines as hosts. These cells allow AcNPV to be cloned by plaque assays (or at limiting dilutions), and grown in monolayer and suspension cultures as well as fermented at densities of $>10^7$ cells/ml in serum-free media.

The fact that the purified baculovirus DNA is infectious indicates that no virion-associated protein is essential for transcription of the initial mRNA species. The transcription of AcNPV genes is temporally regulated (see Fig. 2). Two main classes of mRNA transcripts are recognized, early and late. A subset of early genes are described as 'immediate early' and represent mRNA species that are transcribed in the presence of inhibitors of protein synthesis. Other early genes are trans-activated by the products of the immediate early genes and are described as 'delayed early' genes. The early genes are transcribed by the insect host RNA polymerase II and are sensitive to inhibition by α-amanitin. By contrast, the late genes are transcribed by an α-amanitin-resistant RNA polymerase although whether this is an altered form of an insect RNA polymerase, or a virus-encoded polymerase, is not known. The so-called 'late' genes include genes coding for most of the structural proteins of virions. During the late phase, virus DNA is replicated and budded viruses are produced. At the so-called 'very late' phase of infection, polyhedrin protein and another protein, termed p10, are made. The availability of polyhedrin protein allows occluded viruses to be made. The roles of the p10 protein and the other viral gene products in the occlusion process are not known.

While the above represents the overall scheme of events during an AcNPV infection, some genes are transcribed both in the early and late phases, often involving separate transcription promoters. Transcriptional activity throughout infection frequently results in nested mRNA transcripts. These may have variable 5' and co-terminal 3' ends or co-terminal 5' and variable 3' ends, reflecting the use of different transcription initiation, termination and polyadenylation sites.

Genes are encoded on both DNA strands. While genes usually do not overlap, many of the downstream transcriptional signals (polyadenylation sites) are found in sequences corresponding to the coding domains of adjacent genes coded on the same or opposing DNA strand. Almost all the genes of AcNPV are translated from contiguous sequences in the viral genome. At least one exception is one of the immediate early genes which is translated from a spliced mRNA. Recently, the sequence of the genome of

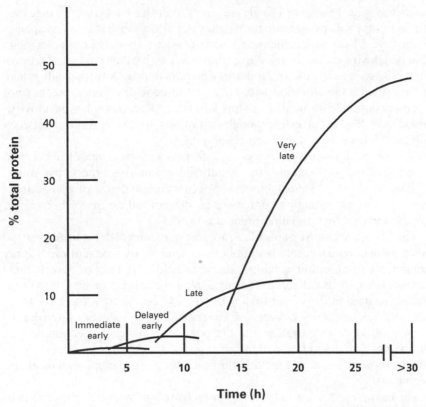

Fig. 2. Schematic of the intracellular infection course of a nucleopolyhedrovirus. The temporal course of synthesis of virus encoded mRNAs and their products is illustrated as a function of the total amount of the cell constituents that such products represent (% total protein). The transcription phases are illustrated as immediate early, delayed early, late and very late (see text). Budded viruses are made during the late phase. Occluded virions are made in the very late phase when polyhedrin and p10 protein are the major protein constituents in the cell.

AcNPV has been reported and some 150 genes and putative genes identified (Ayres *et al.*, 1994).

BACULOVIRUSES AS INSECTICIDES

Baculoviruses have several important advantages as pest control agents which make them obvious candidates for use as insecticides and for further development. Foremost amongst these is the fact that they have only been isolated from invertebrates (specifically arthropods) and so pose no risk to higher organisms when used to control pest infestations. They also produce a high level of pathogenicity in susceptible species, but have no effect on other species, including beneficial insects such as ants, beetles (e.g. lady-birds), wasps, bees, lacewings, etc. Some baculoviruses are effectively restricted in host range to a single species. Others have a broader range (see Table 1) which, in population terms, may not affect more than a few species in the field. The host specificity and lack of effect on beneficial species mean that baculoviruses can be used in conjunction with other biological control methods, such as the use of predators and parasitoids. A further advantage is that, unlike the situation with chemical insecticides, resistance has not been encountered in field situations where repeated use of a baculovirus insecticide has occurred, although variations in susceptibility between different virus strains have been observed.

Natural baculovirus epizootics of infection can be common in certain insect species, for example, the lymantriid moths (e.g. the gypsy moth, *Lymantria dispar*), whereas in other species the incidence of epizootics is low. A number of factors contribute to this, including insect behaviour, population densities and environmental factors.

The history of the use of baculoviruses for pest control dates back over 100 years. Baculoviruses can be applied to a crop by conventional spray equipment using either machine, or hand-held devices. For forest infestations, the spray is delivered from an airplane or helicopter since larvae are usually located in the upper foliage of trees. Other factors that have to be addressed in an aerial delivery of an insecticide include the provision of agents to reduce evaporation and materials to enhance protection against ultraviolet light degradation of the virus on foliage. Other means of delivery that have been employed include the use of baits, or trapping and release regimes.

An example of the use of trapping and release is provided by the control of dynastid beetles such as the rhinoceros beetle, *Oryctes rhinoceros*, a pest of coconut and oil palm plantations in many Pacific Ocean islands and countries (for example, Malaysia), as well as those in and bounding the Indian Ocean (for example, the Seychelles, Oman, East Africa). The infective virus for this pest is a non-occluded baculovirus that, unlike most other baculoviruses, may infect and replicate in both adult and larval forms of the host. Infections in these species appear to be limited to midgut cells and involve the excretion of large numbers of viruses. To provide control, adult beetles are collected by pheromone trapping, infected with adequate amounts of virus to establish an infection, and released in order to spread the infection to female beetles and into the sites (within tree trunks) where larvae breed and feed. Virus transmission during mating also decreases the fecundity of female rhinoceros beetles. The objective of the trapping and release procedure is to allow infections to build up in the natural reservoirs of the host (often outside the palm groves) in order to suppress the populations of beetles that infest the plantations. Such control programmes have been used with success in several of the aforementioned countries.

Inundative release of baculovirus insecticides usually involves spraying suitable quantities of virulent virus on to populations of larvae that have already reached, or which are likely to reach, high densities, and cause an unacceptable damage to a crop. Doses of the order of 10^{10}–10^{12} occlusion bodies per hectare are commonly employed (depending on the virus and the host). The amount of virus that is used varies according to the virus, its infectivity and stability, etc., as well as the host, its instar, location, etc., and other environmental considerations. Ideally, larvae should be infected when they are at an early stage of development in order to reduce any subsequent crop damage. To become lethally infected, first instar larvae require less virus than later instars. However, they also consume less foliage. As a consequence, the likelihood of infection depends on the distribution density of the applied virus. For this reason, ultra-low volume spray regimes are particularly effective. Since later instars consume more foliage, this increases the opportunity of exposure to virus and the establishment of an infection. However, later instars require much more virus to become productively infected. Generally, the dose requirements for lethal infection of particular larval instars are balanced by the greater amounts of contaminated foliage consumed.

While inundative releases usually involve a single application, alternatives that could be employed are the use of lower spray doses, but more frequent application, or the introduction of virus on a lattice regime to take advantage of the dispersive capacity of the virus (virus from primary infections spreading to other larvae by biotic and abiotic means). However, the major costs associated with the use of any insecticide are those associated with labour and equipment (airplane, helicopter, tractor, spray equipment rental, etc.), so that multiple applications are correspondingly more expensive than a single inundative release.

Baculoviruses have been successfully applied against pests of agricultural crops, forests, orchards, greenhouses, as well as stored products and urban pests, such as the brown-tail moth, *Euproctis chrysorrhoea*. The latter poses a health hazard owing to the effects associated with their urticating hairs (Speight *et al.*, 1992).

As noted above, the use of an insecticide has to be geared to the damage tolerance of the crop. For temperate forest pests, where some crop damage can be tolerated with little loss or reduction of yield (tree growth), baculoviruses have found particular favour. Examples include the control of pine sawfly, *Neodiprion sertifer*, the Douglas fir tussock moth, *Orgyia Pseudotsugata*, and the pine beauty moth, *Panolis flammea*, among others. Viruses have been used either singly, or in combination where multiple infestations occur (e.g. pine sawfly and pine beauty moth), although this depends on the availability of effective viruses.

Baculoviruses have also been isolated from a large number of agricultural pests. Many of these viruses have the potential to be developed into viable control agents (Entwistle & Evans, 1985). Field trials have been carried out on some, confirming their potential as control agents. In certain countries they are produced and distributed by national authorities (e.g. China, the former USSR, Brazil). Key target pest species for which baculoviruses have been isolated and developed as control agents include various *Heliothis* species, the *Spodoptera* complex of pests, cutworms such as *Agrotis segetum*, the codling moth, *Cydia pomonella*, the diamondback moth, *Plutella xylostella*, the rice armyworm, *Mythimna separata*, and the cabbage white butterfly, *Pieris rapae*, among others.

One of the most successful agricultural programmes has been the control of the soybean looper, *Anticarsia gemmatalis*, in Brazil where over 1 million hectares of soybean (out of 10 million hectares planted annually) are treated with a nucleopolyhedrovirus (Moscardi & Sosa-Gomez, 1992). This programme is an excellent example of how baculovirus technology can be developed and used by small-scale farmers over a wide area. In several regions of Brazil each year, virus-killed larvae are collected from the field and frozen in large quantities at centralized facilities until they are processed. Processing consists of macerating and filtering the infected larvae, mixing the extracts with kaolin and drying the mixture at ambient temperatures. The final product is then milled to produce a wettable powder before packaging in bags sufficient to treat one hectare of soybeans at a dose of 1–1.5×10^{11} occlusion bodies per hectare. A single application using traditional spray techniques provides sufficient control for one season.

Occluded baculoviruses have the inherent ability to persist in an infectious form and to infect further generations of a target host, provided the virus is available in adequate quantities and is protected *in situ* from exposure to ultraviolet irradiation. Natural or artificially induced baculovirus epizootics often exhibit a biphasic course. The first phase represents the initial round of infections, the second comes from virus resulting from these infections, so that eventually <98–99% of the population are infected.

In time, the applied and progeny viruses degrade, and are otherwise removed from the sites where they may come into contact with further generations of the host. Interestingly, studies conducted with pine beauty moth nucleopolyhedrovirus have shown that, with high pest infestations, a virus infection may produce sufficient progeny virus to cause new epizootics of disease in subsequent seasons (1 or more years later) when new hosts emerge either from the surviving species, or from newly introduced insects.

Ensuring that the pest comes into contact with a virus reservoir requires an intimate knowledge of the dynamics of the virus–host relationships. It has been shown that the lepidopteran pasture pest, *Wiseana cervinata*, in New Zealand causes greater damage in ploughed pastures than in unploughed pastures owing to the occurrence of fewer natural virus epizootics in the former. By instigating a regime of culture involving minimal disturbance of the soil, less of the virus pool is removed from the host's environment and the pest is kept to low levels by the higher frequency of natural epizootics (Kalmakoff & Crawford, 1982).

The commercial use of microbial insecticides only represents *ca*. 2% of the total insecticide market. Most of this reflects sales of the bacterium *Bacillus thuringiensis* to control forest and agricultural pests. The limited use of virus insecticides is due to the lack of suitable viruses, the slow rate at which they produce an effect, and the availability of faster acting, broad spectrum chemical alternatives. Generally, chemical insecticides are cheap to produce, easy to apply and affect a range of pests. The latter is an important issue where a complex of different pests infests a particular crop. The down side to the use of chemicals (and derived residues) is their acquisition and effect on non-target species, their transmission through the food chain (pollution, etc.), and the development of resistance by insect species following prolonged and repeated exposures. Nevertheless, the use of chemicals to control insect pests has provided major economic advantages throughout the world, decreasing the need to overproduce and increasing the quality of life and livelihoods of people across the world.

GENETICALLY IMPROVED BACULOVIRUS INSECTICIDES

With the increasing incidence of pesticide resistance and, for environmental considerations, the need to use less of the available chemical insecticides, baculoviruses offer a viable and attractive control option for many insect pest situations. Viruses are more selective than their chemical counterparts, but suffer from being slow at exerting an effect.

One option for obtaining an improved virus insecticide is to select, from natural isolates, clones of virus that are the most effective. Effectiveness is measured either in terms of the dose required to establish an infection in a susceptible species (described as the average lethal dose causing 50% mortality in members of a particular larval instar, LD_{50}), or in the time taken to kill such species (average lethal time in 50% of members of the exposed instar, LT_{50}). Commonly, different isolates of a virus exhibit different LD_{50} or LT_{50} results. As a consequence, a more virulent species can often be identified among natural isolates (see Hughes, Gettig & McCarthy, 1983). It has been reported for one virus that more virulent isolates can be prepared following exposure to nucleotide analogues (Wood *et al.*, 1981), although whether this applies to other baculoviruses is not known.

Apart from the development of naturally occurring baculoviruses as insecticides, opportunities to improve their speed of action by genetic engineering are being explored. Current improvements under assessment involve recombinants that affect the insect hormonal balance and those that produce insect specific toxins. Successful examples reported in the literature are the expression of an insect juvenile hormone esterase gene (Hammock *et al.*, 1990), an insect specific scorpion toxin gene (Stewart *et al.*, 1991), and an insect specific mite toxin gene (Tomalski & Miller, 1991). Each of these genes has been inserted into the genome of AcNPV, and recombinant viruses are produced which exhibit improved phenotypes in laboratory studies by comparison to the wild-type virus. Only the scorpion toxin modified virus has been tested in field conditions (Cory et al., 1994) and confirmed to be more effective at providing crop protection. In anticipation of such trials, research has been undertaken in the Institute of Virology and Environmental Microbiology, Oxford, UK, over a number of years to establish the conditions and protocols for such field trails, to undertake risk assessment analyses, and to investigate alternative types of modified insecticide.

FIELD STUDIES INVOLVING GENETICALLY MODIFIED BACULOVIRUS INSECTICIDES

In 1986, the first field experiment conducted anywhere in the world with a genetically modified virus (or any other genetically modified organism) was undertaken in a netted enclosure at Wytham, Oxfordshire, using a marked AcNPV in order to monitor the effect of the virus and its persistence on foliage and in soil (Bishop, 1986). The marker consisted of a short, unique and non-coding synthetic oligonucleotide inserted into the 3′ non-coding region of the AcNPV polyhedrin gene. In the laboratory it was shown that this marker did not affect the replication of the virus in terms of virus yields in cultured insect cells, or in insect larvae. The identity of any recovered virus was readily confirmed by hybridizing extracts of infected cells, or larvae, with a radioactive probe specific for the introduced sequence, and thereby differentiating the modified virus from the parent virus, or other viruses, or host material. The modified virus was also shown to be genetically and phenotypically stable during repetitive passage in insect cells and larvae. As with the parent virus, laboratory attempts to infect beneficial insects such as honey bees, ants, ladybirds, lacewings, hoverflies, etc., were unsuccessful. No changes in host range between the parent and modified virus were identified among a large number of Lepidoptera tested in parallel and using species representing a range of families and genera.

The above and other data were reviewed by the appropriate regulatory authorities in the UK as well as by other interested parties prior to granting a licence to conduct a field experiment in the summer of 1986. The site chosen was located in a field of light loam bounded by agricultural land at the University Farm of Oxford University at Wytham, Oxfordshire. A 10 m square, netted, insect- , bird- and vertebrate-proof enclosure was erected and surrounded by a 2 m high wire fence to exclude large herbivores and other animals. Within the netted enclosure, plexiglass-walled sub-enclosures were erected to provide further containment of the insect larvae. Wire netting was placed in the ground to exclude moles (Bishop et al., 1988).

For the field trial, the host insects (small mottled willow caterpillars, Spodoptera exigua) were infected with virus in the laboratory and then released on to sugar beet planted within one of the plexiglass sub-enclosures. This ensured the infection of all the experimental larvae and limited the spread of the virus which might have resulted from a spray application. To serve as controls, uninfected larvae were applied to sugar beet in another sub-enclosure. One week later, all the infected insect larvae succumbed to virus infection. In the control plot, the uninfected insect larvae continued to feed over a three week period and eventually pupated.

When the virus-infected insects were returned to the laboratory, it was demonstrated that they contained the genetically marked baculovirus. The uninfected larvae did not. Sugar beet leaves collected from the sub-enclosure containing the infected larvae (but not those from the control plot) were also shown to be contaminated with the marked AcNPV. The presence or absence of virus was demonstrated by allowing *Trichoplusia ni* larvae to feed on the sugar beet leaves. These larvae subsequently succumbed to infection by the marked virus (as confirmed by hybridization analyses). Also, in the laboratory cabbage seeds were germinated in soil samples collected at the site and the seedlings similarly fed to *T. ni* larvae. For soil samples from the sub-enclosure containing the infected *S. exigua* (but not with samples from the other sub-enclosure, or the surrounds) the *T. ni* larvae died with typical virus symptoms. They were shown to contain the genetically marked virus as demonstrated by hybridization tests. Repeated assays over a six-month period on soil samples recovered from the sub-enclosures after the plants and remaining insects (pupae) had been removed gave the same results.

The first field trial was terminated in February 1987 by disinfecting the site with formalin. Using tests similar to those described above this procedure was shown to inactivate and reduce to undetectable levels any remaining infectious AcNPV in the soil. The results from the first field trial with a genetically marked AcNPV demonstrated that the virus could persist in the environment for at least six months in soil.

This stability of baculovirus infectivity is attributed to the virus occlusion body. The assembly of this structure is dependent on the synthesis of the polyhedrin protein in virus-infected cells. However, deletion of the polyhedrin gene from the AcNPV genome does not affect the production of recombinants nor the formation of non-occluded, budded virus particles (Smith, Fraser & Summers, 1983).

The ability of a polyhedrin-negative or so-called 'crippled' baculovirus to persist in the environment was tested in a field trial conducted in 1987 after obtaining the requisite permissions (Bishop *et al.*, 1988). The aim of the trial was to determine whether the polyhedrin-negative phenotype was a suitable vehicle for a recombinant baculovirus insecticide containing a foreign gene encoding an insecticidal protein. The complete polyhedrin gene promoter and coding sequence were removed from AcNPV and replaced with another synthetic and unique genetic marker. The host range of the modified virus was tested as before. For this purpose, budded virus preparations from *in vitro* cell cultures were fed to insect larvae. It was found that, again as expected, the insertion of the genetic marker did not alter the host range of the virus and that the virus was genetically and phenotypically stable after 50 cycles of replication in insect cells. However, compared to the infectious doses of occluded wild-type virus, some thousand-fold more tissue culture infectious doses of budded viruses were required to establish infections in susceptible lepidopteran larvae. Also it was shown that the polyhedrin-negative virus rapidly lost infectivity when mixed with soil. For example, after 7 days of exposure, less than 0.1% of the initial virus remained infectious. These data suggested that, unlike the wild-type virus, the non-occluded mutant would not persist in the environment after release.

For the 1987 trial the same field enclosure and protocols used in the 1986 study with the occluded marked virus were employed. In this case *S. exigua* larvae were infected with the marked, polyhedrin-negative virus in the laboratory prior to release on to sugar beet plants in one of the sub-enclosures. Uninfected larvae served as controls. After a few days, all the infected larvae died. One week later, and as assayed using *T. ni* larvae, infectious virus could not be recovered from the plant surfaces, soil samples, or the remains of dead larvae. This was in contrast to the first field release with the occluded AcNPV, where infectious virus persisted in the soil for six months and until the site was disinfected. Following the second trial, the site was again disinfected.

Further experiments in 1988 and 1989 utilized a polyhedrin-negative AcNPV with a complete β-galactosidase coding region under the control of a functional polyhedrin gene promoter (AcNPV.lacZ). This marker gene served to test whether a virus that would express a foreign gene could function as an insecticide under field situations. Again, host range, genetic and phenotypic stability tests were conducted giving results similar to those described above. Following permission, and although conducted in exactly the same manner as the previous field trials, the results with the AcNPV.-lacZ virus were unsatisfactory. Few of the insects which had been fed the polyhedrin-negative virus died in the field from virus infection. For the 1988 trial, this was attributed to the fact that the trial was conducted in late summer, when cool temperatures and damp weather conditions precluded normal insect development. However, similar results were obtained in 1989 when the conditions were more favourable. Based on these results it was concluded that a non-occluded form of AcNPV was not a suitable insecticide under field conditions. After these trials in 1991, the site was disinfected, the facilities incinerated on site and, after ploughing, the land returned to its normal use.

A field trial with a genetically modified AcNPV has also been performed in the USA (H. A. Wood, personal communication). In this experiment a 'co-occluded' AcNPV was sprayed onto cabbages infested with *T. ni* larvae.

The co-occluded virus was produced by co-infection of insect cells with a polyhedrin-positive and a genetically modified polyhedrin-negative AcNPV. The polyhedrin-positive virus packaged a proportion of the virus particles produced by the polyhedrin-negative recombinant. The mixture of viruses was sprayed on to a crop in the field. With successive rounds of replication in insect larvae, it was found that the proportion of the polyhedrin-negative genotype decreased although, surprisingly, it did not disappear from the population.

A further field trial was conducted at Wytham in 1993 using new netted facilities to test the effectiveness of an occluded recombinant AcNPV containing a scorpion toxin gene (Cory *et al.*, 1994). This virus (AcST-3) was constructed by inserting a synthetic copy of the *Androctonus australis* Hector insect-specific neurotoxin (AaHIT) coding region under the control of a duplicated AcNPV p10 gene promoter placed upstream of the polyhedrin gene in the virus (Stewart *et al.*, 1991). This arrangement retained the function of the polyhedrin gene and permitted the production of polyhedra. The toxin coding region was downstream and in-frame with a duplicated copy of the coding sequence of the AcNPV gp67 signal peptide to facilitate secretion of the toxin from virus-infected cells. AaHIT acts by causing specific modifications to the sodium conductance of insect neurons, producing a presynaptic excitatory effect leading to paralysis and death. When neonate *T. ni* larvae were infected in the laboratory with AcST-3, they died 25% earlier than insects infected with the parental AcNPV. Furthermore, feeding damage to cabbages by AcST-3-infected larvae was reduced by up to 50% (Stewart *et al.*, 1991).

The host range of AcST-3 has been compared to that of AcNPV using a number of representative lepidopteran larvae (Table 1). Due to the logistical problems associated with obtaining and handling many different insects, LD_{50} assays have only been undertaken with a limited range of species and only with second instar larvae. (For complete LD_{50} assays to obtain statistically valid data at least 30 insect larvae are required per dose with 5–7 doses per bioassay.) Insects were assayed with the available numbers of larvae using 2 virus doses (10^3 or 10^5 PIBs) per second instar larva. For each assay, recombinant virus was compared with the unmodified AcNPV at the same time. With wild-type AcNPV, or the AcST-3 recombinant using *T. ni*, the LD_{50} values were determined to be *ca.* 40 PIBs per second instar caterpillar. Thus, for the other species tested, the two doses employed (10^3, 10^5 PIBS) corresponded to 25- and 2500-fold the LD_{50} values obtained with second instar *T. ni*.

To perform host range assays with the parent and modified virus, various insect species were collected in the field, returned to the laboratory and allowed to produce eggs. These were surface sterilized with formalin to inactivate any associated pathogens (viruses, fungi or bacteria). Neonate larvae were transferred to semi-synthetic diet if this was acceptable to the species. Others were fed the preferred foliage (washed to reduce the effects of other pathogens), as available. When the insects had progressed to their second instar, they were assigned to separate groups and fed small portions of diet contaminated with 10^3 or 10^5 polyhedra (unmodified AcNPV or AcST-3), or water as controls. After 24 h, those insects which had consumed all the contaminated food were transferred to fresh food and incubated until death (or pupation). Deaths were diagnosed as a consequence of virus infection by Giemsa staining of smeared larvae to identify polyhedra. The results are summarized in Table 1. The various insect species are operationally described as 'permissive' for virus replication if the majority (>50%) were infected at doses of $\leqslant 10^3$ polyhedra, 'semi-permissive' if >5% virus deaths occurred after infection with 10^5 polyhedra, and 'non-permissive' if <5% deaths occurred following exposure to this dose. In summary, the data showed that the host ranges of the parental AcNPV and recombinant AcST-3 were comparable. Differences in the percentages of insect deaths in a single species after infection with each virus are well within the normal experimental variation expected for this method of assay. The data presented in Table 1 support the conclusion that the recombinant AcST-3 has a similar host range to the parent AcNPV as may be expected where the host range is determined by viral gene products rather than, those of the introduced gene.

While the host range of the genetically modified AcST-3 was the same as that of the parental AcNPV, there remained the possibility that the recombinant toxin produced by virus that replicated in infected caterpillars would have an effect on non-target insect species. The available data on the mode of action of the AaHIT were reviewed and, while no reason was identified for the expressed toxin exerting an effect when consumed by other species (due to the natural midgut and cuticle 'skin' barriers), confirmatory experiments were performed to verify this point.

The symptomology (paralysis of injected insects), the low doses required, and the rapid onset of symptoms indicate that the site of action of AaHIT is the insect nervous system. Experiments have shown that AaHIT causes repetitive firing of the insect's motor nerves, resulting in massive and uncoordinated stimulation of skeletal muscle. Experiments have been reported with several insect species including *Musca domestica* (Loret *et al.*, 1991), *Locusta migratoria* (Walther, Zlotkin & Rathmeyer, 1976), and *Periplaneta americana* (D'Ajello *et al.*, 1973). The selectivity of AaHIT for the insect nervous system has also been demonstrated by showing that no effect was observed on similar muscle preparations from members representing the Crustacea (Rathmeyer, Walther & Zlotkin, 1977), Arachnida (Ruhland, Zlotkin & Rathmeyer, 1977), and mammals (guinea-pig) (Tintpulver, Zerachia & Zlotkin, 1976). In experiments using isolated single insect nerve fibres, voltage clamp experiments showed that the repetitive firing induced by AaHIT is due to a unique modification of the sodium

Table 2. *Responses of various species to AaHIT*[a]

Test animal (route)	LD$_{50}$ ng/100 mg	95% confid.
Diptera (injection)		
Musca domestica	2	1.3–2.4
Sorcophaga argyrostoma	15	13.7–17.2
Dictyoptera (injection)		
*Blattela germanica*26	24.3–28.0	
Periplaneta americana	46	36.5–57.5
Orthoptera (injection)		
Gryllus domesticus	289	224–373
Gryllus domesticus	375	384–400
Lepidoptera (injection)		
Spodoptera littoralis	1310	1180–1460
Mouse		
(sub-cutaneous injection)	>50 mg/kg	
(intra-cerebral injection)	>2.5 mg/kg	

[a]Data summarized from De Dianous *et al.* (1987).

channel conductance of the insect neuronal membrane. It is manifest as an increase of the sodium current and a slowing of its turn-off (Lester *et al.*, 1982; Gordon, Zlotkin & Catterall, 1985; Zlotkin *et al.*, 1985). In binding experiments using synaptosomes (membrane vesicles prepared from isolated insect nervous tissues), [^{125}I]-labelled AaHIT has been shown to bind to preparations from insects (*L. migratoria*, *Gryllus bimaculatus* and *M. domestica*), but not to those of crustaceans, or mammals (Teitelbaum, Lazarovici & Zlotkin, 1979; Gordon *et al.*, 1984, 1985). The toxin binds to a single class of non-interacting binding site with high affinity ($K_d = 1.2$–3 nM) and low capacity (0.5–2.0) pmol/mg membrane protein). Comparative assays with saxitoxin, which is known to bind to sodium channels owing to its displacement by tetrodotoxin (Walther *et al.*, 1976), have shown that membrane binding capacity for both toxins is the same. Further, the binding site on the sodium channel is different from that of other well-characterized toxins. AaHIT is not displaced by veratridine, tetrodotoxin, sea anemone toxin, or the α and β scorpion toxins which are specific for vertebrates (Gordon *et al.*, 1985).

Injecting purified AaHIT into various insect species has revealed considerable variation in toxicity (De Dianous, Hoarau & Rochat, 1987). These LD$_{50}$ data are summarized in Table 2. Lepidopteran species, such as *S. littoralis*, were about 650-fold less sensitive than a dipteran host (*M. domesticus*). The relative lack of sensitivity (Table 3) of AaHIT by injection in lepidopterous larvae representing some six species has recently been shown to be a consequence of non-specific binding and proteolytic degradation of AaHIT. It is not due to a reduction in its high affinity binding to

Table 3. *Responses of lepidopterous larvae to AaHIT*[a]

Insect	Dose µg/100 mg	Response[b]
Spodoptera littoralis	0.65	0/5
	2.0	5/5
Heliothis peltigera	2.5	0/3
	5.0	1/3
	10.0	3/3
Galleria mellonella	0.8	1/3
Bombyx mori	0.35	2/3
	0.65	5/5
Ocnogyna loewi	0.65	0/3
Cydia pomonella	0.8	0/3

[a]Data summarized from Herrmann *et al.* (1990).
[b]Lethality was determined by inability of the insect to pupate.

neurons since the toxin has been shown to bind efficiently to lepidopterous neuronal preparations (Herrmann, Fishman & Zlotkin, 1990).

The route of application of AaHIT to insects also affects toxicity. It has been shown that the toxin was 500-fold less effective after topical application to *M. domestica* than by injection (De Dianous, Carle & Rochat, 1988). Our studies have shown that feeding 340 ng of toxin to 1 mg second instar *T. ni* larvae produced no obvious toxic effect (unpublished data). All larvae continued to develop normally and eventually pupated. In LD_{50} terms this is equivalent to a value of $>35\,000$ ng/100 mg larva. These results support the conclusion that, for the recombinant toxin to have an effect on insects, it must be synthesized by the virus continuously and *in loco*.

In the environment, caterpillars such as those of *T. ni* are susceptible to predation by beetles. The effect of the scorpion toxin on predatory beetles was assessed by feeding virus-infected larvae to *Pterostichus madidus* in the laboratory. Adult beetles were collected from the field and fed on healthy insect larvae until required for experimental use. Beetles were kept separately to prevent fighting. They were observed to be active hunters, rapidly locating any live material within their containers. Virus-infected larvae were prepared by droplet-feeding neonate *T. ni* larvae with AcST-3, or AcNPV (2 × 10^6 polyhedra/ml). Two days later, 40 fully developed first instar larvae were fed to each beetle. The beetles were monitored 1 and 20 h after feeding. They demonstrated no abnormal behaviour and continued to hunt and feed on larvae. The exposed beetles were kept for a further 10 days and remained unaffected. To quantify the amount of toxin within the virus-infected larvae fed to the beetles, 40 larvae at 75 h post-infection were homogenized, and dilutions assayed for toxin activity by dorso-lateral injection of *M. domestica* (adult flies). This species is highly susceptible to the toxin after injection. The response to the recombinant toxin produced in the AcST-3 infected larvae was compared with that produced after injection of known amounts of the natural toxin purified from total venom using HPLC. It was estimated that each *T. ni* larva contained 2.25–4.5 ng toxin. On the basis of the numbers of larvae consumed, it was calculated that each beetle would have received 225–450 ng toxin. In a similar experiment, eight beetles were fed with approximately 20 mg of moribund AcST-3-infected *T. ni* larvae. The next day, a further 20 mg of dead, AcST-3 infected larvae were fed to the beetles. The beetles accepted the cadavers. The beetles were maintained for a further 10 days, hunted actively and showed no behavioural changes.

Although the safety of baculoviruses in small mammals is well accepted, confirmatory tests were undertaken with the recombinant baculovirus. Three separate trials were undertaken using rats and guinea pigs. Three groups of experimental rats were injected subcutaneously with 0.5 ml of sterile water, AcNPV (unmodified virus), or AcST-3 containing 10^6 polyhedra (equivalent to 10^4 LD_{50} in second instar *T. ni* larvae). In the subsequent 28-day test period, individual body weights, general appearance, feeding and drinking habits were recorded. There was no difference between the three groups, with all animals remaining healthy. At the end of the 28 days, animals were humanely killed (anaesthetic overdose) and necropsies performed. There were no differences observed in the organs in animals from each of the three groups. The sera derived from the animals before and after the toxicity tests did not contain antibodies to AcST-3.

Three further groups of rats were fed 1.0 ml of sterile water, or AcNPV, or AcST-3 containing 10^6 polyhedra. The animals were observed for 28 days as described above, and showed no abnormal symptoms. Subsequent autopsy also confirmed that there were no differences in the appearance of the organs in animals from each experimental group. The sera derived from these animals did not contain virus-specific antibodies.

Guinea pigs were randomly assigned to three experimental groups as described above. Test material was applied to an area of shaved back ($6 \, cm^2$) which was divided into four sub-areas, two of which were abraded using a 26 gauge needle. Each group received either 0.1 ml distilled water, or 0.1 ml AcNPV, or 0.1 ml AcST-3 containing 10^6 polyhedra. The treated areas were re-covered with gauze for a 12 h period. Daily observations were recorded for individual animals over a 14-day period, including evidence for skin reactions – erythema and oedema, body weight, general appearance, feeding and drinking habits. There were no differences between the control and experimental groups of animals, all remained normal. On day 14, the animals were humanely killed, blood samples taken and necropsies performed. The appearance of the organs from animals in all groups was

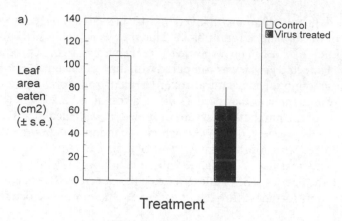

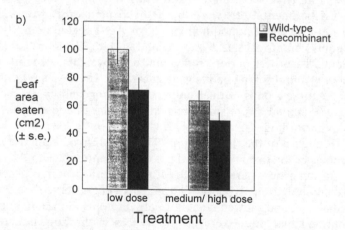

Fig. 3. In (*a*) are shown the mean leaf areas eaten (cm^2 ± s.e.) in infested control and virus (wild-type and recombinant combined) treated cabbages at 16 days after virus application. In (*b*) are shown the mean leaf areas consumed in wild-type AcNPV and recombinant AcST-3 virus treated plots at 16 days after virus application and as a function of the doses applied (see text, Cory *et al.*, 1994, with permission).

normal. The blood samples were analysed for antibodies to the AcST-3 virus, with negative results.

Twenty adult mice were randomly assigned to two groups. Mice in the first group were injected with 0.1 ml of sterile distilled water. Mice in the second group received 0.1 ml containing 1 μg AaHIT. Daily observations were recorded for individual animals over a 14-day period, i.e. skin reactions – erythema and oedema, body weight, general appearance, feeding and drinking habits. There were no differences between the control and experimental groups of animals, all remained normal. At the end of the experiment the animals were humanely killed (anaesthetic overdose) and necropsies performed. The appearance of the organs from animals in all groups was normal.

Finally, adult mice, were exposed to AaHIT ($1\,\mu g$), or water, via the intranasal route. All remained normal throughout the trial period and showed no abnormalities in internal organs.

The above analyses demonstrated that, as expected, the expressed toxin had no effect on vertebrates, or non-target species that acquired the materials by ingestion.

FIRST FIELD TRIAL OF A GENETICALLY IMPROVED VIRUS INSECTICIDE

In the summer of 1993, the efficacy of the AcST-3 recombinant was tested in the field, using *T. ni* as the target pest and cabbages as the crop. A replicated trial was undertaken using 1 m square, totally netted enclosures in a fenced area at the University Farm, Wytham, Oxfordshire. The 2 mm nylon mesh netting was provided to prevent the dispersal of caterpillars from the enclosures and to prevent the entry of other insects, or predators such as birds and rodents. Pitfall traps were placed in the enclosures before the trial so that invertebrate predators, such as beetles, could be removed. A bare earth policy was maintained between the enclosures, with pitfall traps, to limit the presence of other insects. As before, the site was fenced off to keep out other animals.

The AcST-3 recombinant was compared with the parent wild-type virus at three doses (10^6, 10^7 and 10^8 PIBs/m^2) which were equivalent to those which might be used in a pest control situation. Normally, insecticidal sprays would be targeted against the youngest larvae; however, in this trial a high density of third instar larvae was used in order to create a more challenging situation where a large amount of crop damage could occur in a relatively short period of time. The viruses were applied as a spray within each enclosure using ultra low volume equipment in order to give good crop coverage. To limit spray drift between plots, the enclosures were lined with polythene-backed absorbant paper. The paper was removed 30 min after spraying, bagged and autoclaved off-site. The lack of cross-contamination between plots, or detection of infectious virus in soil samples surrounding the enclosures, attested to the efficacy of this regime.

Larvae and cabbages were recovered from the plots at four times after virus application. The larvae were then reared individually in the laboratory to assess viral mortality.

The results of the field trial showed that both viruses caused a significant reduction in crop damage as compared to the uninfected infested controls (Fig. 4(a) and 5). More importantly, plants from the recombinant virus treatments showed significantly less damage than those from the wild-type

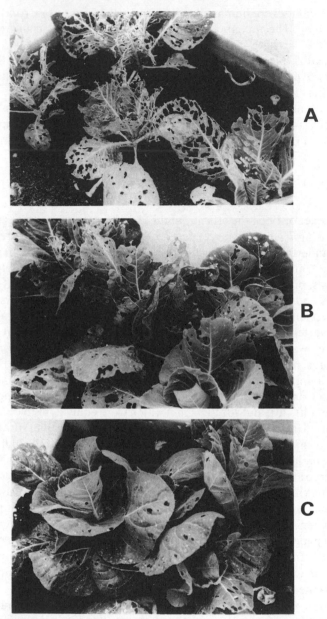

Fig. 4. Damage to cabbage plant leaves in (A) infested control, (B) high dose wild-type AcNPV and (C) high dose recombinant AcST-3 virus treatments. Photographs taken at 16 days post-spray application (Cory *et al.*, 1994, with permission).

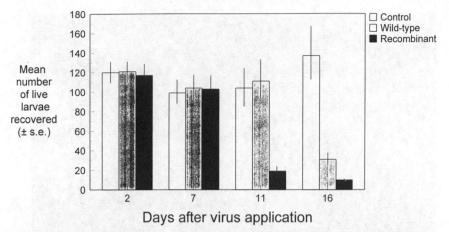

Fig. 5. Mean numbers (± s.e.) of live *T. ni* larvae recovered from infested controls and the high dose virus treatments (at 10^8 polyhedra per m^2) at four times (days after virus application). The control represented untreated larvae, other treatments involved wild-type AcNPV or recombinant AcST-3 virus (Cory *et al.*, 1994, with permission).

virus treated plots (Fig. 4(*b*). This reduction in damage was clearly the result of larvae infected with the recombinant virus dying earlier than those infected with the wild-type virus. This was estimated to be 1.1 days earlier at the first sampling point (2 days after virus application) for larvae infected by equivalent doses, and between 1.3 and 1.7 days for the second sampling time (7 days after virus application). The earlier deaths resulted in the reduced recovery of live larvae in later samples. This effect was most clearly illustrated by the recovery of live larvae from plots which had been treated with the highest virus dose of 10^8 PIBs per m^2 (Fig. 5). There was no difference in the numbers of live *T. ni* larvae recovered on the first two sample dates, but by the third sample, 11 days after virus application, there was a significant fall·in larval recovery from the recombinant virus treated plots compared to both the control and the wild-type virus treatments. However, after a further five days, both virus epizootics appeared to have reached completion, since both showed significant reductions in live larvae compared to the control plots.

Other interesting findings of the trial were the effects both viruses had on their hosts and the consequences on subsequent virus transmission (secondary infections). Larvae infected with the wild-type AcNPV tended to climb up the plants before dying (see Fig. 6(*a*)), after which they lysed *in situ*, providing the opportunity to release virus that could establish infections in any remaining susceptible larvae. However, because the virus expressing the scorpion toxin caused paralysis prior to death, the AcST-3 recombinant infected larvae generally fell off the plants (Fig. 6(*b*)). Also, the earlier deaths of the AcST-3 infected larvae resulted in the production of less virus and reduced breakdown of the larval cadavers in AcST-3 virus infected

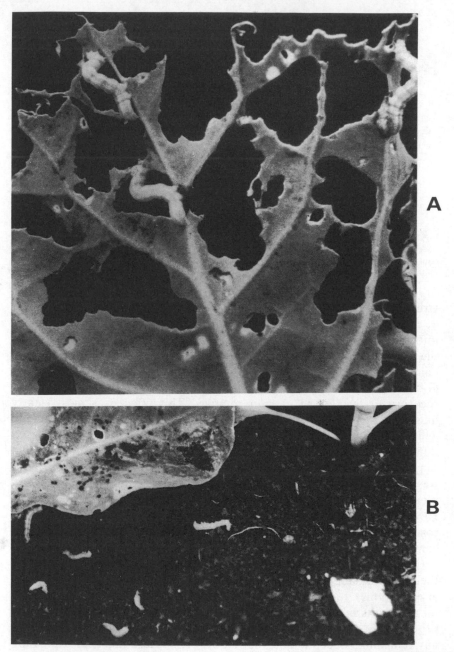

Fig. 6. Positions of foliage located wild-type (A) and soil located recombinant (B) virus infected *T. ni* larvae at death.

insects. Yields from the AcST-3 infections were estimated from polyhedra counts to be only 10–30% of the yields from wild-type AcNPV infections. In the field trial the combination of these effects appeared to produce a reduced secondary wave of infection in the AcST-3 virus treated plots of comparison to the wild-type AcNPV treatment where higher mortalities were evident at the final timepoint. This result, if confirmed by further experiments, and for other recombinants with similar phenotypes, has important implications for risk assessment of genetically modified baculoviruses as it may indicate that such viruses would be less likely to persist in insect populations in Nature.

In summary, the first field trial of a genetically improved virus insecticide encoding an insect-selective scorpion toxin killed plant-feeding larvae more rapidly in the field, resulting in a concomitant reduction in crop damage, as compared to the parent wild-type virus. There was some evidence that in the secondary cycle of infection, mortalities were slightly lower for the recombinant virus-treated insects. However, the data clearly highlight the critical importance of speed of action in achieving reduced crop damage, and showed that the genetically modified AcST-3 virus provided better crop protection than the wild-type AcNPV.

RECOMBINATION BETWEEN BACULOVIRUSES

The transfer of the AaHIT coding region and associated transcription regulatory elements for AcST-3 into another baculovirus might confer improved insecticidal properties to the recipient. This is likely to have little consequence since, as demonstrated above, the host range of the modified species is not changed. The potential for transfer of the viral DNA sequences into other virus species, however, is realistic only for recipients representing other AcNPV strains, or a limited number of other, closely related baculoviruses co-infecting the same cells in the same insect host. Recombination has been demonstrated experimentally between AcNPV and *Galleria mellonella* NPV (Croizier & Quiot, 1981) as well as between AcNPV and *Rachiplusia ou* NPV (Croizier *et al.*, 1988; Smith & Summers, 1980). However, these viruses are considered to be variants of AcNPV, i.e. the same virus species isolated from alternative hosts.

The frequency with which recombination occurs in larvae infected with two strains of a baculovirus is difficult to determine accurately. This is particularly difficult for closely related baculoviruses with similar phenotypes (e.g. host-range) and where characterization of recombinants depends on time-consuming analyses of individual virus genomes using restriction enzymes and electrophoresis. Co-transfecting insect cells with infectious AcNPV DNA and plasmid transfer vectors containing a marker gene (β-galactosidase or polyhedrin) in a context of AcNPV sequences to promote recombination has allowed estimates to be made of recombination frequencies giving values of the order of 1–2% of the total progeny (Kitts,

Ayres & Possee, 1990). In similar experiments, when the AcNPV transfer vector was substituted with one derived from the *M. brassicae* nucleopoly-hedrovirus, which shares only 2% overall sequence similarity with AcNPV, stable recombinant viruses were never detected in the virus progeny (representing less than 1 in 10^9 progeny) although, as expected, there was some evidence for transient gene expression from the MbNPV transfer vector in the AcNPV-infected cells (R. D. Possee, unpublished data). On the basis of these and similar experiments conducted by others (M. Maeda, G. Croizier, personal communication), recombination between different baculovirus species, if it ever occurs in Nature in larvae infected at the same time with different baculovirus species, is likely to be at an extremely low frequency and hence below the limits of detection. This is not a surprising conclusion. Where virus species diverge, their gene products become incompatible, even when the strategies of infection and replication are similar. Examples abound in the literature to support this view, for example, with viruses representing the different genera or serogroups of the Bunyaviridae, or viruses representing different serotypes of influenza virus. While natural recombination between different baculovirus species cannot be excluded, no evidence for ancestral recombination among baculoviruses has yet been obtained. It is likely that the evolution of baculoviruses mostly involves co-evolution with their hosts, as in other parasite–host systems.

In summary, the likelihood of recombination occurring between baculoviruses is restricted not only by the limited opportunities for co-infection of the same cell in the same host but also by the genetic incompatibilities of the species. As with any species, recombination is most likely between members and genetic variants of that species, rather than between members representing different species.

In relation to the question of whether recombination of a genetically modified virus could occur following release, a pertinent question is whether natural variants of the virus exist in Nature. While unambiguous evidence for baculoviruses closely related to AcNPV in the UK is lacking, two observations suggest that strains of the virus are present. Both derive from DNA hybridization and analyses of DNA restriction profiles of viruses in the collection held at the Institute of Virology in Oxford (M. Hirst, D. Goulson, unpublished data). This collection came from the late Kenneth D. Smith FRS and his colleagues and was transferred to Oxford from Cambridge when the Unit of Invertebrate Pathology was established in Oxford University in the late 1960s. Although to-date only a few examples of these viruses have been analysed, and the records of their passage history are incomplete, isolates from *A. gamma* and scarlet tiger moth, *Callimorpha (Panaxia) dominula*, are highly comparable, albeit distinguishable by DNA restriction profile analyses from that of AcNPV. Work on the scarlet tiger moth virus was originally reported by K. D. Smith and colleagues as a new virus in the early 1950s (Smith, 1951; Smith & Wyckoff, 1951), which report antedates the isolation of AcNPV from Irvine, California made in 1967, or the related *Rachiplusia ou* virus (Newland, Indiana) isolated in 1960 (Smith & Summers, 1981). Another isolate of *A. gamma* in the Oxford collection is more distinct from AcNPV, underlining the point made previously about why viruses should not be named after their hosts of origin. While unambiguous evidence for the presence of AcNPV strains in the UK (or elsewhere) would require further field isolations of such viruses, the available data support the view that not only can AcNPV infect species commonly resident in the UK and Europe (see Table 1), but also that it is likely that such viruses are present in the UK.

SUMMARY AND CONCLUSIONS

The results described in this paper support the view that natural and genetically modified baculoviruses may be used in addition to, or as alternatives to chemical insecticides, employed in conjunction with other forms of pest management (integrated pest management programmes) to control insect pests of agriculture as well as other crops. The introduction of an insect-specific scorpion toxin gene into AcNPV provides a marked improvement in the effectiveness of the virus insecticide, at least from the point of view of providing greater protection to the treated crop against pest damage (Cory *et al.*, 1994). The modified virus does not result in any significant alteration to the host range of the virus. Also the recombinant toxin has no effect on small mammals or insect predators of virus-infected larvae. The potential for recombination and transfer of a foreign gene to other distantly related baculoviruses is considered to be potentially extremely small, and most probably limited to related strains of the virus that exist in Nature.

The successful completion of a field trial with a recombinant baculovirus exhibiting improved insecticidal properties has been an essential step in the evaluation of whether such agents offer significant advantages over natural baculoviruses. While to-date the improvements do not provide an insecticide comparable to the available chemical insecticides, where resistance to such chemicals is a problem, the modified insecticide could have a use. In particular, this pertains to the control of the diamondback moth, a major pest of agricultural crops around the world. Further studies are underway to answer questions concerning persistence of the virus in field situations and the long term control of pests on crops, as well as whether the virus is at a competitive advantage or disadvantage in comparison to the wild-type virus in field conditions.

REFERENCES

Ayres, M. D., Howard, S. C., Kuzio, K., Lopez-Ferber & Possee, R. D. (1994). The complete DNA sequence of *Autographa californica* nuclear polyhedrosis virus. *Virology*, **202**, 586–605.

Bishop, D. H. L. (1986). UK release of genetically marked virus. *Nature*, **323**, 496.

Bishop, D. H. L., Entwistle, P. F., Cameron, I. R., Allen, C. J. & Possee, R. D. (1988). Field trials with genetically-engineered baculovirus insecticides. In *The Release of Genetically-engineered Micro-Organisms*, (Sussman, M., Collins, C. H., Skinner, F. A. & Stewart-Tull, D. E., eds.). pp. 143–179. London: Academic Press.

Blissard, G. W. & Rohrmann, G. F. (1990). Baculovirus diversity and molecular biology. *Annual Review of Entomology*, **35**, 127–55.

Cory, J. S., Hirst, M. L., Williams, T., Hails, R. S., Goulson, D., Green, B. M., Carty, T. M., Possee, R. D., Cayley, P. J. & Bishop, D. H. L. (1994). Field trial of a genetically improved baculovirus insecticide. *Nature*, **370**, 138–40.

Croizier, G. & Quiot, J. M. (1981). Obtention and analysis of two genetic recombinants of baculoviruses of Lepidoptera, *Autographa californica* Speyer and *Galleria mellonella* L. *Annals de Virologie*, **132**, 3–18.

Croizier, G., Croizier, L., Quiot, J. M. & Lereclus, D. (1988). Recombination of *Autographa californica* and *Rachiplusia ou* nuclear polyhedrosis viruses in *Galleria mellonella* L. *Journal of General Virology*, **69**, 177–85.

D'Ajello, V., Zlotkin, E., Miranda, F., Lissitzky, S. & Bettini, S. (1973). The effect of scorpion venom and pure toxins on the cockroach nervous system. *Toxicon*, **10**, 399–404.

De Dianous, S., Hoarau, F. & Rochat, H. (1987). Reexamination of the specificity of the scorpion *Androctonus australis* Hector insect toxin toward arthropods. *Toxicon*, **25**, 411–17.

De Dianous, S., Carle, P. R. & Rochat, H. (1988). The effect of the mode of application on the toxicity of *Androctonous australis* Hector insect toxin. *Pesticide Science*, **23**, 35–40.

Entwistle, P. F. & Evans, H. F. (1985). In *Comprehensive Insect Physiology, Biochemistry and Pharmacology*, Vol. 12 (Gilbert, L. I. & Kerkut, G. A., eds.). pp. 347–412, Oxford: Pergamon Press.

Gordon, D., Jover, E., Couraud, F. & Zlotkin, E. (1984). The binding of the insect selective neurotoxin (AaIT) from scorpion venom to locust synaptosomal membranes. *Biochimica et Biophysica Acta*, **778**, 349–58.

Gordon, D., Zlotkin, E. & Catterall, W. A. (1985). The binding of an insect-selective neurotoxin and saxitoxin to insect neuronal membranes. *Biochimica et Biophysica Acta*, **821**, 130–6.

Hammock, B. D., Bonning, B. C., Possee, R. D., Hanzlik, T. N. & Maeda, S. (1990). Expression and effects of the juvenile hormone esterase in a baculovirus vector. *Nature*, **344**, 458–61.

Herrmann, R., Fishman, L. & Zlotkin, E. (1990). The tolerance of lepidopterous larvae to an insect selective neurotoxin. *Insect Biochemistry*, **20**, 625–37.

Hughes, P. R., Gettig, R. R. & McCarthy, W. J. (1983). Comparison of the time-mortality response of *Heliothis zea* to 14 isolates of *Heliothis* nuclear polyhedrosis virus. *Journal of Invertebrate Pathology*, **41**, 246–61.

Kalmakoff, J. & Crawford, A. M. (1982). Enzootic virus control of *Wiseana* spp. in a pasture environment. In *Microbial and Viral Pesticides* (Kurstak, E., ed.). pp. 435–448, New York: Marcel Dekker.

Kitts, P. A., Ayres, D. & Possee, R. D. (1990). Linearization of baculovirus DNA enhances the recovery of recombinant expression vectors. *Nucleic Acids Research*, **18**, 5667–72.

Lester, D., Lazarovici, P., Pelhate, M. & Zlotkin, E. (1982). Purification, characterization and action of 2 insect toxins from the venom of the *Buthotus judaicus*. *Biochimica et Biophysica Acta*, **701**, 370–81.

Loret, E. P., Martin-Eauclaire, M.-F., Mansuelle, P., Sampieri, F., Granier, C. & Rochat, H. (1991). An anti-insect toxin purified from the scorpion Androctonus australis Hector also acts on the α- and β-sites of the mammalian sodium channel sequence and circular dichoism studies. Biochemistry, 30, 633–40.

Moscardi, F. & Sosa-Gomez, D. R. (1992). Use of viruses against soybean caterpillars in Brazil. In Pest Management in Soybean (Copping, L. G., Green, M. B. & Rees, R. T., eds.). pp. 98–109, SCI.

Rathmeyer, M., Walther, C. H. & Zlotkin, E. (1977). The effect of different toxins from scorpion venom on neuromuscular transmission and nerve action potential in the crayfish. Comprehensive Biochemistry and Physiology, 56c, 35–8.

Ruhland, M., Zlotkin, E. & Rathmeyer, W. (1977). The effect of toxins from the venom of the scorpion Androctonus australis on a spider nerve-muscle preparation. Toxicon, 15, 157–60.

Smith, K. M. (1951). The polyhedral diseases of insects. Endeavour, Vol. X, No. 40.

Smith, K. M. & Wyckoff, R. W. G. (1951). Electron microscopy of insect viruses. Research, 4, 148–55.

Smith, G. E. & Summers, M. D. (1980). Restriction map of Rachiplusia ou and Rachiplusia ou-Autographa californica baculovirus recombinants. Journal of Virology, 33, 311–19.

Smith, G. E. & Summers, M. D. (1981). Application of a novel radioimmunoassay to identify baculovirus structural proteins that share interspecies antigenic determinants. Journal of Virology, 39, 125–37.

Smith, G. E., Fraser, M. J. & Summers, M. D. (1983). Molecular engineering of the Autographa californica nuclear polyhedrosis virus genome: deletion mutations within the polyhedrin gene. Journal of Virology, 46, 584–93.

Speight, M. R., Kelly, P. M., Sterling, P. H. & Entwistle, P. F. (1992). Field application of a nuclear polyhedrosis virus against the Brown-tail moth, Euproctis chrysorrhoea (L.). Journal of Applied Entomology, 113, 295–306.

Stewart, L. M. D., Hirst, M., Ferber, M. L., Merryweather, A. T., Cayley, P. J. & Possee, R. D. (1991). Construction of an improved baculovirus insecticide containing an insect-specific toxin gene. Nature, 352, 85–8.

Teitelbaum, Z., Lazarovici, P. & Zlotkin, E. (1979). Selective binding of the scorpion venom insect toxin to insect nervous tissue. Insect Biochemistry, 9, 343–6.

Tintpulver, M., Zerachia, T. & Zlotkin, E. (1976). The actions of toxins derived from scorpion venom on the ileal smooth muscle preparation. Toxicon, 14, 371–7.

Tomalski, M. D. & Miller, L. K. (1991). Insect paralysis by baculovirus-mediated expression of a mite neurotoxin gene. Nature, 352, 82–5.

Van Regenmortel, M. H. V. (1989). Virus species, a much overlooked but essential concept in virus classification. Intervirology, 31, 241–54.

Walther, C., Zlotkin, E. & Rathmeyer, W. (1976). Action of different toxins from the scorpion Androctonus australis on a locust nerve-muscle preparation. Journal of Insect Physiology, 22, 1187–94.

Wood, H. A., Hughes, P. R., Johnston, L. B. & Langridge, W. H. R. (1981). Increased virulence of Autographa californica nuclear polyhedrosis virus by mutagenesis. Journal of Invertebrate Pathology, 38, 236–41.

Zlotkin, E., Kadouri, D., Gordon, D., Palhate, M., Martin, M. F. & Rochat, H. (1985). An excitatory and a depressant insect toxin from scorpion venom both affect sodium conductance and possess a common binding site. Archives of Biochemistry and Biophysics, 240, 877–87.

ONLY 35 YEARS OF ANTIVIRAL NUCLEOSIDE ANALOGUES!

GRAHAM DARBY

Wellcome Research Laboratories, Langley Court, Beckenham, Kent BR3 3BS, UK

INTRODUCTION

We tend to think of antiviral research as a relatively young discipline. Nevertheless, the first description of a small molecule which could block the replication of a virus appeared in the literature almost 50 years ago (Hamre, Bernstein & Donovick, 1950). The molecule was a thiosemicarbazone and it inhibited the replication of viruses from several diverse families. However, whilst being an efficient inhibitor of virus replication, this compound also had undesirable effects on the host cell, a problem which has recurred time and again with antiviral compounds over the years. Viruses multiply inside the cells of their host organism making extensive use of the machinery of the host cell, and the difficulty, therefore, is to find a molecule which will inhibit virus replication without impacting unfavourably on normal cellular processes. Ideally inhibitors should interfere specifically with the function of a protein encoded by the virus and essential for its replication, and it should have no interactions with cellular proteins. This is a goal which has rarely been achieved but, as will be shown later, it is one which clearly can be attained.

The most fertile field for the discovery of antiviral molecules has been the area of nucleoside analogues, and the remainder of this chapter will review the progress made in this exciting area. Both the successes and failures will be discussed, and the chapter will close with an attempt to predict likely future trends. It will not be a comprehensive review but will be an attempt to outline the principles, using as examples those molecules which have achieved a degree of success in the clinic.

EARLY NUCLEOSIDE ANALOGUES

Fifty years ago, as now, one of the most serious unsolved medical problems was that of cancer. A characteristic of this disease is the local uncontrolled growth of cells to generate tumours which ultimately disrupt the functions of the affected organism. It was known at that time that genetic information is both maintained in an organism, and transferred from parent to progeny in DNA molecules. The double helix provided a molecular explanation of heredity, explaining how the informational content of DNA molecules could be replicated and passed down from cell to cell and generation to generation.

DNA is a polymeric molecule made up of four different nucleosides linked through phosphodiester bonds, and it is the sequence of nucleosides in the DNA strand which encodes the genetic information. Prior to cell division the DNA must be replicated, but replicated in such a way that the primary sequence of nucleosides in the polymer is maintained. It was argued that, if the DNA replication process could be disrupted, this would block cell division and therefore arrest tumour growth. Of course, it would also be expected that any other rapidly dividing cells in the body would be inhibited. Attention focused on analogues of nucleosides, the building blocks for DNA synthesis, and in the ensuing years many potential anti-cancer drugs from this class of molecule were synthesized. Thus an extensive library of nucleoside analogues was established, any one of which was potentially capable of blocking DNA synthesis. It was among these potential anti-cancer agents that the earliest antiviral nucleosides were discovered.

The rationale for looking at these molecules was simple. Many viruses have DNA genomes which must be replicated in the infected cell if infectious progeny are to be produced. Any compound which has the potential to inhibit DNA synthesis also therefore has the potential to inhibit the multiplication of DNA viruses. It was this idea which led to many of the early analogues being tested against DNA-containing viruses, particularly poxviruses and herpesviruses. There is less interest today in poxviruses because the scourge of smallpox has been eliminated. However, herpesviruses remain a serious public health problem with genital herpes infection, for example, increasing in most of the developed world. It has been estimated that around 40 million individuals suffer from genital herpes in the USA alone.

One of the first compounds to be tested and shown to be active against DNA viruses was 5-iodo-2'-deoxyuridine (IDU; Prusoff, 1959). It was soon shown to be effective against herpes simplex virus infection in a rabbit eye model, a model which mimics closely the infection in humans, and this led to the compound being licensed in the USA for the treatment of herpes keratitis (Kaufman, 1962; Kaufman, Martola & Dohlman, 1962). The drug is used topically but cannot be used systemically because of the wide range of serious side-effects that it causes. These include bone marrow suppression, nausea, vomiting and loss of hair. In addition, the compound is tumourigenic and teratogenic. It is clear from this catalogue of unfortunate side-effects that IDU is not a selective antiviral. However, it has an important place in history as the first licensed antiviral and the first for which there was a clear demonstration of clinical efficacy.

MODE OF ACTION OF IDU

Since IDU has been known as an antiviral for the past 35 years, it is perhaps surprising that its mode of action remains poorly understood. As described above, it was tested initially against herpesviruses because of its potential to block DNA synthesis. Is there any evidence that this is in fact how it works? The answer is that this is almost certainly not how it works. When IDU penetrates the cell it is phosphorylated by kinases to give successively the monophosphate, diphosphate and finally, triphosphate form. It is the nucleoside triphosphates which are the direct substrates of the DNA-synthesizing enzymes, the polymerases, and IDU-triphosphate is mistaken by the virus (or cellular) polymerase for the natural substrate, thymidine triphosphate. It is incorporated into the DNA in place of the natural substrate.

There have been suggestions that, once incorporated into the DNA, it may slow down DNA synthesis, destabilize the DNA or disrupt the transcription process (Fischer, Chen & Prusoff, 1980; Otto, Lee & Prusoff, 1982). There is little direct evidence for any of these mechanisms, and it is quite possible that there is an, as yet, unrecognized mode of action. However it works, this compound clearly inhibits herpesvirus growth, but it is its lack of selectivity that limits its use. Since selectivity is a key issue with all nucleoside analogues, and it is their effects on the host cell which often limit their usefulness, it is worth considering how these compounds might interfere with normal cellular processes.

THE PROBLEM OF SELECTIVITY

Acute toxicity

As discussed above, nucleoside analogues were initially investigated because of their potential to interfere with cellular DNA synthesis and function and thus inhibit the growth of dividing cells. It was anticipated therefore that they would be cytotoxic. Although chromosomal DNA replication is the major target for many of these compounds, it is by no means the only one. They have the potential to disrupt many other aspects of nucleic acid metabolism such as mitochondrial DNA synthesis, transcription and DNA repair, and they may also interfere at an earlier stage, in purine or pyrimidine metabolism, affecting the synthesis of nucleic acid precursors through either *de novo* or salvage pathways (Fig. 1). Interference with many of these important cellular processes can lead to cytotoxicity. Generally, the analogues must be phosphorylated to some degree, but once phosphorylated there are many potential sites of interference.

Cytotoxicity can be manifested in many different ways when translated to the whole organism, although usually the toxicity observed can be explained by inhibitory effects on rapidly dividing cells. Thus certain organs such as

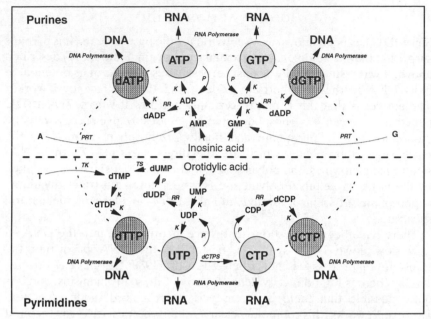

Fig. 1. Enzymic reactions involved in the synthesis of purine and pyrimidine nucleotides and nucleic acids. Potential sites of interference by nucleoside and nucleotide analogues. Abbreviations: K, kinase; P, phosphatase; PRT, phosphoribosyl transferase; RR, ribonucleotide reductase; TK, thymidine kinase; TS, thymidylate synthase.

bone marrow, testes, liver and gut epithelium are particularly sensitive to the effects of cytotoxic compounds.

Chronic toxicity

Although nucleoside analogues may cause acute toxicity in the animal host, because of the potential of many to be incorporated into DNA we must also consider the possibility of longer-term damage to the organism. The danger is that they may be incorporated into mature DNA molecules causing errors in subsequent rounds of DNA replication. Such mutational events may contribute to the development of metastatic disease, or in the embryo they can result in developmental defects. Again, in order to exert these effects, an analogue must be phosphorylated, in this case through the triphosphate.

Interestingly, if we look at the side-effects of systemic exposure to IDU, we see many of the problems outlined above, acute toxicity caused by the cytotoxicity of the compound resulting in bone marrow suppression and loss of hair, and the chronic toxicities of tumourigenicity and teratogenicity.

THE DISCOVERY OF ACYCLOVIR

The real breakthrough in antiviral drug discovery came with the synthesis of a series of acyclic purine nucleoside analogues, and in particular with the synthesis of 9-(2-hydroxyethoxymethyl)guanine (acyclovir or ACV; Elion *et al.*, 1977). This molecule is illustrated in Fig. 2. It was shown by Bauer and colleagues (Schaeffer *et al.*, 1978) to be a potent inhibitor of herpes simplex virus in culture and to have little effect on the replication of uninfected cells. Later work showed that it also inhibits several other human herpesviruses including varicella-zoster, Epstein–Barr and to a lesser extent human cytomegalovirus (Collins, 1983), although it does not inhibit viruses in other families. The compound has become one of the world's most successful drugs. It has now been used to treat in excess of 30 million patients, and despite early justifiable concerns about the wisdom of using a nucleoside analogue to treat non-life-threatening conditions, it has an excellent safety profile. This is illustrated by studies in which the drug has been used to suppress genital herpes in individuals who suffer frequent recurrences of disease. The formal study has completed 5 years (Goldberg *et al.*, 1993), but some individuals have taken oral acyclovir twice daily for 10 years with no apparent serious side-effects.

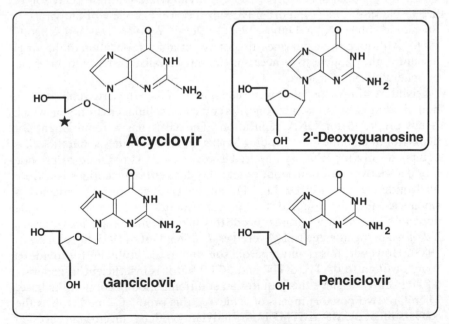

Fig. 2. Structures of acyclic guanosine analogues compared with the natural nucleoside 2'-deoxyguanosine (inset). Bonds shown in bold are those involved in the formation of the sugar–phosphate backbone of DNA. * Indicates the deleted portion of the sugar–phosphate linkage in acyclovir.

In order to understand why this nucleoside is so specific and selective for the virus we need to consider its mode of action.

Mode of action of acyclovir

As we saw earlier, nucleoside analogues require phosphorylation, both to exert antiviral effects and also to exert their undesirable effects on uninfected cells. The remarkable property of acyclovir, which makes a major contribution to its safety profile and lack of side-effects, is that this compound remains unphosphorylated in uninfected cells. There are no cellular kinases capable of recognizing this molecule as a substrate, and so it remains almost entirely in the innocuous form of the free nucleoside. What then is the difference in the infected cell?

Herpesviruses are relatively large and complex, and they encode a set of enzymes that are required for synthesis of new copies of the virus DNA. In addition to the proteins involved directly in DNA replication such as DNA polymerase, DNA helicase, DNase, DNA-binding protein and a protein which binds to the replication origin (Challberg, 1986; Olivo et al., 1989), herpesviruses also encode several enzymes involved in nucleoside and nucleotide metabolism, notably a ribonucleotide reductase (Averett et al., 1983; Dutia, 1983; Spector et al., 1983), thymidylate synthase (Davison & Scott, 1986) and most importantly from the point of view of antivirals, a thymidine kinase (Kit & Dubbs, 1963; Littler et al., 1986; Davison & Scott, 1986). Although these enzymes are not essential for replication of the virus in culture, they appear to be necessary in the specialized environment of the peripheral nervous system.

It is interesting to speculate why this might be. When the virus replicates in a non-dividing cell such as a neurone, it is entering a cellular environment which is not prepared for DNA replication. The virus has a requirement for deoxyribonucleotides, the building blocks for DNA synthesis, but the cell is equipped only for RNA synthesis. However, what is required to provide a bridge between ribonucleotides and the deoxyribonucleotides is a ribonucleotide reductase (see Fig. 1), an enzyme capable of reducing a ribonucleotide diphosphate to the corresponding deoxyribonucleotide derivative. Most herpesviruses encode this enzyme and are able to synthesize it de novo in the infected cell (Averett et al., 1983; Dutia, 1983; Spector et al., 1983). However, this is only a partial solution to the problem. It provides a route through to dATP, dGTP and dCTP but it is insufficient to generate dTTP from UTP since this requires an additional modification of the base. There are two possible means of achieving this product. One involves the conversion of dUMP to dTMP by thymidylate synthase, an enzyme encoded by varicella zoster virus (Davison & Scott, 1986). The other involves induction of the salvage enzyme, thymidine kinase, which enables the use of exogenous thymidine. This is seen with herpes simplex (Kit & Dubbs,

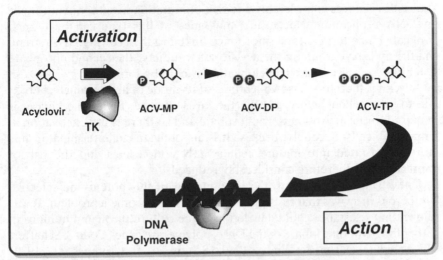

Fig. 3. Mode of action of acyclovir. Phosphorylation to monophosphate is carried out by the virus-coded thymidine kinase (TK). Other activation steps are carried out by cellular enzymes. The target of activated drug is the virus-coded DNA polymerase.

1963), varicella zoster (Davison & Scott, 1986) and Epstein–Barr (Littler *et al.*, 1986) viruses, each of which encodes a new thymidine kinase.

It is the latter enzyme which provides the explanation for the antiviral activity of acyclovir since it can recognize acyclovir as a substrate for phosphorylation (see Fig. 3). Acyclovir is converted to its monophosphate which is then recognized by the cellular guanylate kinase and other cellular kinases and is ultimately converted to its triphosphate (Fyfe *et al.*, 1978). This mode of activation ensures that acyclovir is free from acute toxicities since it is only activated in infected cells. However, this is not the end of the story because the target of acyclovir triphosphate is DNA polymerase and here again acyclovir had characteristics which enhance its safety potential.

In creating the acyclic sugar, a nucleoside analogue was produced which cannot be incorporated into mature DNA molecules (see Fig. 2) since it lacks a hydroxyl moiety equivalent to the 3′ hydroxyl group on the natural sugar. It is, however, incorporated into the nascent DNA strand at its growing point and there causes chain termination (Furman *et al.*, 1979). This is not especially important as far as blocking replication of the virus is concerned since it does not matter whether the compound is incorporated or not into virus DNA as long as the infection is aborted. However, when considering the potential for causing chronic toxicity such as neoplastic disease or teratogenic effects, this becomes an extremely important consideration.

Very low levels of acyclovir triphosphate are generated in uninfected cells, but the failure of the analogue to be incorporated into mature DNA molecules reduces considerably any risk of mutation in subsequent rounds of DNA replication. Nucleoside analogues of this type can be termed 'obligate chain terminators' since, once linked to the growing DNA strand via the hydroxyl group on the acyclic sugar moiety, the second phosphate ester bond required for chain elongation cannot be formed.

Acyclovir therefore has two features of its mode of action which contribute to its excellent safety. Firstly, the virus enzyme, thymidine kinase, is required for its activation to triphosphate and so there is little activation in uninfected cells. Secondly, because it is an 'obligate chain terminator' it is not incorporated into mature cellular DNA molecules and the risk of mutagenesis is therefore considerably reduced.

One of the concerns with the introduction of this potent and selective antiherpes drug was that resistance might rapidly become a problem. It was shown that resistance could develop in tissue culture through mutations in either the thymidine kinase or the DNA polymerase gene (Coen & Schaffer, 1980; Field, Darby & Wildy, 1980; Schnipper & Crumpacker, 1980) although the isolated variants were often somewhat attenuated in animal model systems (Field et al., 1980; Darby, Field & Salisbury, 1981; Larder & Darby, 1985). However, resistance remains rare in treated immunocompetent individuals, although it is seen in a small proportion of immunocompromised patients (see Collins & Darby, 1991).

Thus the picture which emerged with acyclovir clearly represents a successful formula, and it is therefore worth asking whether it is applicable to other systems.

Is the ACV paradigm applicable to other targets?

The difficulty in applying the ACV formula to other virus groups is that a key aspect of the mode of action is the restricted activation of the drug which only occurs in virus-infected cells. The only other human viruses which encode a thymidine kinase gene are the poxviruses and there is little interest in developing antivirals against members of this group. This may change in the future should widespread use of recombinant vaccinia virus vaccines develop.

It is, however, feasible for viruses to encode other enzymes capable of performing the activation function. In fact, this has been shown with another member of the herpesvirus group, cytomegalovirus (CMV). Human CMV is only weakly inhibited by acyclovir but it is more sensitive to the related acyclic nucleoside, 9-[{2-hydroxy-1-(hydroxymethyl)-ethoxy}methyl]-guanine (ganciclovir, or GCV; Field et al., 1983). This virus does not encode thymidine kinase (Chee et al., 1990) and yet studies on drug resistance had suggested that activation of the drug involves a virus-specific event (Biron et al., 1986). Subsequently it was shown that it is the product of the CMV gene UL97 which is responsible for the initial phosphorylation of GCV (Sullivan

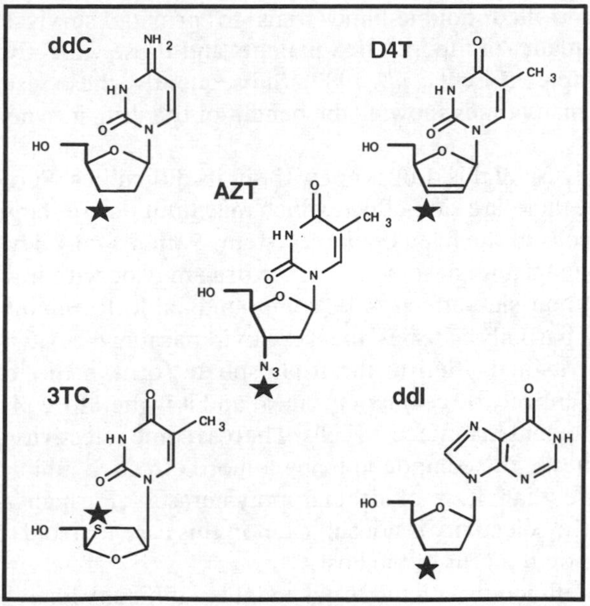

Fig. 4. Dideoxynucleotide inhibitors of HIV. The site of the 3' hydroxyl group in normal nucleosides is indicated by the symbol, *.

et al., 1992; Littler, Stuart & Chee, 1992). Although the normal function of this protein is not known, comparison of the amino acid sequence with those of known protein kinases shows that they share conserved domains (Chee *et al.*, 1990), and this suggests a role in protein phosphorylation for the virus gene product. This illustrates the point that enzymes other than thymidine kinase can phosphorylate nucleoside analogues, and systems should be scrutinized to determine whether they encode such functions.

The second important aspect of acyclovir action is its obligate chain termination, and this is an aspect which has certainly been exploited in other drugs. With the advent of AIDS, the hunt was on for effective anti-retroviral agents, and it was in the area of obligate chain terminators that the early successes were obtained. The first drug shown to be effective for the treatment of AIDS (Fischl *et al.*, 1987) was the nucleoside analogue 3'-azido-2',3'-dideoxythymidine (zidovudine or AZT; see Fig. 4). Although the sugar moiety of this molecule is cyclic, the 3'hydroxyl group, which provides the attachment point for the next nucleotide in the growing DNA chain during reverse transcription of the genome, is absent and in its place is an azido group. Thus this molecule is also an obligate chain terminator (Furman *et al.*, 1986).

AZT acts as an alternative substrate for the viral reverse transcriptase and is incorporated in place of thymidine. It is an extremely potent inhibitor of the virus with an IC_{50} of approximately 10 nM. This drug was shown in early placebo-controlled double-blind trials to provide survival benefit and improved quality of life in AIDS patients and those with advanced AIDS-related complex (Fischl et al., 1987). Subsequently, these results have been confirmed many times showing the benefit of the drug in symptomatic HIV disease.

However, use of this drug is not without its difficulties. Particularly in the severely ill, there are side-effects which may limit its use, largely inhibitory effects on cells of the haemopoietic system. Patients may develop anaemia or neutropaenia and, in some cases, the drug may be withdrawn. These are not uncommon side-effects with compounds able to inhibit cell growth. AZT only partially satisfies the acyclovir paradigm. Although it chain terminates, its activation to the triphosphate form in the infected cell is carried out entirely by cellular enzymes, and it is therefore phosphorylated to a similar extent in uninfected cells. There is some selectivity at the level of DNA synthesis, the compound being a more effective inhibitor of reverse transcriptase than it is of cellular polymerases (Furman et al., 1986). However, the selectivity is not sufficient in this case to avoid some undesirable effects on the cells of the host.

A further difficulty with this drug, and it is a difficulty likely to apply to all specific antiretroviral drugs, is that treatment has to be over a prolonged period and this allows the development of drug-resistant variants (Larder, Darby & Richman, 1989). In view of the development of drug resistance, many drugs are now being tested in combination.

Subsequently, two further analogues, 2',3'-dideoxycytidine (ddC), and 2',3'-dideoxyinosine (ddI) (see Fig. 4; Mitsuya & Broder, 1987), which again are both obligate chain terminators, were introduced for the management of HIV. Unfortunately, these also suffer from a lack of selectivity and cause side-effects in man, namely peripheral neuropathy and pancreatitis. There are other potent dideoxy analogues in development such as 2'-deoxy-3'-thiacytidine (3TC; Belleau et al., 1989) and 2',3'-dideoxy-2',3'-didehydrothymidine (D4T; De Clercq, 1989) and it is hoped that they may provide more selectivity (see Fig. 4).

Because of the requirement for activation, the structural features of nucleosides which can be exploited in antivirals are limited. Those analogues which are not recognized by cellular enzymes can only be useful if there are virus enzymes capable of phosphorylating them. The block in phosphorylation is often at the first stage of activation, conversion to the nucleoside monophosphate, and therefore one way around this is to make nucleoside monophosphate analogues. Several series of compounds of this type have been investigated, most notably the phosphonates which have a modified sugar phosphate linkage (C–P instead of C–O–P), and they have a broad spectrum of antiviral activities. Perhaps the most widely studied of these compounds is the cytidine derivative (5)-9-(3-hydroxy-2-phosphonylmethoxypropyl)-cytosine (HPMPC), which is currently being investigated as a possible treatment for human CMV infection (Flores-Aguilar, Besen & Freeman, 1994). This drug has the advantage of an extremely long half-life *in vivo* which permits infrequent dosing, but its use has been associated with nephrotoxicity. The toxicity can be avoided if probenicid is administered concomitantly (Jaffe, 1994).

Experience with the compounds described above has underscored the importance of both aspects of the mode of action of acyclovir. This is emphasized further by some recent experiences with other antiviral nucleosides.

RECENT PROBLEMS WITH NEWER ANTIVIRAL NUCLEOSIDES

When acyclovir was in its early development, there was great scepticism about the likely utility of nucleoside analogues for the treatment of virus disease. Many felt that a molecule with the potential to disrupt the transfer of genetic information could not have a safety profile appropriate for the treatment of non-life-threatening diseases. We now know that this scepticism was misplaced in the case of acyclovir, but it may be that experience with this drug has lulled us into a false sense of security. Certainly, recent experiences with a number of newer molecules suggests that this may be the case. Two examples are discussed below:

Sorivudine

Sorivudine (E-5-{bromovinyl} arabinofuranosyluracil or BVaraU; Yokota *et al.*, 1989) is the most potent inhibitor of varicella zoster virus (VZV) described to date. In tests in culture it has a large therapeutic window (the ratio of the IC_{50} against the uninfected cell to the antiviral IC_{50}). The compound is activated by the virus thymidine kinase to both its monophosphate and diphosphate forms (Yokota *et al.*, 1989), but the precise mechanism of inhibition of DNA synthesis is not known. Since this molecule is phosphorylated specifically by a viral enzyme, it would not be expected to exhibit adverse effects on the cell. Indeed, it is not the compound itself nor any of its phosphorylated derivatives which have caused problems in the clinic, but rather it is a metabolite formed from its degradation.

All nucleosides and their analogues have the potential to be cleaved by phosphorylases to yield the sugar and the free base. In the case of BVaraU this cleavage is seen to a degree in treated patients and it is one of the products of this cleavage, 5-bromovinyl uracil, which may under some circumstances lead to serious clinical problems.

Following trials in Japan, the drug was approved in that country for use in the treatment of shingles. However, in the months following approval there were several drug-related fatalities. The common thread linking these

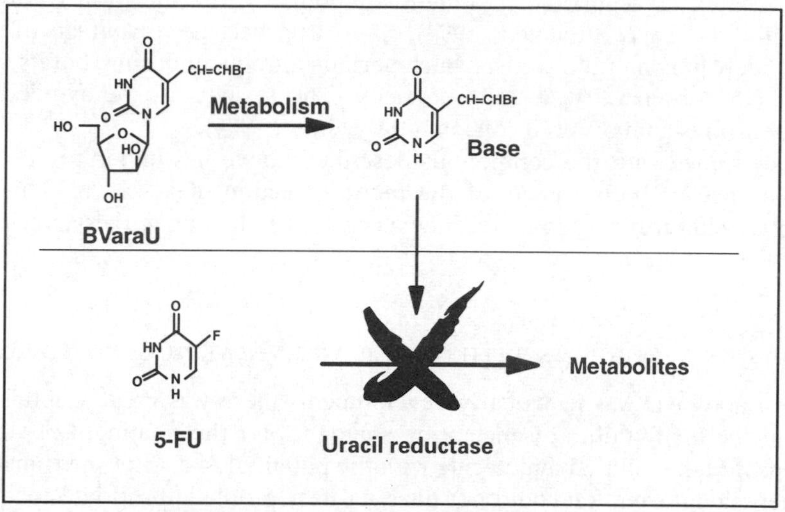

Fig. 5. Proposed mechanism for the interaction between BVaraU and 5-FU *in vivo*. Uracil reductase, which is responsible for the clearance of 5-FU, is inhibited by the base metabolite derived from BVaraU.

patients was that they were all being treated for neoplastic disease with a base analogue, 5-fluorouracil or 5-FU.

In the case of 5-FU, a cytotoxic compound, there is a delicate balance between destroying tumours and producing unacceptable toxicity to the patient. Even in the absence of other drugs, this balance is difficult to achieve because patients have somewhat variable levels of an enzyme, uracil reductase, which is able to metabolize 5-FU and which thus modulates the amount of 5-FU in the plasma. If patients are treated concurrently with BVaraU, it is believed that the metabolite, bromovinyluracil, inhibits uracil reductase and this results in elevated levels of 5-FU in the plasma, sometimes tipping the balance towards lethal consequences (see Fig. 5).

Fialuridine

In another recent unfortunate episode, fialuridine (2'-fluoro-5-iodo-β-D-arabinofuranosyl uracil or FIAU; Watanabe *et al.*, 1979) was responsible for the deaths of several patients enrolled in a trial of this compound against chronic hepatitis B infection (Anon., 1993). This drug had appeared to be satisfactory in pre-clinical studies and also in phase-1 and early phase-2 clinical trials. Unfortunately, the organ affected by the disease was also the major target organ for toxicity, and this may explain why the toxic effects were not recognized earlier. In any case, when patients were given prolonged treatment with the drug many developed a range of serious side-effects and approximately one-third died from liver failure.

FUTURE PROSPECTS FOR NUCLEOSIDE ANALOGUES

There have been numerous attempts to improve on acyclovir through synthesis of analogues of the compound. An example is the compound ganciclovir which was discussed earlier in this chapter. This molecule lacks some of the specificity of acyclovir, but it has proven to be useful in the management of human CMV infection in the immunocompromised patient (Whitley, 1988).

Penciclovir

More recently another acyclic nucleoside, penciclovir (PCV or 9-[4-hydroxy-3-hydroxymethylbut-1-yl]guanine; Boyd *et al.*, 1987), has been licensed in several countries for use against VZV (Fig. 2), or more accurately, the pro-drug of penciclovir, famciclovir (Harnden *et al.*, 1989; Vere-Hodge *et al.*, 1989), has been licensed (Fig. 6). Nucleoside analogues are often poorly bioavailable if given by the oral route, and in the case of penciclovir the bioavailability is so low as to preclude use of the drug orally. This problem was overcome by producing a pro-drug of the compound which, on oral administration, is converted to penciclovir. The pro-drug of penciclovir, famciclovir (FCV) is complex (Fig. 6). The hydroxyl groups on the sugar moiety are protected by acetyl groups which must be removed by

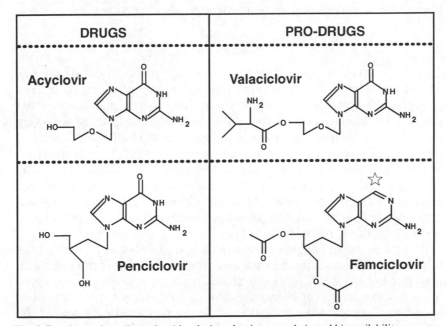

Fig. 6. Pro-drugs of acyclic nucleosides designed to improve their oral bioavailability.

hydrolysis. This is readily achieved in the case of the first hydroxyl, but is less efficient in the case of the second. In addition, the base also must be converted to guanine by the action of xanthine oxidase. Using this pro-drug, good plasma levels of penciclovir can be achieved (Pue & Benet, 1993; Vere-Hodge, 1993).

The mode of action of penciclovir is similar to that of acyclovir in that it is phosphorylated in the infected cell by herpesvirus thymidine kinase and following activation to the triphosphate form it works by shutting down viral DNA synthesis (Earnshaw *et al.*, 1992; Vere-Hodge & Cheng, 1993). However, in contrast to acyclovir it is not an obligate chain terminator because the acyclic sugar moiety has two hydroxyl groups equivalent to the 2'- and 3'-hydroxyl groups of normal sugars (see Fig. 2) and it is therefore possible that it could be incorporated into cellular DNA. In addition, it is now clear that the triphosphate form of penciclovir is a relatively weak inhibitor of HSV (and VZV) DNA polymerase with a K_i value that is 100-fold lower than that of acyclovir. This is compensated for by a long triphosphate half-life in the cell and the achievement of higher levels of PCV-triphosphate (Vere-Hodge & Cheng, 1993).

These factors result in the two molecules having similar antiviral activities against HSV and VZV. In clinical trials where the two drugs have been compared (so far only for zoster treatment) famciclovir has not shown any advantage. Because of its long intra-cellular half-life, dosing with famciclovir can be less frequent than with acyclovir, but it remains to be seen whether famciclovir can match the excellent safety profile of acyclovir.

Valaciclovir

Despite the excellent antiviral profile of acyclovir, its use is limited by its low oral bioavailability. Thus, although impressive clinical results have been achieved with oral acyclovir over the years, there have been indications that the higher plasma levels of the drug which can be achieved following intravenous administration result in improvements in efficacy. In view of this, a pro-drug of acyclovir has recently been developed (Beauchamp *et al.*, 1992). This is the valine ester (valaciclovir; see Fig. 6). When given orally this compound breaks down rapidly and almost quantitatively to produce acyclovir. The only other product made is the natural amino acid L-valine, a component of our normal diet. Thus it is expected that the safety profile of the compound will resemble that of acyclovir (Jacobson, 1993).

It has been shown using this compound that the levels of acyclovir that can be achieved are indeed three to five times higher than those achievable with oral drug (Weller *et al.*, 1991) and clinical trials are in progress to compare the two compounds. Valaciclovir has already been shown to be superior to acyclovir in speeding the resolution of zoster-associated pain, and in the treatment of genital herpes it reduces the severity of the disease as well as

Table 1. *Properties of nucleoside analogues in comparison with acyclovir. The ideal antiviral would be phosphorylated specifically in infected cells and would be an obligate chain terminator. Only acyclovir fulfils both criteria.*

DRUG	Specific Phosphorylation	Obligate Chain Termination	Target Virus
ACV	✓	✓	Herpes
GCV	✓	✗	Herpes
PCV	✓	✗	Herpes
BVaraU	✓	✗	VZV
IDU	✗	✗	Herpes
AZT	✗	✓	HIV
ddI	✗	✓	HIV
ddC	✗	✓	HIV
3TC	✗	✓	HIV/Hep B
FIAU	✗	✗	Hep B

acyclovir but with the added convenience of twice-daily dosing (Crooks & Murray, 1994).

Finally, the work on nucleoside analogues has, to date, been largely hit and miss with little other than intuition guiding the synthetic chemistry. This may be about to change as we develop a better understanding of the interactions between nucleoside analogues and their target enzymes. Considerable progress has been made recently in the area of protein structure and one of the important enzymes to be structured is the HIV reverse transcriptase, the target of several dideoxynucleosides including AZT (Kohlstaedt *et al.*, 1992; Jacobo-Molina *et al.*, 1993). It is expected that a detailed understanding of protein-drug interactions will generate new ideas on the structural characteristics of potentially effective inhibitors. However, because of their complex modes of action involving activation mediated by several different enzymes, it is unlikely that we will ever be able to develop truly rational drug design programmes in this area.

CONCLUSIONS

Early experience with antiviral nucleosides showed that they can have clinical efficacy but they were uniformly cytotoxic compounds producing a variety of unwanted side-effects. The discovery and development of acyclovir showed that not all antiviral nucleosides are cytotoxic, since this compound was shown to be highly selective for the virus, and subsequently it was shown to be an extremely well-tolerated molecule.

There are two features of its mode of action which appear to make major contributions to its safety profile. One is the specific activation of the drug only in virus-infected cells, and the second is its capacity for obligate chain termination which ensures that the molecule cannot be incorporated into mature cellular DNA. Only with ACV do we see both features, and although one or other alone can be sufficient for selectivity, both are preferable. Unfortunately, to date this is the only antiviral to have both characteristics (Table 1).

More recently, there have been further reminders of the potential risks associated with nucleoside analogues in the experiences with BVdU and FIAU. Such molecules have multiple opportunities for interactions with cellular enzymes, and the often dramatic changes in biological properties brought about by seemingly simple chemical modifications mean that there are no 'class effects' with these compounds and each must be judged on its own merits. For this reason, the introduction of valaciclovir, the pro-drug of acyclovir, is an exciting development, since it builds on the established safety profile of acyclovir, incorporating the safety features of the parent molecule.

REFERENCES

Anon. (1993). FIAU trial for hepatitis B: a catastrophe with multiple deaths. *Antiviral Agents Bulletin*, **6**, 193–5.

Averett, D. R., Lubbers, C., Elion, G. B. & Spector, T. (1983). Ribonucleotide reductase induced by herpes simplex type 1 virus. Characterisation of a distinct enzyme. *Journal of Biological Chemistry*, **258**, 9831–8.

Beauchamp, L. M., Orr, G. F., de Miranda, P., Burnette, T. & Krenitsky, T. A. (1992). Amino acid ester prodrugs of acyclovir. *Antiviral Chemistry and Chemotherapy*, **3**, 157–64.

Belleau, B., Dixit, D., Nguyen-Ba, N. & Krus, J. L. (1989). Design and activity of a novel class of nucleoside analogues effective against HIV. *Abstract, 5th International AIDS Conference, Montreal*, p515.

Biron, K. K., Fyfe, J. A., Stanat, S. C., Leslie, L. K., Sorrell, J. B., Lambe, C. U. & Coen, D. M. (1986). A human cytomegalovirus mutant resistant to the nucleoside analog 9{[2-hydroxy-1-(hydroxymethyl)-ethoxy]methyl}guanine (BW B759U) induces reduced levels of BW B759U triphosphate. *Proceedings of the National Academy of Sciences, USA*, **83**, 8769–73.

Boyd, M. R., Bacon, T. H., Sutton, D. & Cole M. (1987). Antiherpes virus activity of 9-(4-hydroxy-3-hydroxymethylbut-1-yl)guanine BRL-39123 in cell culture. *Antimicrobial Agents and Chemotherapy*, **31**, 1238–42.

Challberg, M. D. (1986). A method for identifying the viral genes required for herpesvirus DNA replication. *Proceedings of the National Academy of Sciences, USA*, **83**, 9094–8.

Chee, M. S., Bankier, A. T., Beck, S., Bohni, R., Brown, C. M., Cerny, R., Horsnell, T., Hutchinson, C. A. III, Kouzarides, T., Martignetti, J. A., Preddie,

E., Satchwell, S. C., Tomlinson, P., Weston, K. M. & Barrell, B. G. (1990). Analysis of the protein coding content of the sequence of human cytomegalovirus strain AD169. *Current Topics in Microbiology and Immunology*, **154**, 125–69.

Coen, D. M. & Schaffer, P. A. (1980). Two distinct loci confer resistance to acycloguanosine in herpes simplex virus type 1. *Proceedings of the National Academy of Sciences, USA*, **77**, 2265–9.

Collins, P. (1983). The spectrum of antiviral activities of acyclovir *in vitro* and *in vivo*. *Journal of Antimicrobial Chemotherapy*, **12** (*Suppl. B*), 19–27.

Collins, P. & Darby, G. (1991). Laboratory studies of herpes simplex virus strains resistant to acyclovir. *Reviews in Medical Virology*, **1**, 19–28.

Crooks, R. J. & Murray, A. (1994). Valaciclovir – a review of a promising new antiherpes agent. *Antiviral Chemistry and Chemotherapy*, **5**, (*Suppl. 1*), 31–7.

Darby, G., Field, H. J. & Salisbury, S. A. (1981). Altered substrate specificity of herpes simplex virus thymidine kinase confers acyclovir-resistance. *Nature*, **289**, 81–3.

Davison, A. J. & Scott, J. E. (1986). The complete DNA sequence of varicella zoster virus. *Journal of General Virology*, **67**, 1759–816.

De Clercq, E. (1989). Potential drugs for the treatment of AIDS. *Journal of Antimicrobial Chemotherapy*, **23** (*Suppl. A*), 35–46.

De Clercq, E., Sakuma, T., Baba, M., Pauwels, R., Balzarini, J., Rosenberg, I. & Holy, A. (1987). Antiviral activity of phosphonylmethoxyalkyl derivatives of purine and pyrimidines. *Antiviral Research*, **8**, 261–72.

Dutia, B. M. (1983). Ribonucleotide reductase induced by herpes simplex virus has a virus-specified constituent. *Journal of General Virology*, **64**, 513–21.

Earnshaw, D. L., Bacon, T. H., Darlison, S. J., Edmonds, K., Perkins, R. M. & Vere-Hodge, R. A. (1992). Mode of antiviral action of penciclovir in MRC-5 cells infected with herpes simplex virus type 1 (HSV-1), HSV-2, and varicella-zoster virus. *Antimicrobial Agents and Chemotherapy*, **36**, 2747–57.

Elion, G. B., Furman, P. A., Fyfe, J. A., de Miranda, P., Beauchamp, L. & Schaeffer, H. J. (1977). Selectivity of action of an antiherpetic agent 9-(2-hydroxyethoxymethyl)guanine. *Proceedings of the National Academy of Sciences, USA*, **74**, 5716–720.

Field, A. K., Davies, M. E., DeWitt, C., Perry, H. C., Liou, R. Germershausen, J., Karkas, J. D., Ashton, W. T., Johnston, D. B. and Tolman, R. L. (1983). 9[{2-Hydroxy-1-(hydroxymethyl)ethoxy}methyl]guanine: a selective inhibitor of herpes group virus replication. *Proceedings of the National Academy of Sciences, USA*, **80**, 4139–43.

Field, H. J., Darby, G. & Wildy, P. (1980). Isolation and characterisation of acyclovir-resistant mutants of herpes simplex virus. *Journal of General Virology*, **49**, 115–24.

Fischer, P. H., Chen, M. S. & Prusoff, W. H. (1980). The incorporation of 5-iodo-5′-amino-2′,5-dideoxyuridine and 5-iodo-2′-deoxyuridine into herpes simplex virus DNA. Relationship between antiviral activity and effects on DNA structure. *Biochimica et Biophysica Acta*, **606**, 236–45.

Fischl, M. A., Richman, D. D., Grieco, M. H., Gottlieb, M. S., Volberding, P. A., Laskin, O. L., Leedon, J. M., Groopman, J. E., Mildvan, D., Schooley, R. T., Jackson, G. G., Dirack, D. T., King, D. & the AZT collaborative working group (1987). The efficacy of azidothymidine (AZT) in the treatment of patients with AIDS and AIDS-related complex. *New England Journal of Medicine*, **317**, 185–91.

Flores-Aguilar, M., Besen, G. & Freeman, W. R. (1994). Treatment of viral retinitis with high dose HPMPC liposomes. *Antiviral Research*, **23** (*Suppl. 1*), 42.

Furman, P. A., Fyfe, J. A., St. Clair, M. H., Weinhold, K., Rideout, J. L., Freeman, G. A., Lehrmann, S. N., Bolognesi, D. P., Broder, S., Mitsuya, H. & Barry, D. W. (1986). Phosphorylation of 3'-azido-3'-deoxythymidine and selective interaction of the 5'-triphosphate with human immunodeficiency virus reverse transcriptase. *Proceedings of the National Academy of Sciences, USA*, **83**, 8333–7.

Furman, P. A., St. Clair, M. H., Fyfe, J. A., Rideout, J. L., Keller, P. M. & Elion, G. B. (1979). Inhibition of herpes simplex virus-induced DNA polymerase activity and viral DNA replication by 9-(2-hydroxyethoxymethyl)guanine and its triphosphate. *Journal of Virology*, **32**, 72–7.

Fyfe, J. A., Keller, P. M., Furman, P. A., Miller, R. L. & Elion, G. B. (1978). Thymidine kinase from herpes simplex virus phosphorylates the new antiviral compound, 9-(2-hydroxyethoxymethyl)guanine. *Journal of Biological Chemistry*, **253**, 8721–7.

Goldberg, L. H., Kaufman, R. H., Kurtz, T. O., Conant, M. A., Eron, L. J., Batenhorst, R. L., Boone, G. S. and the Acyclovir Study Group (1993). Continuous five-year treatment of patients with frequently recurring genital herpes simplex virus infection with acyclovir. *Journal of Medical Virology, (Suppl. 1)*, 45–50.

Hamre, D., Bernstein, J. & Donovick, R. (1950). Activity of *p*-aminobenzaldehyde, 3-thiosemicarbazone in the chick embryo and in the mouse. *Proceedings of the Society for Experimental Biology and Medicine*, **73**, 275–8.

Harnden, M. R., Jarvest, R. J., Boyd, M. R., Sutton, D. & Vere-Hodge, R. A. (1989). Prodrugs of the selective antiherpesvirus agent 9-(4-hydroxy-3-[hydroxymethyl]but-1-yl)guanine (BRL 39123) with improved gastrointestinal absorption properties. *Journal of Medicinal Chemistry*, **32**, 1738–43.

Jacobo-Molina, A., Ding, J., Nanni, R. G., Clarke, A. D., Jr, Lu, X., Tantillo, C., Williams, R. L., Kamer, G., Ferris, A. L., Clark, P., Hizi, A., Hughes, S. H. & Arnold, E. (1993). Crystal structure of human immunodeficiency virus type 1 reverse transcriptase complexed with double stranded DNA at 3.0 Å resolution shows bent DNA. *Proceedings of the National Academy of Sciences, USA*, **90**, 6320–4.

Jacobson, M. A. (1993). Valaciclovir (BW256U87): the L-valyl ester of acyclovir. *Journal of Medical Virology, (Suppl. 1)*, 150–3.

Jaffe, H. S. (1994). Clinical trials of (*S*)-1-[3-hydroxy-2-{phosphonylmethoxy}-propyl]cytosine (HPMPC). *Antiviral Research*, **23**, (*Suppl. 1*), 43.

Kaufman, H. E. (1962). Clinical cure of herpes simplex keratitis by 5-iodo-2'-deoxyuridine. *Proceedings of the Society for Experimental Biology and Medicine*, **109**, 251–2.

Kaufman, H. E., Martola, F. & Dohlman, C. (1962). The use of 5-iodo-2'deoxyuridine (IDU) in the treatment of herpes keratitis. *Archive of Ophthalmology (New York)*, **68**, 235–9.

Kit, S. & Dubbs, D. R. (1963). Acquisition of thymidine kinase activity by herpes simplex virus infected mouse fibroblast cells. *Biochemical and Biophysical Research Communications*, **11**, 55–59.

Kohlstaedt, L. A., Wang, J., Friedman, J. M., Rice, P. A. & Steitz, T. A. (1992). Crystal structure at 3.5 Å resolution of HIV-1 reverse transcriptase complexed with an inhibitor. *Science*, **256**, 1783–90.

Larder, B. A. & Darby, G. (1985). Selection and characterisation of acyclovir-resistant herpes simplex virus type 1 mutants inducing altered DNA polymerase activities. *Virology*, **146**, 262–71.

Larder, B. A., Darby, G. & Richman, D. D. (1989). HIV with reduced sensitivity to zidovudine (AZT) isolated during prolonged therapy. *Science*, **243**, 1731–4.

Littler, E., Stuart, A. D. & Chee, M. S. (1992). Human cytomegalovirus UL97 open reading frame encodes a protein that phosphorylates the antiviral nucleoside analogue ganciclovir. *Nature*, **358**, 160–2.

Littler, E., Zeuthen, J., McBride, A. A., Trost Sorensen, E., Powell, K. L., Walsh-Arrand, J. E. & Arrand, J. R. (1986). Identification of an Epstein–Barr virus-coded thymidine kinase. *EMBO Journal*, **5**, 1959–66.

Mitsuya, H. & Broder, S. (1987). Strategies for antiviral therapy in AIDS. *Nature*, **325**, 773–8.

Olivo, P. D., Nelson, N. J. & Challberg, M. D. (1989). Herpes simplex virus type 1 gene products required for DNA replication: identification and overexpression. *Journal of Virology*, **63**, 196–204.

Otto, M. J., Lee, J. J. & Prusoff, W. H. (1982). Effects of nucleoside analogues on the expression of herpes simplex type 1-induced proteins. *Antiviral Research*, **2**, 267–81.

Prusoff, W. H. (1959). Synthesis and biological activities of iododeoxyuridine, an analog of thymidine. *Biochimica et Biophysica Acta*, **32**, 295–6.

Pue, M. A. & Benet, L. Z. (1993). Pharmacokinetics of famciclovir in man. *Antiviral Chemistry and Chemotherapy*, **4**, *(Suppl. 1)*, 47–55.

Schaeffer, H. J., Beauchamp, L., de Miranda, P., Elion, G. B., Bauer, D. J. & Collins, P. (1978). 9-(2-Hydroxyethoxymethyl)guanine activity against viruses of the herpes group. *Nature*, **272**, 583–5.

Schnipper, L. E. & Crumpacker, C. S. (1980). Resistance of herpes simplex virus to acycloguanosine: role of viral thymidine kinase and DNA polymerase loci. *Proceedings of the National Academy of Science, USA*, **77**, 2270–3.

Spector, T., Stonehuerner, J. G., Biron, K. K. & Averett, D. R. (1983). Ribonucleotide reductase induced by varicella zoster virus. Characterisation, and potentiation of acyclovir by its inhibition. *Biochemical Pharmacology*, **36**, 4341–6.

Sullivan, V., Talarico, C. L., Stanat, S. C., Davis, M., Coen, D. M. & Biron, K. K. (1992). A protein kinase homologue controls phosphorylation of ganciclovir in human cytomegalovirus-infected cells. *Nature*, **358**, 162–4.

Vere-Hodge, R. A. (1993). Famciclovir and penciclovir. The mode of action of famciclovir including its conversion to penciclovir. *Antiviral Chemistry and Chemotherapy*, **4**, 67–84.

Vere-Hodge, R. A. and Cheng, Y.-C. (1993). Mode of action of penciclovir. *Antiviral Chemistry and Chemotherapy*, **4**, *(Suppl. 1)*, 13–24.

Vere-Hodge R. A., Sutton, D., Boyd, M. R., Harnden, M. R. & Jarvest, R. L. (1989). Selection of an oral prodrug (BRL 42810; famciclovir) for the antiherpes agent BRL 39123 (0-{4-hydroxy-3-hydroxymethylbut-1-yl}guanine; penciclovir). *Antimicrobial Agents and Chemotherapy*, **33**, 1765–73.

Watanabe, K. A., Reichman, U., Hirota, K., Lopez, C. & Fox, J. J. (1979). Synthesis and antiherpes virus activity of some 2'fluoro-2'-deoxyarabino-furanosyl-pyrimidine nucleosides. *Journal of Medicinal Chemistry*, **22**, 21–4.

Weller, S., Blum, M. R., Doucette, M., Smiley, M. L., Burnette, T. & de Miranda, P. (1991). Multiple dose pharmacokinetics (PK) of 256U87, a new acyclovir (ACV) prodrug in normal volunteers. *Pharmaceutical Research*, **8** (*Suppl.*), 314.

Whitley, R. J. (1988). Ganciclovir – have we established clinical value in the treatment of cytomegalovirus infections? *Annals of Clinical Medicine*, **108**, 452–4.

Yokota, T., Konno, K., Mori, S., Shigeta, S., Kunagai, M., Watanabe, Y. & Machida, H. (1989). Mechanism of selective inhibition of varicella zoster virus replication by 1-β-D-arabinofuranosyl-E-5-(2-bromovinyl)uracil. *Molecular Pharmacology*, **36**, 312–16.

ANTIPROTOZOAL DRUGS: SOME ECHOES, SOME SHADOWS

SIMON L. CROFT

Department of Medical Parasitology, London School of Hygiene and Tropical Medicine, Keppel Street, London WC1E 7HT, UK

INTRODUCTION

Parasitic protozoa remain a major threat to the health of human populations throughout the world. Despite this, there are few effective drugs for the treatment of many protozoal diseases. The chemotherapy of malaria, a disease causing 250 million infections and 2 million deaths per annum, has been undermined by the widespread development of resistance by *Plasmodium falciparum* to chloroquine, as well as quinine, mefloquine and pyrimethamine (Wernsdorfer & Payne, 1991). The first line drugs for human African trypanosomiasis (sleeping sickness), South American trypanosomiasis (Chagas' disease) and leishmaniasis, are arsenical, nitroheterocyclic and antimonial compounds, respectively (De Castro, 1993; Olliaro & Bryceson, 1993; Van Nieuwenhove, 1992). These therapies have variable efficacy, toxic side effects and require long courses of administration. Opportunistic protozoa, some unknown in humans only a decade ago, have emerged as a major cause of mortality and morbidity in immunocompromised patients. No drugs have been identified for the treatment of cryptosporidiosis caused by *Cryptosporidium parvum*, and treatments for microsporidiosis and some forms of amoebiasis are limited. Some drugs are available for the treatment of toxoplasmosis and others are on clinical trial (Araujo & Remington, 1992); however, the main problem with this infection is that maintenance therapy is required throughout the period of immunosuppression with relapses and side effects occurring commonly for the standard pyrimethamine/sulphadiazine combination. Appropriate drugs for protozoa of veterinary importance remain effective in the Western countries although development of resistance, for example, of coccidia to the ionophore drugs, is a frequent occurrence (Chapman, 1993). In Africa, the control of cattle trypanosomiasis (nagana) is threatened by the spread of resistance to the phenanthridinium drug, isometamidium chloride and the diamidine drug, diminazene aceturate (Peregrine & Mamman, 1993).

Protozoa display a remarkable diversity of morphology and of life cycles as well as showing enormous genomic and biochemical variation within and between species and this taxonomic grouping does not imply a close evolutionary relationship (Sogin *et al.*, 1989). Some protozoa, for example,

Plasmodium and *Trypanosoma cruzi*, have more than one stage of their life cycle in the mammalian host, and these stages have different sites of development, different biochemistry and, not surprisingly, different drug sensitivities. Protozoan parasites have adopted a wide range of strategies to adapt to the immunologically hostile but nourishing environment of the mammalian host. Biochemical and molecular studies of the past decade have elucidated many metabolic differences between protozoa and mammals, which offer potential targets for new and improved chemotherapeutics (Coombs & North, 1991). In addition, adaptation to life in the mammal appears to have led to a simplification of many biochemical pathways as parasites evolved to exploit and become more reliant on metabolites from the host; such pathways are obvious targets for chemotherapeutic intervention (Fairlamb, 1989).

In this review I will focus on drugs for human pathogenic protozoa, in particular on current areas of development and themes important in the search for more effective drugs. A discussion of all antiprotozoal drugs in use (Table 1) and biochemical targets is beyond the scope of this review. The excellent compendium of Coombs & North (1991) covers many aspects of protozoan biochemistry germane to this review and Campbell & Rew (1986) provide a general text that lists the uses and structures of standard antiprotozoal drugs. Sneader (1985) provides a lively account of the history of antiprotozoal chemotherapy. Before discussing the different groups of drugs, it is worth emphasizing that for most of the past century antiprotozoal drugs have been used without any idea of their mechanism of action or of their biochemical targets. The sections within this review therefore reflect a history of empirical and structure–activity studies as well as the current search for specific metabolic inhibitors; there is no neat classification for these drugs.

QUINOLINES AND TRICYCLIC COMPOUNDS

Paul Ehrlich provided an early rational basis for antiprotozoal chemotherapy when, in 1891, he and Guttmann showed that the dye methylene blue, known to stain the blood stages of the recently discovered *Plasmodium* parasites at concentrations that did not affect leukocytes, could be used to cure human malaria. It took 30 years to progress to identifying more active and less toxic antimalarials because models of *Plasmodium* infection were not available for testing new drugs. When an avian model of malaria became available, researchers in Germany were able to increase the antimalarial activity of methylene blue by the addition of a basic diethylaminoethyl sidechain and then to remove the dye properties by the use of a quinoline nucleus. This approach produced the first synthetic antimalarial, pamaquine (plasmoquine), in 1925, a compound active against the exoerythrocytic liver stages. Further work soon led to primaquine.

Table 1. *Drugs recommended for treatment of protozoal infections*[a]

Disease	Parasite	Drugs	Class
Malaria	*Plasmodium falciparum*	quinine, quinidine[b]	Cinchona alkaloids
		chloroquine	4-aminoquinoline
		mefloquine	4-quinoline meth-anol
		halofantrine[b]	9-phenanthrene methanol
		tetracycline	tetracycline
		clindamycin	lincosamide
		proguanil	biguanide
		pyrimethamine	diaminopyrimidine
		sulphadoxine	sulphonamide
	Plasmodium vivax	chloroquine	4-aminoquinoline
		primaquine[b]	8-aminoquinoline
Toxoplasmosis	*Toxoplasma gondii*	pyrimethamine	diaminopyrimidine
		sulphadiazine	sulphonamide
		spiramycin	macrolide
Leishmaniasis	*Leishmania donovani, L. major, L. braziliensis, L. tropica*, etc.	sodium stibogluco-nate	pentavalent antimo-nials
		meglumine antimo-nate	
		pentamidine	diamidine
		amphotericin B	polyene antibiotic
South American Trypanosomiasis (Chagas' disease)	*Trypanosoma cruzi*	benznidazole	nitroimidazole
		nifurtimox	nitrofuran
Human African Trypanosomiasis (sleeping sicke-ness)	*T.b. gambiense*	melarsoprol	trivalent arsenical
		eflornithine	amine analogue
		pentamidine	diamidine
	T.b. rhodesiense	melarsoprol	trivalent arsenical
		suramin	sulphated naphthy-lamine
Amoebiasis	*Entamoeba histolytica*	metronidazole	nitroimidazole
		tinidazole	nitroimidazole
		paromomycin	aminoglycoside
		diloxanide	acetamide
		iodoquinol	hydroxyquinoline
	Naegleria sp.	amphotericin B	polyene antibiotic
	Acanthamoeba (ocular)	propamidine	diamidine[c]
		polyhexamethyline-guanide	biguanide[c]
Giardiasis	*Giardia intestinalis*	metronidazole	nitroimidazoles
		tinidazole	
		mepacrine	acridine
		albendazole	benzimidazole
Cryptosporidiosis	*Cryptosporidium parvum*	?	
Trichomoniasis	*Trichomonas vaginalis*	metronidazole	nitroimidazoles
		tinidazole	
Microsporidiosis	*Septata intestinalis Enterocytozoon bieneusi*	albendazole	benzimidazole
	Encephalitozoon hellem	?	

[a] This list does not include all human protozoal infections, nor does it include all drugs on trial or used in different regions of the world.
[b] Therapeutic use. All other antimalarials can be used as prophylactics.
[c] Drugs reported to be effective in limited trials.

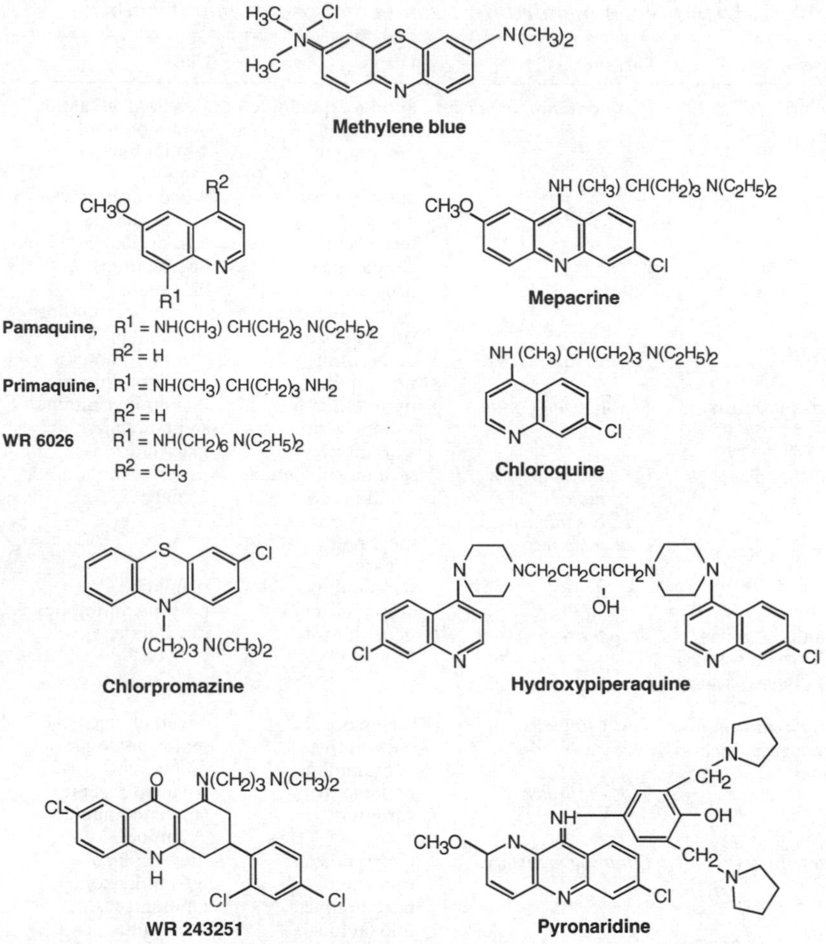

Fig. 1. Quinolines and other synthetic tricyclic antiprotozoals.

In the subsequent search to find compounds active against the erythrocytic stages of malaria, mepacrine was synthesized in 1932 by replacing the quinoline nucleus with acridine, retaining the diethylaminoethyl group on the 9-position, and adding chlorine to the 3-position (Fig. 1). Mepacrine soon replaced quinine as the main antimalarial drug. Subsequently, the removal of a benzene ring from mepacrine led to the 4-aminoquinolines; the most notable being chloroquine, which although first synthesized in 1934, only came into widespread use after 1945.

There are three reasons for this historical digression. First, chloroquine (despite current resistance problems) and primaquine (despite toxicity problems) are still widely used for the treatment of malaria. Although surpassed by chloroquine, in its day mepacrine was considered to be better than quinine for malaria, and it is still recommended for the treatment of giardiasis and cutaneous leishmaniasis. Secondly, the structures essential to these activities remain a starting point for the design and synthesis of new drugs. Bis-quinolines have been synthesized in an attempt to circumvent the problems of chloroquine resistance. For example, piperaquine and hydroxypiperaquine (Fig. 1), in which a pair of piperazine rings bridge two quinoline nuclei, showed good activity in experimental models, and hydroxypiperaquine has been reported to cure chloroquine resistant malaria from patients in China (see Slater, 1993). A series of alkane-bridged bisquinolines have also been found to be highly active against chloroquine-resistant *Plasmodium* in experimental models (Vennerstrom *et al.*, 1992). Other research groups in China started from the acridine nucleus and developed pyronaridine (Fig. 1) and pyracrine, which have been used in that country for the treatment of both *P. falciparum* and *P. vivax* malaria as well as of multidrug resistant cases (Fu & Xiao, 1991). Other acridine derivatives synthesized for malaria include floxacrine, which was not effective enough for clinical studies (Schmidt, 1979) and the more promising WR 243251 (Fig. 1) (Berman *et al.*, 1994). A series of 9-analino-acridines is active against *Plasmodium*, *Leishmania* and *Trypanosoma* (Figgitt *et al.*, 1992; Mauel *et al.*, 1993). Thirdly, quinoline and acridine structures have appeared in active leads from screening programmes. Over the past 30 years, two 8-aminoquinolines have emerged from extensive studies by the Walter Reed Army Institute for Research; WR 6026 (Fig. 1) is on clinical trial for visceral leishmaniasis and WR 238,605 for *P. vivax* malaria (Peters, Robinson & Milhous, 1993). From 300,000 compounds screened in the search for antimalarials between 1963 and 1986 (Peters, 1987) mefloquine, a quinoline methanol with many of the characteristics of quinine, and halofantrine, a phenanthrene methanol, emerged (see Pudney, this volume). In natural product screens, acridone alkaloids, for example atalphillinine, have been identified which have *in vitro* and *in vivo* activity against *Plasmodium* (Basco *et al.*, 1994).

Phenothiazine, synthesized in the late nineteenth century with aniline dyes like methylene blue, was used as an anthelmintic, and antihistaminic and neuroleptic derivatives have shown concordant antiprotozoal activity. Chlorpromazine (Fig. 1) and other phenothiazines have activity against *Plasmodium* (Kristiansen & Jepsen, 1985), *Leishmania donovani* (Pearson *et al.*, 1982), *Trypanosoma brucei* (Seebeck & Gehr, 1983) and *Trypanosoma cruzi* (De Castro, 1993). Although the mechanisms of action have been related to calmodulin function, microtubule binding and inhibition of phosphodiesterases, it seems more likely that the antiprotozoal activity of phenothiazines results from nonspecific effects common to amphiphilic cationic drugs (Hammond, Hogg & Gutteridge, 1985), predictable from the lipophilic and steric components of their structures (Evans & Croft, 1994). Tricyclic antidepressant drugs, for example, clomipramine and amitriptyline, are structurally similar to phenothiazines but are stereochemically distinct and derived from the iminodibenzyl ring. These drugs also have concordant activity against trypanosomes and leishmanias (Hammond *et al.*, 1985; Zilberstein, Liveanu & Gepstein, 1990). Although this activity has been related to the specific inhibition of trypanothione reductase (Benson *et al.*, 1992) and ATPase proton pumps(Zilberstein & Dwyer, 1984) non-specific effects are again more probable (Hammond *et al.*, 1985; Evans & Croft, 1994). The tricyclic drugs imipramine and amitryptiline also have anti-*P. falciparum* activity, possibly through the inhibition of riboflavin metabolism (Dutta, Pinto & Rivlin, 1990). In addition, desipramine reverses chloroquine resistance in *P. falciparum* (Bitonti *et al.*, 1988).

NITROHETEROCYCLIC COMPOUNDS

Five-membered nitroheterocyclic compounds have also made a major contribution to antiprotozoal chemotherapy. Antiprotozoal nitrofurans were first developed in the 1950s and nitrofurazone and furazolidone have been used widely against these parasites (see Townson *et al.*, 1994). One nitrofuran derivative, nifurtimox (Fig. 2), although no longer manufactured, has been used extensively since 1976 for the treatment of Chagas' disease; one of the only two recommended drugs. Both nitrofurazone and nifurtimox have also been used to treat late stage African trypanosomiasis and the latter is still recommended (Van Nieuwenhove, 1992). Nifurtimox is believed to be active against trypanosomes through the generation of nitroanion radicals (see Townson *et al.*, 1994). More recently, nitrofuran derivatives have been used in a rational approach to design 'subversive' substrates for the unique trypanosomatid enzyme trypanothione reductase (TR); derivatives with basic side chains inhibited TR, underwent a one-electron reduction and were trypanocidal (Henderson *et al.*, 1988).

The screening of synthetic derivatives of a 2-nitroimidazole antibiotic, azomycin isolated in 1953, led to metronidazole and other 5-nitroimidazoles (Fig. 2) which remain the first line drugs for the treatment of trichomoniasis, giardiasis, amoebiasis and balantidiasis caused by anaerobic protozoa (see Townson *et al.*, 1994). The antiprotozoal activity of metronidazole and related compounds is dependent upon their reduction to toxic unstable reactive intermediates by reduced ferridoxin on specific anaerobic pathways of carbohydrate metabolism. This pathway is found in either a special organelle, the hydrogenosome, in *Trichomonas vaginalis* or in the cytoplasm of *Giardia* and *Entamoeba*. There have been frequent reports of resistance of *T. vaginalis* to metronidazole and cross-resistance to other nitroimidazoles in this parasite and also in *Giardia intestinalis* (Meingassner & Thurner, 1979; Boreham, Phillips & Shepherd, 1988). Although Quon, D'Oliveira & Johnson (1992) reported reduced transcription and expression of ferridoxin in resistant organisms, resistance is probably dependent upon a

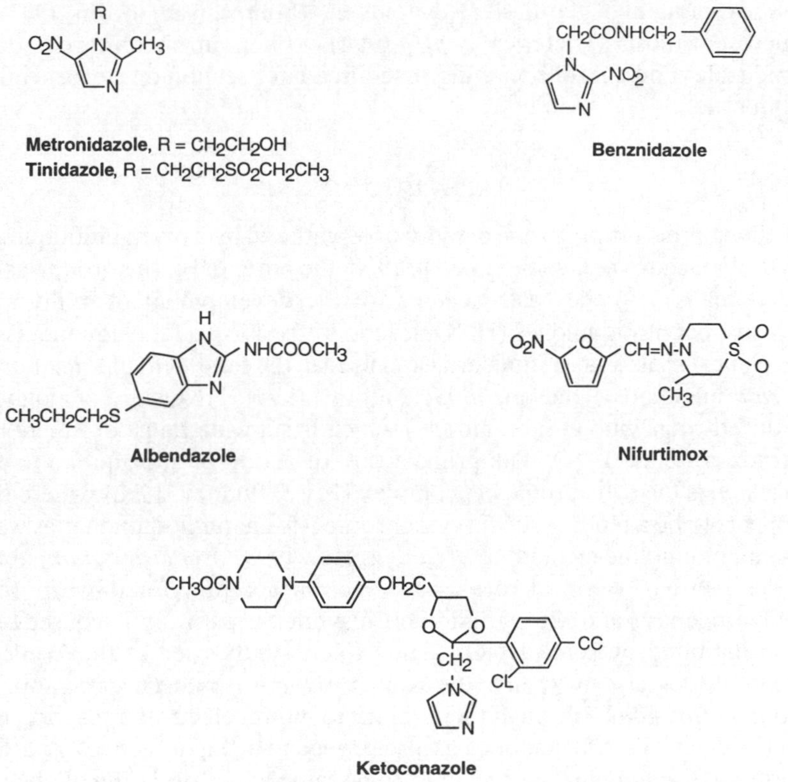

Metronidazole, R = CH_2CH_2OH
Tinidazole, R = $CH_2CH_2SO_2CH_2CH_3$

Benznidazole

Albendazole

Nifurtimox

Ketoconazole

Fig. 2. Nitroheterocyclic antiprotozoals.

variety of mechanisms (Townson et al., 1994). New 5-nitroimidazole derivatives have been tested, and surprisingly, two compounds with β-lactam substitution at the 2-position on the nitroimidazole were 100- and 50-fold more active than metronidazole against T. vaginalis and Entamoeba histolytica, respectively (Vanelle et al., 1991). Nitroimidazoles are also important in the treatment of Chagas' disease. Benznidazole (Fig. 2) was introduced for treatment in 1979 and remains at the forefront of treatment despite its variable efficacy in the treatment of chronic infections and toxicity. Although other nitroimidazoles have shown good experimental activity against T. cruzi, for example, megazole (CL-64,855) and MK-436 (see De Castro, 1993), and against T. brucei CNS infections in the mouse model (Jennings, 1993), development of new drugs of this class is unlikely owing to toxicity problems.

Another group of nitroheterocyclic compounds, the benzimidazoles, were developed principally as anthelmintics. These compounds disrupt microtubule function by the inhibition of the polymerization of tubulin subunits. One benzimidazole, albendazole (Fig. 2) is proving to be effective in the treatment of giardiasis (Reynoldson, Thompson & Horton, 1992) and microsporidiosis (Dieterich *et al.*, 1994). Other nitroheterocyclic drugs, imidazoles and triazoles, are discussed in a later section on sterol synthesis inhibitors.

NAPHTHOQUINONES

Although the antiprotozoal activity of a synthetic hydroxynaphthoquinone, hydrolapachol was first identified in 1938, the potential of this group has only been realised in the past decade with the development of orally active hydroxynaphthoquinones (HNQs). One of the HNQs, atovaquone, is now on clinical trials as an antimalarial and used in the treatment of *Pneumocystis carinii* infections (Hudson, 1993; Pudney, 1995). HNQs are analogues of reduced coenzyme Q (ubiquinone) which has a wide range of functions in protozoa (Ellis, 1994). The primary site of action of atovaquone in *Plasmodium* is the cytochrome bc_1 complex (Fry & Pudney, 1992), where there appears to be a binding site on cytochrome b for naphthoquinones as well as 8-aminoquinoline metabolites (Vaidya *et al.*, 1993). In *Plasmodium*, ubiquinone also has a critical role as an electron acceptor for dihydroorotate dehydrogenase and the pyrimidine biosynthetic pathway is consequently also inhibited by some HNQs. The HNQs synthesized at the Wellcome Laboratories also have activity against *Eimeria* spp., *Theileria* spp. and *Toxoplasma gondii* through their ability to inhibit electron transport (Hudson *et al.*, 1985). Atovaquone has also proved to be highly active against cyst forms of *T. gondii* in the brain of mice (Araujo *et al.*, 1992) and has been on trial for toxoplasmosis in AIDS patients (Kovacs *et al.*, 1992). Two other HNQs, parvaquone and buparvaquone, have been developed for the treatment of theileriosis in cattle and other domestic animals (McHardy, 1992) and buparvaquone is also highly active against *L. donovani in vitro* (Croft *et al.*, 1992).

NATURAL PRODUCTS

Plant products have many apocryphal uses in traditional medicine for the treatment of parasitic diseases and some products and their derivatives continue to make a major impact on antiprotozoal chemotherapy. Quinine and emetine were isolated, respectively, from the bark of *Cinchona* and the rhizome of *Cephaelis ipecacuanha* in the 1820s, and are still used in the treatment of malaria and amoebiasis. The renewed interest in plant products (Phillipson & Wright, 1991; Iwu, Jackson & Schuster, 1994) has been stimulated in part by the identification in the 1970s of the antimalarial activity of the sesquiterpene lactone peroxide artemisinin (Qinghaosu) (Fig. 3), isolated from *Artemisia annua*, which was used in traditional Chinese medicine for the treatment of fevers and malaria. Artemisinin has

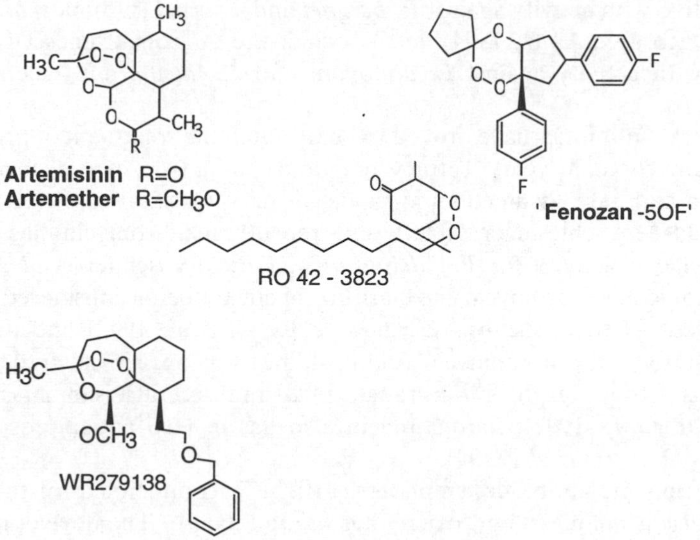

Fig. 3. Artemisinin and synthetic derivatives containing an endoperoxide moiety.

proved to be a rapidly acting blood schizonticide that is effective against strains of *P. falciparum* resistant to standard drugs. Since the isolation of artemisinin in the 1970s and the synthesis of semisynthetic derivatives, sodium artesunate and artemether with improved bioavailability, over one million people have been treated with these compounds (White, 1994). Another derivative, artether is also under development (Davidson, 1994). The identification of the novel 1, 2, 4-trioxane ring of artemisinin with an endoperoxide bridge which is essential for antimalarial activity, has prompted synthetic studies. Tricyclic trioxane derivatives, for example, WR 279138 (Posner *et al.*, 1992) and cyclopenteno-trioxanes, for example, Fenozan-50F (Peters *et al.*, 1992), have shown promising experimental activity. Another sesquiterpene lactone, yingzhaosu, isolated from *Artabotrys uncinatus* and synthetic derivatives of this bicyclic peroxide, Ro 42-3823 (Fig. 3) and Ro 42-1611 also have marked antimalarial activity (Stohler, Jacquet & Peters, 1988). Artemisinin and trioxanes have also shown activity against intracellular, although not extracellular, *Toxoplasma gondii* (Chang, Jefford & Pechere, 1989; Ou-Yang *et al.*, 1990) and *Leishmania major* (Yang & Liew, 1993) but not against *L. donovani* (Croft, unpublished data). These compounds appear to be activated by iron and haem to generate free radicals which act as alkylating agents (Meshnick, 1994). However, recent reports of the neurotoxicity of these compounds are of concern (Brewer *et al.*, 1994).

Other plant derived compounds with notable antiprotozoal activity include the naphthoquinones, plumbagin and β-lapachone that have antileishmanial and antitrypanosomal activity (Iwu *et al.*, 1994), ajoene a garlic derivative with activity against *P. berghei* and *T. cruzi* (Urbina *et al.*, 1991; Perez, Rosa & Apitz, 1994) and lipochalcone A from Chinese liquorice roots with activity against *L. donovani* and *P. falciparum* (Chen *et al.*, 1994*a, b*).

Several antibiotics have proved to be useful in the treatment of protozoal infections for many years. Tetracyclines and clindamycin have been used for malaria and have assumed importance for drug resistant cases (Krogstad, Herwaldt & Schlesinger, 1988) and recently azithromycin has shown prophylactic activity for *P. falciparum* malaria (Kuschner *et al.*, 1994). Clindamycin, azithromycin and clarithromycin are being considered for the treatment of toxoplasmosis (Araujo & Remington, 1992) and different formulations of paromomycin (aminosidine) are on clinical trial for cutaneous (El-On, Jacobs & Weinrauch, 1988) and visceral leishmaniasis (VL) (Thakur *et al.*, 1992). Paromomycin also has *in vivo* anticryptosporidial activity (Verdon *et al.*, 1994).

The polyene antibiotic, amphotericin B, is recommended for the treatment of leishmaniasis but toxicity has limited its use. The development of amphotericin B-lipid formulations for the treatment of systemic mycoses in immunocompromised patients has renewed interest in the use of this polyene for visceral leishmaniasis (VL). The lipid formulations, which have greatly reduced toxicity over the free drug, may also improve activity through drug targeting as the *L. donovani* parasites which cause visceral leishmaniasis live in the macrophages of the liver, spleen and bone marrow, cells that remove drug carriers from the circulation after parenteral administration. One liposome formulation, 'AmBisome', showed a four-fold improvement over free drug against *L. donovani* in mice (Croft *et al.*, 1991) and has now been used to treat over 30 cases of VL, including some resistant to antimonials (Davidson *et al.*, 1994), although some VL patients with HIV co-infection relapsed. Two other formulations have been evaluated: an amphotericin colloidal dispersion (Amphocil) was effective against visceral leishmaniasis in Brazil (Dietze *et al.*, 1993), while an amphotericin B lipid complex was not effective against mucocutaneous leishmaniasis (Llanos-Cuentas *et al.*, 1991).

ARSENICALS AND ANTIMONIALS

The organic arsenical and antimonial compounds are currently the recommended drugs for the treatment of human African trypanosomiasis and leishmaniasis (Table 1). The trivalent arsenical melarsoprol, introduced in 1949, has for many years been used for the treatment of late stage CNS infections of *T. brucei gambiense* and *T. brucei rhodesiense*. Reactive encephalopathy is a major side effect in up to 10% of patients treated with this drug. In the absence of alternative drugs, attempts have been made to find dose regimens that improve efficacy and reduce toxicity. The recent development of assays for melarsoprol has enabled the pharmacokinetics of this drug to be analysed in patients in West Africa and, consequently, a new dose schedule using single doses for 10 consecutive days has been recommended (Burri et al., 1993). The causes of the encephalopathy are disputed (Jennings, 1993). Approaches to reduce the severity of this inflammatory side effect have been to use co-administration of prednisolone in patients (Pepin et al., 1989) and azathioprine in mouse models of the CNS infection (Jennings, 1993). It is unlikely that two recently synthesized and highly effective arsenicals, melarsen cysteamine, which is being used to treat T. evansi in camels and IMOL 881 (Maes, Songa & Hamers, 1993), will be used in the treatment of humans.

Some advances have also been made in understanding the mode of action of arsenical drugs. Trypanothione (N^1, N^8-bis(glutathionyl)spermidine, the trypanosomatid equivalent of glutathione, binds to melarsoprol, and the thioarsane adduct formed by these two molecules prevents trypanothione from performing its cellular function and is a competitive inhibitor of trypanothione reductase (Fairlamb, Henderson & Cerami, 1989).

The use of pentavalent antimonials for the treatment of leishmaniasis has also benefited from a reappraisal of dose regimens. This followed pharmacokinetics studies which showed that 80% of intramuscular injections was excreted in 6 hours (Rees et al., 1980). Increased and more frequent doses are now recommended for both visceral and cutaneous forms of the disease (Olliaro & Bryceson, 1993). In another approach, liposomal encapsulation of the standard antimonials increased activity 200- to 700-fold in rodent models of visceral leishmaniasis through targeting to the infected cells and reduced antimony excretion. Unfortunately, development of a commercial liposomal-sodium stibogluconate formulation (Croft, Neal & Rao, 1989) was halted owing to toxicity problems.

ANTI-FOLATES

Folate derivatives are essential donors in purine and pyrimidine biosynthesis, in particular the transfer of a methyl group from tetrahydrofolate to uridylate to form thymidylate in the pyrimidine pathway. Protozoan parasites synthesize folate co-factors from pteridine, glutamate and p-aminobenzoic acid. Two enzymes on the biosynthetic pathway, dihydropteroate synthetase (DPS) and dihydrofolate reductase (DHFR) have proved to be important targets: DPS for sulphomanides and sulphones and DHFR for pyrimidine analogues. The biguanide proguanil and its triazene metabolite cycloguanil (Fig. 4), were originally developed from pyrimidine analogues and shown to be useful antimalarials in the 1940s. Subsequently, folic acid analogues with a structural similarity to cycloguanil were tested and the diaminopyrimidine pyrimethamine emerged as the most potent antimalarial (Fig. 4). For many years these two drugs, both specific DHFR inhibitors,

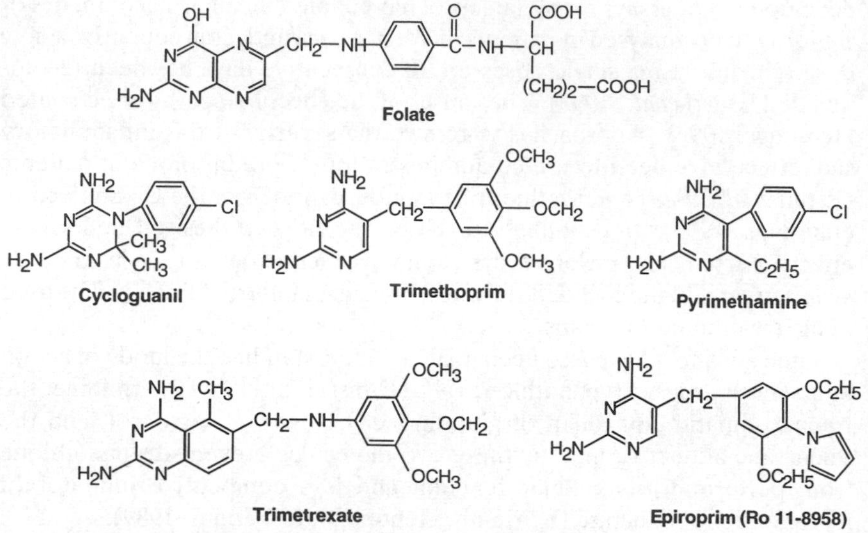

Fig. 4. Antiprotozoal dihydrofolate reductase inhibitors.

used either in combination with chloroquine in the case of proguanil and sulphonamides or dapsone in the case of pyrimethamine, have been the backbone of malaria prophylaxis. The combination of pyrimethamine and sulphadoxine (Fansidar) is also an important therapeutic for chloroquine-resistant malaria. Resistance to pyrimethamine and proguanil is now wide-spread (Wernsdorfer & Payne, 1991) owing to specific amino acid point mutations in *P. falciparum* DHFR. These two drugs are often not cross-resistant as resistance may be conferred by different point mutations on DHFR (Peterson, Milhous & Wellems, 1990).

   The design of new DHFR inhibitors has been greatly aided by studies that revealed that the structure of DHFR in protozoa is different from that in other eukaryotes and bacteria, being present as a bifunctional enzyme with thymidylate synthase (TS) in a 110–140 kD dimer of two subunits (Ivanetich & Santi, 1990). For *Plasmodium* there is a need for alternative DHFR inhibitors without cross-resistance to established drugs; other diaminopyrimidines, biguanides, triazines and quinazolines (Canfield *et al.*, 1993) and compounds more specific for TS, including 5-fluororotate and a quinazoline, ICI D1694, are being investigated (Rathod & Reshimi, 1994). In *T. gondii* the folate pathway enzymes, DPS and DHFR, are also inhibited by sulphonamides and pyrimethamine, although the enzymes of this organism are less sensitive than those of *Plasmodium*. A pyrimethamine/sulphadiazine combination has been used to treat toxoplasmosis for years despite concern over toxicity (Araujo & Remington, 1992). The increase in toxoplasmosis cases, mainly due to infections in the immunosuppressed, has stimulated the search for more effective DHFR inhibitors, including a re-examination of the use of trimethoprim and studies on more lipid soluble piritrexim and trimetrexate (Fig. 4) which have greater DHFR inhibiting activity (see Araujo & Remington, 1992) and epiroprim (Fig. 4) (Chang *et al.*, 1994). The distinct DHFR/TS enzyme has also been identified in *Leishmania* and *Trypanosoma* and the potential of this enzyme target in these species, which were previously considered to be relatively insensitive to DHFR inhibitors, is illustrated by identification of antileishmanial activity in 5-substituted 2, 4-diaminopyrimidines (Sirawaraporn *et al.*, 1988) and triazines (Booth *et al.*, 1987).

PURINE BIOSYNTHESIS INHIBITORS

Protozoa are unable to synthesize purines *de novo* and therefore rely upon salvage pathways containing unique enzymes to obtain purine bases and nucleosides from their hosts. Although these differences have been studied in the search for new antimalarials, for example thio-purine analogues inhibit hypoxanthine-guanine-xanthine phosphoribosyltransferase and growth of *P. falciparum in vitro* (Queen, Van der Jagt & Reyes, 1990), this approach has made most impact upon the chemotherapy of leishmaniasis and Chagas' disease. In *Leishmania* and *T. cruzi* the salvage pathway enzymes hypoxanthine–guanine phosphoribosyl-transferase and adeno-succinate synthetase/lyase, normally phosphorylate and aminate purine bases or nucleosides to nucleotides; these enzymes also have a high affinity for pyrazolopyrimidines (for example allopurinol) and C-nucleosides (for example, formycin B). Incorporation of the metabolized analogues into RNA inhibits protein synthesis and some intermediates also have activity against guanosine monophosphate reductase (Marr, 1991).

Allopurinol has been on trial against both visceral and cutaneous leishma-niasis but results have been equivocal probably due to the rapid metabolism of this drug and steady state serum levels remaining below the sensitivity of the parasite *in vitro*. However, combinations with antimonials have been more promising against the visceral disease (Di Martino *et al.*, 1990) and an 80% cure rate was reported in a trial against cutaneous leishmaniasis caused by *L. panamensis* (Martinez & Marr, 1992). Allopurinol riboside (Fig. 5) was also ineffective against South American cutaneous leishmaniasis, de-spite co-administration with probenicid to elevate plasma levels of the drug (Guederian *et al.*, 1991).

Reports on the treatment of Chagas' disease with allopurinol have also been equivocal; it proved ineffective against acute phase infections but was as effective, with a 75% cure rate, and less toxic than the standard drugs nifurtimox and benznidazole against the chronic form of the disease when used at high doses (Gallerano, Marr & Sosa, 1990). A multi-centre study on the treatment of chronic Chagas' disease with allopurinol is in progress.

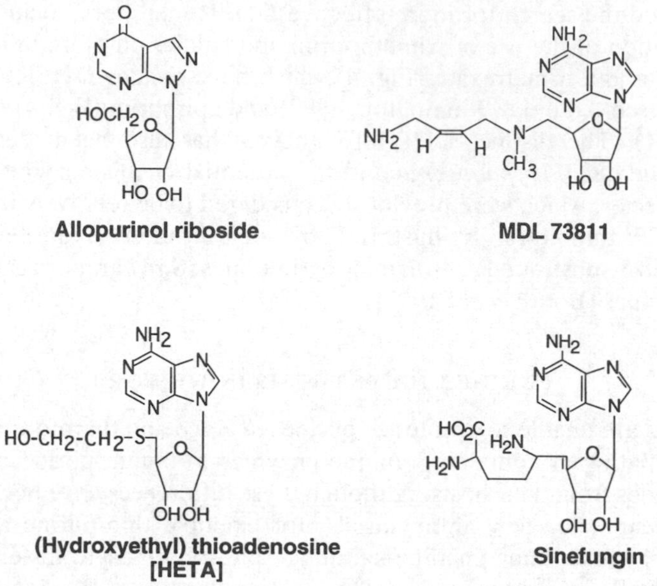

Allopurinol riboside

MDL 73811

(Hydroxyethyl) thioadenosine [HETA]

Sinefungin

Fig. 5. Purine nucleoside analogues with antiprotozoal activity.

POLYAMINE INHIBITORS

The polyamines, putrescine, spermidine and spermine, are co-factors for macromolecular synthesis and membrane stabilization, and are essential for cell multiplication. Polyamines are formed in most eukaryotes by decarboxylation of ornithine to putrescine by ornithine decarboxylase (ODC) and the subsequent addition of aminopropyl groups from decarboxylated S-adenosylmethionine (AdoMet). The byproduct of this aminopropyl donation, methylthioadenosine (MTA) is broken down further to adenine and methylthioribose by MTA phosphorylase. In 1980, the ODC inhibitor eflornithine (DL-α-difluoromethylornithine) was shown to deplete levels of putrescine and spermidine in trypanosomes and cure experimental rodent infections of *T. brucei* (Bacchi *et al.*, 1980). Eflornithine, which had been tested in humans previously in anticancer studies, was soon on trial for the treatment of African trypanosomiasis. It was registered for the treatment of African trypanosomiasis in the USA in 1990 and in Europe in 1991 (Kuzoe, 1993). By 1992, eflornithine was reported to have cured over 90% of 711 patients infected with *T.b.gambiense*, many of whom were refractory to treatment with the standard arsenical drugs (Van Nieuwenhove, 1992). However, there are three major drawbacks to the use of eflornithine: first, the recommended intravenous dosage of 100 mg/kg every 6 hours for 14 days, secondly, that strains of *T.b.rhodesiense* are unresponsive to eflornithine alone (Bacchi *et al.*, 1990), probably owing to differences in AdoMet metabolism (Bacchi *et al.*, 1993), and thirdly, the high costs of the drug and administration (Kuzoe, 1993). Currently clinical trials of eflornithine plus suramin are in progress for the treatment of *T.b.rhodesiense* sleeping sickness (see Kuzoe, 1993). The sensitivity of *T. brucei* to eflornithine could be due to the slow turnover of ODC in these organisms compared to mammalian cells (Ghoda *et al.*, 1990). Other protozoa are affected to a lesser extent by eflornithine (see McCann & Pegg, 1992).

The link between polyamine, AdoMet and MTA metabolism has been mentioned above. In trypanosomatids the spermidine produced in this pathway is required for the synthesis of trypanothione which protects these parasites against toxic metabolites and maintains intracellular thiol balance. Depletion of spermidine levels through inhibition of ODC or limiting the supply of AdoMet through the inhibition of AdoMet decarboxylase, results in parasite death. One AdoMet analogue, MDL 73811 (Fig. 5) inhibits AdoMet decarboxylase, decreases spermidine levels and has proved to be highly active against *T. brucei*, including those strains of *T.b.rhodesiense* which do not respond to eflornithine (Bitonti *et al.*, 1990). MTA analogues, for example 5'-deoxy-5'-(hydroxyethyl) thioadenosine (Fig. 5), are also active against *T. brucei* in experimental models (Bacchi *et al.*, 1991). The nucleoside antibiotic, sinefungin, another AdoMet analogue (Fig. 5) is a broad-spectrum anti-protozoal agent (*Plasmodium, Trypanosoma, Leishmania, Cryptosporidium*, Microsporidia) but it appears to be primarily a competitive inhibitor of protein-O-methyltransferases (Lawrence & Robert-Gero, 1993).

Another polyamine analogue, the *bis*(benzyl)tetramine MDL 27695 has also shown promising antiprotozoal activity, against *Plasmodium* (Bitonti *et al.*, 1989) but in particular against *L. donovani* infections in mice following oral administration (Baumann *et al.*, 1991). Two other compounds of this class, MDL 27695 and 27696, were active against *T. cruzi* (Majumder & Kierszenbaum, 1993).

STEROL AND LIPID BIOSYNTHESIS INHIBITORS

The presence of ergosterol, rather than cholesterol, as the principal sterol in the plasmamembranes of *Leishmania* and *T. cruzi* is the basis for the selective toxicity of polyene antibiotics and azoles to these parasites. Since 1981 a number of azoles, including ketoconazole, miconazole and itraconazole, developed as antifungal drugs due to their ability to inhibit cytochrome p450-dependent C-14 demethylation of lanosterol in the ergosterol biosynthetic pathway, have shown experimental activity against these two parasites. While biochemists have dissected the sterol pathway in *Leishmania* and *T. cruzi* to find parallels with fungal metabolism (see Haughan & Goad, 1991), the clinicians, delighted to test the potential of cheaper, licensed, orally active drugs have used the azoles, in particular ketoconazole and itraconazole for the treatment of cutaneous leishmaniasis. Until recently many of the clinical trials were uncontrolled and produced inconclusive results. However, more recently a controlled trial with 120 cases suggested that ketoconazole was more effective against cutaneous lesions caused by *L. mexicana* than against cutaneous lesions caused by *L. braziliensis* (Navin *et al.*, 1992). Although the ketoconazole cure rate was not statistically significant in either group, this clinical study and earlier biochemical studies (Beach, Goad & Holz, 1988) have shown that there is a variation in sensitivity of *Leishmania* spp. to these compounds. The possibility that sterols might be salvaged from host cell macrophages (Berman *et al.*, 1986) is another factor which might contribute to the disappointing results from the use of azoles alone in the treatment of leishmaniasis. Ketoconazole and itraconazole were also effective against acute but not against chronic experimental infections of *T. cruzi* in mice (McCabe, Remington & Araujo, 1987); ketoconazole also failed to cure patients with chronic Chagas' disease at doses used to treat deep mycoses (Brener *et al.*, 1993).

Although new azoles are under development, and some, for example, the *bis*-triazole ICI 195,739, have high anti-*T. cruzi* activity (Ryley, McGregor & Wilson, 1988), further studies are concentrating upon combinations of inhibitors with synergistic activity and with matching pharmacokinetic properties. Inhibitors of other enzymes of the sterol biosynthesis pathway, for example the allylamines which inhibit squalene epoxidase, and inhibitors of hydroxy-methyl-glutaryl coenzyme A (HMGCoA) reductase also affect growth of these parasites (Berman & Gallalee, 1987; Urbina *et al.*, 1993). In experimental studies, potentiation of activity against *T. cruzi* has been shown for the allylamine terbinafine in combination with ketoconazole or ICI 195,739 (Maldonado *et al.*, 1993) and for lovastatin, an inhibitor of HMGCoA reductase, in combination with ketoconazole or ketoconazole plus terbinafine (Urbina *et al.*, 1993). Lovastatin also inhibits *P. falciparum in vitro* (Grellier *et al.*, 1994), but this may be due to inhibition of the synthesis of the polyprenyl sidechain for ubiquinone (Ellis, 1994).

Lipid biosynthesis is a novel target in protozoa. The glycosyl phosphatidyl inositol that anchors the variable surface glycoprotein of *T. brucei* contains myristate, a fatty acid that is rare in mammalian cell anchors. This difference provided a rational approach which led to the identification of a myristate analogue 10-(propoxy)decanoic acid with activity against *T. brucei in vitro* (Doering *et al.*, 1991). Several alkyl lysophospholipids, some on clinical trial as anticancer drugs, have shown selective activity against *Leishmania* and two of these miltefosine and ilmofosine are active against *L. donovani* in animals following oral administration (Kuhlencord *et al.*, 1992; Croft *et al.*, 1993). Phospholipid biosynthesis is inhibited in *Leishmania* by ether lipids (Achterberg & Gercken, 1987), in *T. cruzi* by ajoene (Urbina *et al.*, 1991) and in *Plasmodium* by choline analogues (Vial & Ancelin, 1992).

PROTEIN SYNTHESIS INHIBITORS

Protein synthesis has been largely overlooked as a target in protozoa, despite the clinical use of some antibiotics (see above). Edlind (1991), however, has provided a valuable re-appraisal of the antiprotozoal potential of protein synthesis inhibiting antibiotics. Although there are fundamental differences in prokaryotic and eukaryotic protein synthesis, from the structure of ribosomal subunits to the mechanism of mRNA binding to SSU rRNA, the precise mechanisms by which most antibiotics work against protozoa is unclear. Sequence analysis of SSU rRNA has shown that base pair sequences of *P. falciparum* (McCutchan *et al.*, (1988) and *L. donovani* (Looker *et al.*, 1988) could explain the selective toxicity of some antibiotics, for example paromomycin. Biochemical studies on the mechanisms of action of some antibiotics have shown tetracycline to affect the mitochondria of *P. falciparum* (Kiatfuengfoo *et al.*, 1989), sparsomycins the peptidyl transferase of *T. brucei* (Bitonti *et al.*, 1985) and aminoglycosides ribosomal dissociation in *Leishmania* (Marouf, Lawrence, Croft & Robert-Gero, unpublished data). Putative elongation factors have been identified in some protozoa and might be the target of fusidic acid, a drug which binds to bacterial EG-F and is known to be active against both *G. intestinalis* (Farthing & Inge, 1986) and *P. falciparum* (Black, Wildfang & Borgbjerg, 1985).

Antibiotics are not the only source of inhibitors of protein synthesis. Emetine, still used for the treatment of amoebiasis, is a specific inhibitor of eukaryotic protein synthesis. More recently the anticancer agent homoharringtonine from *Cephalotaxus harringtonia*, which binds to the ribosomal A site, and several quassinoids have shown activity against *P. falciparum* (Ekong *et al.*, 1990).

Protein synthesis can also be inhibited selectively through the design of complementary 'antisense' oligonucleotide sequences, targeted to mRNA sequences which control splicing, initiation or elongation (Sartorius & Franklin, 1991). The presence of a common 35-nucleotide leader sequence in *T. brucei* was used in the design of a 9-mer acridine linked oligonucleotide which rapidly killed these organisms *in vitro* (Vespieren *et al.*, 1987). More recently, 18–21 oligodeoxynucleotide phosphorothioates complementary to DHFR-TS mRNA sequences (Rapaport *et al.*, 1992) and 24-mer oligonucleotide phosphoramidites complementary to hypoxanthine-guanine phosphoribosyltransferase mRNA (Dawson *et al.*, 1993) have shown activity against *Plasmodium falciparum in vitro*.

DNA-DEPENDENT PROCESSES

One step further down the pathway of molecular design from mRNA inhibitors are antigene products. DNA nucleotide sequence selectivity is important in relation to the activity of certain groups of antitumour drugs, for example the saframycins and bleomycins. Bleomycin is also a trypanocide (Clarkson *et al.*, 1983). Two oligopeptide antibiotics, netropsin and distamycin, which bind to the DNA minor groove in AT-rich regions, can be modified by the addition of heterocyclic moieties (imidazole, furan, thiazole) to allow binding to GC sites. These 'lexitropsins', which allow the reading of sequences, have been identified as potential antiviral and anticancer drugs. The potential of this approach to protozoa, where there are unique sequences in the nuclear and mitochondrial DNA, is unexplored although the selectivity of distamycin A for *P. falciparum* at submicromolar concentrations has been reported (Ginsburg *et al.*, 1993).

DNA gyrase in prokaryotes and the equivalent, but dimeric, enzyme DNA topoisomerase II, in eukaryotes are important targets in the development of antibacterial and anticancer drugs. The antibacterial fluoroquinolones, classic DNA gyrase inhibitors, have shown poor activity against trypanosomes and leishmanias (see Douc-Rasy *et al.*, 1988; Shapiro, 1993) and in the treatment of malaria (McClean, Hitchman & Shafran, 1992). However, topoisomerase II remains a potential target in trypanosomatids where it has important functions in the replication of kinetoplast DNA, a mass of mitochondrial DNA network made of circular DNA monomers in mini- and maxicircles, through the unwinding of the helix for the replication fork, decatenation and reattachment of minicircles to the DNA network and topology during reorganization for division. Genes for DNA topoisomerase II have already been sequenced and expressed for *T. brucei* and *T. cruzi* (Fragoso & Goldenberg, 1992) and show extensive homology with other eukaryotic enzymes. Isolated topoisomerase II from *T. equiperdum* and *T. cruzi* was inhibited by anticancer drugs, such as etoposide, teniposide, ellipiticine and mAMSA, as well as standard antitrypanosomal drugs, for example the intercalators ethidium bromide and isometamidium chloride or the diamidine minor groove binders diminazene (berenil) and pentamidine (Douc-Rasy *et al.*, 1988; Shapiro, 1993). 9-anilino-acridines which have shown activity *in vitro* against *Plasmodium* and trypanosomes (Figgitt *et al.*, 1992) and *Leishmania* (Mauel *et al.*, 1993) are also putative topoisomerase inhibitors.

DNA polymerases have also been identified as targets in rapidly dividing protozoa. Phosphonylmethoxyalkylpurines, which are also antiviral compounds, for example, HPMPA and its 3-deaza analogue, were active against *P. falciparum in vitro* at nanomolar concentrations and also against rodent malaria (De Vries *et al.*, 1991). RNA polymerases are also different in protozoa, and one inhibitor of this enzyme complex, rifampicin, has been used in several studies in the treatment of cutaneous leishmaniasis with conflicting results (Evans *et al.*, 1989). The presence of circular DNA in *Plasmodium* which encodes the B-subunit of a prokaryote like RNA polymerase has been suggested to explain sensitivity of *P. vivax* to rifampin

(Pukrittayakamee *et al.*, 1994). Another rifampicin derivative, rifabutin, is active against *T. gondii* in a rodent model (Araujo, Slifer & Remington, 1994).

CONCLUSIONS

Despite the growing threat of protozoal infections to human populations, there have been few major advances in drug treatment for these diseases. Most recent advances derive either from the use of drugs developed for other microbial diseases and cancer or from traditional medicines. This is not due to the absence of targets as there are sufficient unique and highly accessible differences between protozoan parasites and mammalian cells to produce selective inhibitors. Given the power of molecular techniques to define targets and the availability of novel compounds, it will be disappointing if new drugs are not discovered soon. Cytokines will probably also have a role in the treatment of some protozoal infections, especially those caused by opportunistic parasites in immunodeficient patients. The benefits of cytokine–drug combinations have already been demonstrated in treatments for leishmaniasis and toxoplasmosis (Murray, 1993). The conundrum for antiprotozoal chemotherapy is that current research on the identification and characterization of specific targets could lead to highly specific and effective drugs; however, these drugs are likely to be expensive to develop and manufacture, have a limited market and may not be produced.

ACKNOWLEDGEMENTS

Thanks to Caroline Ash, Geoffrey Kirby, Keith Smith and David Warhurst for reading the manuscript and making valuable comments.

REFERENCES

Achterberg, V. & Gercken, G. (1987). Metabolism of ether lysophospholipids in *Leishmania donovani* promastigotes. *Molecular and Biochemical Parasitology*, **26**, 277–87.

Araujo, F. G., Huskinson-Mark, J., Gutteridge, W. E. & Remington, J. S. (1992). *In vitro* and *in vivo* activities of the hydroxynaphthoquinone 566C80 against the cyst form of *Toxoplasma gondii*. *Antimicrobial Agents and Chemotherapy*, **36**, 326–30.

Araujo, F. G. & Remington, J. S. (1992). Recent advances in the search for new drugs for treatment of toxoplasmosis. *International Journal of Antimicrobial Agents*, **1**, 153–64.

Araujo, F. G., Slifer, T. & Remington, J. S. (1994). Rifabutin is active in murine models of toxoplasmosis. *Antimicrobial Agents and Chemotherapy*, **38**, 570–5.

Bacchi, C. J., Garofalo, J., Ciminelli, M., Rattendi, D., Goldberg, B., McCann,

P. P. & Yarlett, N. (1993). Resistance to DL-α-difluoromethylornithine by clinical isolates of *Trypanosoma brucei rhodesiense*. Role of S-adenosylmethionine. *Biochemical Pharmacology*, **46**, 471–81.

Bacchi, C. J., Nathan, H. C., Hunter, S. H., McCann, P. P. & Sjoerdsma, A. (1980). Polyamine metabolism: a potential therapeutic target in trypanosomes. *Science*, **210**, 332–4.

Bacchi, C. J., Nathan, H. C., Livingston, T., Valladares, G., Saric, M., Sayer, P. D., Njogu, A. R. & Clarkson, A. B. Jr. (1990). Differential susceptibility to DL-α-difluoromethylornithine in clinical isolates of *Trypanosoma brucei rhodesiense*. *Antimicrobial Agents and Chemotherapy*, **34**, 1183–8.

Bacchi, C. J., Sufrin, J. R., Nathan, H. C., Spiess, A. J., Hannan, T., Garofalo, J., Alecia, K., Katz, L. & Yarlett, N. (1991). 5′-alkyl-substituted analogs of 5′-methylthioadenosine as trypanocides. *Antimicrobial Agents and Chemotherapy*, **35**, 1315–20.

Basco, L. K., Mitaku, S., Skaltsounis, A.-S., Ravelomanantsoa, N., Tillequin, F., Koch, M. & Le Bras, J. (1994). *In vitro* activities of furoquinoline and acridone alkaloids against *Plasmodium falciparum*. *Antimicrobial Agents and Chemotherapy*, **38**, 1169–71.

Baumann, R. J., McCann, P. P. & Bitonti, A. J. (1991). Suppression of *Leishmania donovani* by oral administration of a *bis*(benzyl)polyamine analog. *Antimicrobial Agents and Chemotherapy*, **31**, 1403–7.

Beach, D. H., Goad, L. J. & Holz, G. G. Jr. (1988). Effects of antimycotic azoles on growth and sterol biosynthesis of *Leishmania* promastigotes. *Molecular and Biochemical Parasitology*, **31**, 149–62.

Benson, T. J., McKie, J. H., Garforth, J., Borges, A., Fairlamb, A. H. & Douglas, K. (1992). Rationally designed selective inhibitors of trypanothione reductase. *Biochemical Journal*, **286**, 9–11.

Berman, J., Brown, L., Miller, R., Andersen, S. L., McGreevy, P., Schuster, B. G., Ellis, W., Age, R. A. & Rossan, R. (1994). Antimalarial activity of WR 243251, a dihydroacridinedione. *Antimicrobial Agents and Chemotherapy*, **38**, 1753–6.

Berman, J. D., Goad, L. J., Beach, D. H. & Holz, G. G. Jr. (1986). Effects of ketoconazole on sterol biosynthesis by *Leishmania mexicana mexicana* amastigotes in murine macrophage tumor cells. *Molecular and Biochemical Parasitology*, **20**, 85–92.

Berman, J. & Gallalee, J. V. (1987). *In vitro* antileishmanial activity of inhibitors of steroid biosynthesis and combinations of antileishmanial agents. *Journal of Parasitology*, **73**, 671–3.

Bitonti, A. J., Byers, T., Bush, T. L., Casara, P. J., Bacchi, C. J., Clarkson, A. B. Jr., McCann, P. P. & Sjoerdsma, A. (1990). Cure of *Trypanosoma brucei brucei* and *Trypanosoma brucei rhodesiense* infections in mice with an irreversible inhibitor of S-adenosylmethione decarboxylase. *Antimicrobial Agents and Chemotherapy*, **34**, 1485–90.

Bitonti, A. J., Dumont, J. A., Bush, T. M., Edwards, M. L., Stemerick, D. M., McCann, P. P. & Sjoerdsma, A. (1989). *Bis*(benzyl)polyamine analogs inhibit the growth of chloroquine resistant human malaria parasites (*Plasmodium falciparum*) *in vitro* and in combination with α-difluormethylornithine cure murine malaria. *Proceedings of the National Academy of Sciences, USA*, **86**, 651–5.

Bitonti, A. J., Kelly, S. E., Flynn, G. A. & McCann, P. P. (1985). Inhibition of *Trypanosoma brucei brucei* peptidyl transferase activity by sparsomycin analogs and effects on trypanosome protein synthesis and proliferation. *Biochemical Pharmacology*, **34**, 3055–60.

Bitonti, A. J., Sjoerdsma, A., McCann, P. P., Kyle, D. E., Oduola, A. M., Rossan,

R. N., Milhous, W. K. & Davidson, D. E. J. (1988). Reversal of chloroquine resistance in the malaria parasite *Plasmodium falciparum*. *Science*, **242**, 1301–3.

Black, F. T., Wildfang, I. L. & Borgbjerg, K. (1985). Activity of fusidic acid against *Plasmodium falciparum in vitro*. *Lancet*, **i**, 578–9.

Booth, R. G., Selassie, C. D., Hansch, C. & Santi, D. V. (1987). Quantitative structure-activity relationship of triazene-antifolate inhibition of *Leishmania* dihydrofolate reductase and cell growth. *Journal of Medicinal Chemistry*, **30**, 1218–24.

Boreham, P. F. L., Phillips, R. E. & Shepherd, R. W. (1988). Altered uptake of metronidazole *in vitro* by stocks of *Giardia intestinalis* with different drug sensitivities. *Transactions of the Royal Society of Tropical Medicine and Hygiene*, **82**, 104–6.

Brener, Z., Cancado, J. R., Galvao, L. M. D. C., Luz, M. P. D., Filardi, L. D. S., Pereira, M. E. S., Santos, L. M. T. & Cancado, C. B. (1993). An experimental and clinical assay with ketoconazole in the treatment of Chagas' disease. *Memorias do Instituto Oswaldo Cruz*, **88**, 149–53.

Brewer, T. G., Peggins, J. O., Grate, S. J., Petras, J. M., Levine, B. S., Weina, P. J., Swearenge N. J., Heiffer, M. H. & Schuster, B. G. (1994). Neurotoxicity in animals due to artether and artemether. *Transactions of the Royal Society of Tropical Medicine and Hygiene*, **88 Suppl. 1**, 33–6.

Burri, C., Baltz, T., Giroud, C., Doua, F., Welker, H. A. & Brun, R. (1993). Pharmacokinetic properties of the trypanocidal drug melarsoprol. *Chemotherapy*, **39**, 225–34.

Campbell, W. C. & Rew, R. S. (1986). *Chemotherapy of Parasitic Diseases*. New York: Plenum Press.

Canfield, C. J., Milhous, W. K., Ager, A. L., Rossan, R. N., Sweeney, T. R., Lewis, N. J. & Jacobus, D. P. (1993). PS-15: a potent, orally active antimalarial from a new class of folic acid antagonists. *American Journal of Tropical Medicine and Hygiene*, **49**, 121–6.

Chang, H. R., Arsenijevic, D., Comte, R., Polak, A., Then, R. L. & Pechere, J.-C. (1994). Activity of epiprim (Ro 11-8958), a dihydrofolate reductase inhibitor, alone and in combination with dapsone against *Toxoplasma gondii*. *Antimicrobial Agents and Chemotherapy*, **38**, 1803–7.

Chang, H. R., Jefford, C. W. & Pechere, J.-C. (1989). *In vitro* effects of three new 1, 2, 4-trioxanes (pentatroxane, thiahexatroxane and hexatroxane) on *Toxoplasma gondii*. *Antimicrobial Agents and Chemotherapy*, **33**, 1748–52.

Chapman, H. D. (1993). Resistance to anticoccidial drugs in fowl. *Parasitology Today*, **9**, 159–62.

Chen, M., Christensen, S. B., Theander, T. G. & Kharazami, A. (1994a). Antileishmanial activity of lipochalcone A in mice infected with *Leishmania major* and in hamsters infected with *Leishmania donovani*. *Antimicrobial Agents and Chemotherapy*, **38**, 1339–44.

Chen, M., Theander, T. G., Christensen, S. B., Hviid, L., Zhai, L. & Kharazami, A. (1994b). Lipochalcone A, a new antimalarial agent, inhibits *in vitro* growth of the human malaria parasite *Plasmodium falciparum* and protects mice from *P. yoelii* infection. *Antimicrobial Agents and Chemotherapy*, **38**, 1470–5.

Clarkson, A. B. Jr., Bacchi, C. J., Mellow, G. H., Nathan, H. C., McCann, P. P. & Sjoerdsma, A. (1983). Efficacy of combinations of difluoromethylornithine and bleomycin in a mouse model of central nervous system African trypanosomiasis. *Proceedings of the National Academy of Sciences, USA*, **80**, 5729–33.

Coombs, G. & North, M. (1991). *Biochemical Protozoology*. London: Taylor & Francis.

Croft, S. L., Davidson, R. N. & Thornton, E. A. (1991). Liposomal amphotericin B in the treatment of visceral leishmaniasis. *Journal of Antimicrobial Chemotherapy*, **28 Suppl. B**, 111–18.

Croft, S. L., Hogg, J., Gutteridge, W. E., Hudson, A. T. & Randall, A. W. (1992). The activity of hydroxynaphthoquinones against *Leishmania donovani*. *Journal of Antimicrobial Chemotherapy*, **30**, 827–32.

Croft, S. L., Neal, R. A. & Rao, L. S. (1989). Liposomes and other drug delivery systems in the treatment of leishmaniasis. In (Hart, D. T., ed.) *Leishmaniasis. The Current Status and New Strategies for Control*. pp. 783–792. New York: Plenum Press.

Croft, S. L., Neal, R. A., Thornton, E. A. & Herrmann, D. B. J. (1993). Antileishmanial activity of the ether phospholipid ilmofosine. *Transactions of the Royal Society of Tropical Medicine and Hygiene*, **87**, 217–19.

Davidson, D. E. Jr. (1994). Role of artether in the treatment of malaria and plans for further development. *Transactions of the Royal Society for Tropical Medicine and Hygiene*, **88 Suppl. 1**, 51–2.

Davidson, R. N., DiMartino, L., Gradoni, L., Giacchino, R., Russo, R., Gaeta, G. B., Pempinell, O. R., Scott, S., Raimondi, F., Cascio, A., Prestileo, T., Caldeira, L., Wilkinson, R. J. & Bryceson, A. D. M. (1994). Liposomal amphotericin B (AmBisome) in Mediterranean visceral leishmaniasis: a multi-centre trial. *Quarterly Journal of Medicine*, **87**, 75–81.

Dawson, P. A., Cochran, D. A. E., Emmerson, B. T. & Gordon, R. B. (1993). Inhibition of *Plasmodium falciparum* hypoxanthine–guanine phosphoribosyltransferase mRNA by antisense oligodeoxynucleotide sequence. *Molecular and Biochemical Parasitology*, **60**, 153–6.

De Castro, S. L. (1993). The challenge of Chagas' disease chemotherapy: an update of drugs assayed against *Trypanosoma cruzi*. *Acta Tropica*, **53**, 83–98.

De Vries, E. D., Stam, J. G., Franssen, F. F. J., Nieuwenhuijs, H., Chavalitshewinkoon, P., Clercq, E. D., Overdulve, J. P. & Vliet, P. C. V. D. (1991). Inhibition of the growth of *Plasmodium falciparum* and *Plasmodium berghei* by the DNA polymerase inhibitor HPMPA. *Molecular and Biochemical Parasitology*, **47**, 43–50.

Dieterich, D. T., Lew, E. A., Kotler, D. P., Poles, M. A. & Orenstein, J. M. (1994). Treatment with albendazole for intestinal disease due to *Enterocytozoon bieneusi* in patients with AIDS. *Journal of Infectious Diseases*, **169**, 178–83.

Dietze, R., Milan, E. P., Berman, J. D., Grogl, M., Falqueto, A., Feitosa, T. F., Luz, K. G., Suassuna, F. A. B., Marinho, L. A. C. & Ksionski, G. (1993). Treatment of Brazilian kala-azar with a short course of Amphocil (amphotericin B cholesterol dispersion). *Clinical Infectious Diseases*, **17**, 981–6.

DiMartino, L., Mantovani, M. P., Gradoni, L., Gramiccia, M. & Guandalini, S. (1990). Low dosage combination of meglumine antimoniate plus allopurinol as first choice treatment of infantile visceral leishmaniasis in Italy. *Transactions of the Royal Society of Tropical Medicine and Hygiene*, **84**, 534–5.

Doering, T. L., Raper, J., Buxbaum, L. U., Adams, S. P., Gordon, J. I., Hart, G. W. & Englund, P. T. (1991). An analog of myristic acid with selective toxicity for African trypanosomes. *Science*, **252**, 1851–4.

Douc-Rasy, S., Riou, J-R., Ahomadegbe, J-C. & Riou, G. (1988). ATP-independent DNA topoisomerase II as potential drug target in trypanosomes. *Biology of the Cell*, **64**, 145–56.

Dutta, P., Pinto, J. & Rivlin, R. (1990). Antimalarial properties of imipramine and amitryptyline. *Journal of Protozoology*, **37**, 54–8.

Edlind, T. D. (1991). Protein synthesis as a target for antiprotozoal drugs. In

(Coombs, G. & North, M., eds.) *Biochemical Protozoology*. pp. 569–586. London: Taylor & Francis.

Ekong, R. M., Kirby, G. C., Patel, G., Phillipson, J. D. & Warhurst, D. C. (1990). Comparison of the *in vitro* activities of quassinoids with activity against *Plasmodium falciparum*, anisomycin and some other inhibitors of eukaryotic protein synthesis. *Biochemical Pharmacology*, **40**, 297–301.

Ellis, J. E. (1994). Coenzyme Q homologs in parasitic protozoa as targets for chemotherapeutic attack. *Parasitology Today*, **10**, 296–301.

El-On, J., Jacobs, G. P. & Weinrauch, L. (1988). Topical chemotherapy of cutaneous leishmaniasis. *Parasitology Today*, **4**, 76–81.

Evans, A. T. & Croft, S. L. (1994). Antileishmanial actions of tricyclic neuroleptics appear to lack structural specificity. *Biochemical Pharmacology*, **48**, 613–16.

Evans, A. T., Croft, S. L., Peters, W. & Neal, R. A. (1989). Antileishmanial effects of clofazimine and other antimycobacterial agents. *Annals of Tropical Medicine and Parasitology*, **83**, 447–54.

Fairlamb, A. H. (1989). Novel biochemical pathways in parasitic protozoa. *Parasitology*, **99**, S93–112.

Fairlamb, A. H., Henderson, G. B. & Cerami, A. (1989). Trypanothione is the primary target for arsenical drugs against African trypanosomes. *Proceedings of the National Academy of Sciences*, *USA*, **86**, 2607–11.

Farthing, M. J. G. & Inge, P. M. G. (1986). Antigiardial activity of the bile salt-like antibiotic sodium fusidate. *Journal of Antimicrobial Chemotherapy*, **17**, 165–71.

Figgitt, D., Denny, W., Chavalitshewinkoon, P., Wilairat, P. & Ralph, R. (1992). *In vitro* study of anticancer acridines as potential antitrypanosomal and antimalarial agents. *Antimicrobial Agents and Chemotherapy*, **36**, 1644–7.

Fragoso, S. P. & Goldenberg, S. (1992). Cloning and characterization of the gene encoding *Trypanosoma cruzi* DNA topoisomerase II. *Molecular and Biochemical Parasitology*, **55**, 127–34.

Fry, M. & Pudney, M. (1992). Site of action of the antimalarial hydroxynaphthoquinone,2-[*trans*-4-(chlorophenyl)cyclohexyl]-3-hydroxy-1,4-naphthoquinone (566C80). *Biochemical Pharmacology*, **43**, 1545–53.

Fu, S. & Xiao, S-H. (1991). Pyronaridine: a new antimalarial drug. *Parasitology Today*, **7**, 310–13.

Gallerano, R. H., Marr, J. J. & Sosa, R. R. (1990). Therapeutic efficacy of allopurinol in patients with chronic Chagas' disease. *American Journal of Tropical Medicine and Hygiene*, **43**, 159–66.

Ghoda, L., Phillips, M. A., Bass, K. E., Wang, C. C. & Coffino, P. (1990). Trypanosome ornithine decarboxylase is stable because it lacks sequences found in the carboxyl terminus of the mouse enzyme which target the latter for intracellular degradation. *Journal of Biological Chemistry*, **265**, 11823–6.

Ginsburg, H., Nissani, E., Krugliak, M. & Williamson, D. H. (1993). Selective toxicity to malaria parasites by non-intercalating DNA-binding ligands. *Molecular and Biochemical Parasitology*, **58**, 7–16.

Grellier, P., Valentin, A., Millerioux, V., Schrevel, J. & Rigomier, D. (1994). 3-Hydroxy-3-methylglutaryl coenzyme A reductase inhibitors lovastatin and simvastatin inhibit *in vitro* development of *Plasmodium falciparum* and *Babesia divergens* in human erythrocytes. *Antimicrobial Agents and Chemotherapy*, **38**, 1144–8.

Guederian, R. H., Chico, M. E., Rogers, M. D., Pattishall, K. M., Grogl, M. & Berman, J. (1991). Placebo controlled treatment of Ecuadorian cutaneous leishmaniasis. *American Journal of Tropical Medicine and Hygiene*, **45**, 92–7.

Hammond, D. J., Hogg, J. & Gutteridge, W. E. (1985). *Trypanosoma cruzi*:

possible control of parasite transmission by blood transfusion using amphiphilic cationic drugs. *Experimental Parasitology*, **60**, 32–42.

Haughan, P. A. & Goad, L. J. (1991). Lipid biochemistry of trypanosomatids. In *Biochemical Protozoology* (Coombs, G. & North, M., eds.). pp. 312–328. London: Taylor & Francis.

Henderson, G. B., Ulrich, P., Fairlamb, A. H., Rosenberg, I., Periera, M., Sela, M. & Cerami, A. (1988). 'Subversive' substrates for the enzyme trypanothione disulfide reductase: alternative approach to chemotherapy of Chagas' disease. *Proceedings of the National Academy of Sciences, USA*, **85**, 5374–8.

Hudson, A. (1993). Atovaquone – a novel broad-spectrum anti-infective drug. *Parasitology Today*, **9**, 66–8.

Hudson, A. T., Randall, A. W., Fry, M., Ginger, C. D., Hill, B., Latter, V. S., McHardy, N. & Williams, R. B. (1985). Novel anti-malarial hydroxynaphthoquinones with potent broad spectrum anti-protozoal activity. *Parasitology*, **90**, 45–55.

Ivanetich, K. M. & Santi, D. V. (1990). Bifunctional thymidylate synthase-dihydrofolate reductase in protozoa. *The FASEB Journal*, **4**, 1591–7.

Iwu, M. M., Jackson, J. E. & Schuster, B. G. (1994). Medicinal plants in the fight against leishmaniasis. *Parasitology Today*, **10**, 65–8.

Jennings, F. W. (1993). Combination chemotherapy of CNS trypanosomiasis. *Acta Tropica*, **54**, 205–13.

Kiatfuengfoo, R., Suthiphongchai, T., Prapunwattana, P. & Yuthavong, Y. (1989). Mitochondria as the site of action of tetracycline on *Plasmodium falciparum*. *Molecular and Biochemical Parasitology*, **34**, 109–16.

Kovacs, J. & the NIAID – Clinical Centre Intramural AIDS Program. (1992). Efficacy of atovaquone in treatment of toxoplasmosis in patients with AIDS. *Lancet*, **340**, 637–8.

Kristiansen, J. E. & Jepsen, S. (1985). The susceptibility of *Plasmodium falciparum in vitro* to chlorpromazine and the stereo-isomeric compounds *cis*(Z)- and *trans*(E)-clopenthixol. *Acta Pathologica Microbiologica Immunologica Scandanavica, Section B*, **93**, 249–51.

Krogstad, D. J., Herwaldt, B. J. & Schlesinger, P. H. (1988). Antimalarial agents: specific treatment regimes. *Antimicrobial Agents and Chemotherapy*, **32**, 957–61.

Kuhlencord, A., Maniera, T., Eibl, H. & Unger, C. (1992). Hexadecylphosphocholine: oral treatment of visceral leishmaniasis in mice. *Antimicrobial Agents and Chemotherapy*, **36**, 1630–4.

Kuschner, R. A., Heppner, D. G., Andersen, S. L., Wellde, B. T., Hall, T., Schneider, I., Ballou, W. R., Foulds, G., Sadoff, J. C., Schuster, B. & Taylor, D. N. (1994). Azithromycin prophylaxis against a chloroquine-resistant strain of *Plasmodium falciparum*. *Lancet*, **343**, 1396–7.

Kuzoe, F. A. S. (1993). Current situation of African trypanosomiasis. *Acta Tropica*, **54**, 153–62.

Lawrence, F. & Robert-Gero, M. (1993). Distribution of macromolecular methylations in promastigotes of *Leishmania donovani* and the impact of sinefungin. *Journal of Eukaryotic Microbiology*, **40**, 581–9.

Llanos-Cuentas, A., Echevarria, J., Chang, J., Campos, P., Cruz, M., Cieza, J., Lentnek, D., Robbins, M. & Williams, R. (1991). Randomized trial of amphotericin B complex (ABLC) versus Fungizone in patients with mucocutaneous leishmaniasis. *Memorias do Instituto Oswaldo Cruz, Rio de Janeiro*, **86 Suppl. 1**, 238.

Looker, D., Miller, L. A., Elwood, H. J., Stickel, S. & Sogin, M. M. L. (1988). Primary structure of the *Leishmania donovani* small ribosomal RNA coding region. *Nucleic Acids Research*, **16**, 7198.

McCabe, R. E., Remington, J. S. & Araujo, F. G. (1987). Ketoconazole promotes parasitological cure of mice infected with *Trypanosoma cruzi*. *Transactions of the Royal Society of Tropical Medicine and Hygiene*, **81**, 613–15.

McCann, P. P. & Pegg, A. E. (1992). Ornithine decarboxylase as an enzyme target for therapy. *Pharmacology and Therapeutics*, **54**, 195–215.

McClean, K. L., Hitchman, D. & Shafran, S. D. (1992). Norfloxacin is inferior to chloroquine for falciparum malaria in Northwestern Zambia: a comparative trial. *Journal of Infectious Diseases*, **165**, 904–7.

McCutchan, T. F., Cruz, V. F. D. L., Lal, A. A., Gunderson, J. H., Elwood, H. J. & Sogin, M. L. (1988). Primary sequences of two small subunit ribosomal RNA genes from *Plasmodium falciparum*. *Molecular and Biochemical Parasitology*, **28**, 63–8.

McHardy, N. (1992). Butalex (buparvaquone): a new therapeutic for theileriosis. In (Dolan, T. T., ed.) *Recent Developments in the Research and Control of Theileria annulata*. pp. 59–66. Nairobi: The International Laboratory for Research on Animal Diseases.

Maes, L., Songa, E. B., & Hamers, R. (1993). IMOL 881, a new trypanocidal compound. *Acta Tropica*, **54**, 261–9.

Majumder, S. & Kierszenbaum, F. (1993). Inhibition of host cell invasion and intracellular replication of *Trypanosoma cruzi* by *N,N-bis*(benzyl)-substituted polyamine analogs. *Antimicrobial Agents and Chemotherapy*, **37**, 2235–8.

Maldonado, R. A., Molina, J., Payares, G. & Urbina, J. A. (1993). Experimental chemotherapy with combinations of ergosterol biosynthesis inhibitors in murine models of Chagas' disease. *Antimicrobial Agents and Chemotherapy*, **37**, 1353–9.

Marr, J. J. (1991). Purine metabolism in parasitic protozoa and its relationship to chemotherapy. In (Coombs, G. & North, M., eds.) *Biochemical Protozoology*. pp. 524–536. London: Taylor & Francis.

Martinez, S. & Marr, J. J. (1992). Allopurinol in the treatment of American cutaneous leishmaniasis. *New England Journal of Medicine*, **326**, 741–4.

Mauel, J., Denny, W., Gamage, S., Ransijn, A., Wojcik, S., Figgitt, D. & Ralph, R. (1993). 9-anilinoacridines as potential antileishmanial agents. *Antimicrobial Agents and Chemotherapy*, **37**, 991–6.

Meingassner, J. G. & Thurner, J. (1979). Strain of *Trichomonas vaginalis* resistant to metronidazole and other 5-nitroimidazoles. *Antimicrobial Agents and Chemotherapy*, **15**, 2254–27.

Meshnick, S. R. (1994). The mode of action of antimalarial endoperoxides. *Transactions of the Royal Society of Tropical Medicine and Hygiene*, **88 Suppl. 1**, 31–2.

Murray, H. W. (1993). Cytokines as antimicrobial therapy for the T cell-deficient patient: prospects for treatment of nonviral opportunistic infections. *Clinical Infectious Diseases*, **17 Suppl. 2**, S407–413.

Navin, T. R., Arana, B. A., Arana, F. E., Berman, J. & Chajon, J. F. (1992). Placebo-controlled clinical trial of sodium stibogluconate (Pentostam) versus ketoconazole for treating cutaneous leishmaniasis in Guatemala. *Journal of Infectious Diseases*, **165**, 528–34.

Olliaro, P. & Bryceson, A. D. M. (1993). Practical progress and new drugs for changing patterns of leishmaniasis. *Parasitology Today*, **9**, 323–8.

Ou-Yang, K., Krug, E. C., Marr, J. J. & Berens, R. L. (1990). Inhibition of growth of *Toxoplasma gondii* by qinghaosu and derivatives. *Antimicrobial Agents and Chemotherapy*, **34**, 1961–5.

Pearson, R. D., Manian, A. A., Harcus, J. L., Hall, D. & Hewlett, E. L. (1982). Lethal effects of phenothiazine neuroleptics on the pathogenic protozoan *Leishmania donovani*. *Science*, **217**, 369–71.

Pepin, J., Milord, F., Guern, C., Mpia, B., Ethier, L., & Mansina, D. (1989). Trial of prednisolone for the prevention of melarsoprol-induced encephalopathy in gambiense sleeping sickness. *Lancet*, **i** 1246–50.

Peregrine, A. S. & Mamman, M. (1993). Pharmacology of diminazene: a review. *Acta Tropica*, **54**, 185–203.

Perez, H. A., Rosa, M. D. L. & Apitz, R. (1994). *In vitro* activity of ajoene against rodent malaria. *Antimicrobial Agents and Chemotherapy*, **38**, 337–339.

Peters, W. (1987). *Chemotherapy and Drug Resistance in Malaria. Volume 1.* London: Academic Press.

Peters, W., Robinson, B. L. & Milhous, W. K. (1993). The chemotherapy of rodent malaria. LI. Studies on a new 8-aminoquinoline, WR238605. *Annals of Tropical Medicine and Parasitology*, **87**, 547–52.

Peters, W., Robinson, B. L., Tovey, D. G., Rossier, J. C. & Jefford, C. W. (1992). The chemotherapy of rodent malaria. L. The activities of some synthetic 1, 2, 4-trioxanes against chloroquine-sensitive and chloroquine-resistant parasites. Part 3: Observations on 'Fenozan-50F', a difluorinated 3,3-spirocyclopentane 1, 2, 4-trioxane. *Annals of Tropical Medicine and Parasitology*, **87**, 111–23.

Peterson, D. S., Milhous, W. K. & Wellems, T. E. (1990). Molecular basis of differential resistance to cycloguanil and pyrimethamine in *Plasmodium falciparum* malaria. *Proceedings of the National Academy of Sciences, USA*, **87**, 3018–22.

Phillipson, J. D. & Wright, C. W. (1991). Antiprotozoal agents from plant sources. *Plant Medica*, **57 Suppl. 1**, S53–9.

Posner, G. H., Oh, C. H., Gerena, L. & Milhous, W. K. (1992). Extraordinarily potent antimalarial compounds: new structurally simple, easily synthesized, tricyclic 1, 2, 4-trioxanes. *Journal of Medicinal Chemistry*, **35**, 2459–67.

Pudney, M. (1995). Antimalarials: from quinine to atovaquone. (This volume).

Pukrittayakamee, S., Viravan, C., Charoenlarp, P., Yeamput, C., Wilson, R. J. M. & White, N. J. (1994). Antimalarial effects of rifampin in *Plasmodium vivax* malaria. *Antimicrobial Agents and Chemotherapy*, **38**, 511–14.

Queen, S. A., Van der Jagt, D. L. & Reyes, P. (1990). *In vitro* susceptibilities of *Plasmodium falciparum* to compounds which inhibit nucleotide metabolism. *Antimicrobial Agents and Chemotherapy*, **34**, 1393–8.

Quon, D. V. K., D'Oliveira, C. E. & Johnson, P. J. (1992). Reduced transcription of the ferrodoxin gene in metronidazole-resistant *Trichomonas vaginalis*. *Proceedings of the National Academy of Sciences, USA*, **89**, 4402–6.

Rapaport, E., Misiura, K., Agrawal, S. & Zamecnik, P. (1992). Antimalarial activities of oligodeoxynucleotide phosphorothioates in chloroquine-resistant *Plasmodium falciparum*. *Proceedings of the National Academy of Sciences, USA*, **89**, 8577–80.

Rathod, P. K. & Reshimi, S. (1994). Susceptibility of *Plasmodium falciparum* to a combination of thymidine and ICI D1694, a quinazoline antifolate directed at thymidylate synthase. *Antimicrobial Agents and Chemotherapy*, **38**, 476–80.

Rees, P. H., Keating, M. I., Kager, P. & Hockmeyer, W. T. (1980). Renal clearance of pentavalent antimony (sodium stibogluconate). *Lancet*, **ii**, 226–9.

Reynoldson, J. A., Thompson, R. C. A. & Horton, R. J. (1992). Albendazole as a future antigiardial agent. *Parasitology Today*, **8**, 412–14.

Ryley, J. F., McGregor, S. & Wilson, R. D. (1988). Activity of ICI 195, 729 – a novel, orally active bistriazole in rodent models of fungal and protozoal infections. *Annals of the New York Academy of Sciences*, **544** 310–8.

Sartorius, C. & Franklin, R. M. (1991). The use of antisense oligonucleotides as chemotherapeutic agents for parasites. *Parasitology Today*, 7, 90–3.

Schmidt, L. H. (1979). Antimalarial properties of floxacrine, a dihydroacridine-dione derivative. *Antimicrobial Agents and Chemotherapy*, 16, 475–85.

Seebeck, T. & Gehr, P. (1983). Trypanocidal action of neuroleptic phenothiazines in *Trypanosoma brucei*. *Molecular and Biochemical Parasitology*, 9, 197–208.

Shapiro, T. (1993). Inhibition of topoisomerases in African trypanosomes. *Acta Tropica*, 54, 251–60.

Sirawaraporn, W., Sertsrivanich, R., Booth, R. G., Hansch, C., Neal, R. A. & Santi, D. V. (1988). Selective inhibition of *Leishmania* dihydrofolate reductase and *Leishmania* growth by 5-benzyl-2, 4-diaminopyrimidines. *Molecular and Biochemical Parasitology*, 31, 79–86.

Slater, A. F. G. (1993). Chloroquine: mechanism of drug action and resistance in *Plasmodium falciparum*. *Pharmacology and Therapeutics*, 47, 2203–35.

Sneader, W. (1985). *Drug Discovery: The Evolution of Modern Medicines*. Chichester: John Wiley & Sons Ltd.

Sogin, M. L., Gundeerson, J. H., Elwood, H. J., Alonso, R. A. & Peattie, D. A. (1989). Phylogenetic meaning of the kingdom concept: an unusual ribosomal RNA from *Giardia lanblia*. *Science*, 243, 75–7.

Stohler, H., Jacquet, C. & Peters, W. (1988). Biological characterization of novel bicyclic peroxides as antimalarial agents. *XIIth International Congress for Tropical Medicine and Malaria*, Amsterdam. Abstract TuP-1-3.

Thakur, C. P., Olliaro, P., Bhowmick, S., Choudhury, B. K., Kumar, M. & Verma, B. B. (1992). Treatment of visceral leishmaniasis (kalazar) with aminosidine (= paromomycin)-antimonial combinations, a pilot study in Bihar, India. *Transactions of the Royal Society of Tropical Medicine and Hygiene*, 86, 615–16.

Townson, S. M., Boreham, P. F. L., Upcroft, P. & Upcroft, J. (1994). Resistance to the nitroheterocyclic drugs. *Acta Tropica*, 56, 173–94.

Urbina, J. A., Lazardi, K., Marchan, E., Visbal, G., Aguirre, T., Piras, M. M., Piras, R., Maldonado, R. A., Payares, G. & De Souza, W. (1991). Mevinolin (lovastatin) potentiates the antiproliferative effects of ketoconazole and terbina-fine against *Trypanosoma* (*Schizotrypanum*) *cruzi*: *in vitro* and *in vivo* series. *Antimicrobial Agents and Chemotherapy*, 37, 580–91.

Urbina, J. A., Marchan, E., Lazardi, K., Visbal, G., Apitz-Castro, R., Gil, F., Aguirre, T., Piras, M. M & Piras, R. (1993). Inhibition of phosphatidylcholine biosynthesis and cell proliferation in *Trypanosoma cruzi* by ajoene, an antiplatelet compound isolated from garlic. *Biochemical Pharmacology*, 45, 2381–7.

Vaidya, A. B., Lashgari, M. S., Pologe, L. G. & Morrisey, J. (1993). Structural features of *Plasmodium* cytochrome *b* that underlie susceptibility to 8-aminoquinolines and hydroxynaphthoquinones. *Molecular and Biochemical Parasitology*, 58, 33–42.

Vanelle, P., Maldonado, J., Gasquet, M., Delmas, F., Timon-David, P., Jentzer, O. & Crozet, M. P. (1991). Studies on antiparasitic agents: effect of the lactam nucleus substitution in the 2-position on the *in-vitro* activity of new 5-nitroimidazoles. *Journal of Pharmacy and Pharmacology*, 43, 735–736.

Van Nieuwenhove, S. (1992). Advances in sleeping sickness therapy. *Annales de la Société Belge de Medicine Tropicale*, 72 Suppl. 1, 39–51.

Vennerstrom, J. L., Ellis, W. Y., Ager, A. L., Andersen, S. L., Gerena, L. & Milhous, W. K. (1992). Bisquinolines. 1. *N*, *N-bis*(7-choroquinolin-4-yl) alkane-diamines with potential against chloroquine resistant malaria. *Journal of Medicinal Chemistry*, 35, 2129–34.

Verdon, R., Polianski, J., Gaudebout, C., Marche, C., Garry, L. & Pocidalo, J-J. (1994). Evaluation of curative anticryptosporidial activity of paromomycin in a dexamethasone-treated rat model. *Antimicrobial Agents and Chemotherapy*, **38**, 1681–2.

Vespieren, P., Cornelissen, A. W. C. A., Thuong, N. T., Helene, C. & Toulme, J. J. (1987). An acridine-linked-oligodeoxynucleotide targeted to the common 5 end of trypanosome mRNAs kills cultured parasites. *Gene*, **61**, 307–15.

Vial, H. J. & Ancelin, M. L. (1992). Malarial lipids: an overview. In (Avila, J. L. & Harris, J. R. eds.) *Subcellular Biochemistry. Volume 18. Intracellular Parasites.* pp. 259–306. New York: Plenum Press.

Wernsdorfer, W. H. & Payne, D. (1991). The dynamics of drug resistance in *Plasmodium falciparum. Pharmacology and Therapeutics*, **50**, 95–121.

White, N. J. (1994). Artemisinin: current status. *Transactions of the Royal Society for Tropical Medicine and Parasitology*, **88 Suppl. 1**, 3–4.

Yang, D. M. & Liew, F. Y. (1993). Effects of qinghaosu (artemisinin) and its derivatives on experimental cutaneous leishmaniasis. *Parasitology*, **106**, 7–11.

Zilberstein, D. & Dwyer, D. M. (1984). Antidepressants cause lethal disruption of membrane function in the human protozoan parasite *Leishmania. Science*, **226**, 977–9.

Zilberstein, D., Liveanu, V. & Gepstein, A. (1990). Tricyclic drugs reduce proton motive force in *Leishmania donovani. Biochemical Pharmacology*, **39**, 935–40.

BIOCIDES: ACTIVITY, ACTION AND RESISTANCE

A. DENVER RUSSELL[1] and NICHOLAS J. RUSSELL[2]

[1]*Welsh School of Pharmacy and* [2]*School of Molecular and Medical Biosciences, Biochemistry Unit, University of Wales, Cardiff*

INTRODUCTION

Early history

Antimicrobial agents have been used in various guises for many centuries. Early concepts, involving an empirical approach, employed copper and silver vessels for storing potable water; wine, vinegar and honey as cleansing agents for wounds; balsams as natural preservatives in aiding mummification; drying, salting and spices for preserving meat and fish; and the use of burning juniper branches in rooms where sufferers of plague had lain (for review see Block, 1991*a*; Hugo, 1991, 1992*a*). Later, in the fifteenth century, came Fracastoro's concept of 'seeds of disease' as possible aetiological agents, and later still the 'animalcules' of van Leeuwenhoek. In the eighteenth century, Pringle devised a method of quantifying chemical preservation, in which he compared solutions of various salts with his standard (sea salt) in terms of their relative ability to preserve lean meat.

Other significant discoveries, which are well described in the reviews cited above, were (i) the use by Davies of iodine (although toxic) as a wound disinfectant; of chlorine water by Semmelweis in obstetrics, and of phenol (again, despite toxicity) by Lemair, Küchenmeister and Lister as wound dressings and in antiseptic surgery; (ii) the studies of microbes and diseases by Agostino Bassi, Pasteur and Koch; (iii) inhibition of bacterial growth, the forerunner of today's minimum inhibitory concentrations, in investigations by Bucholz; (iv) the claimed sporicidal activity (later shown to be incorrect) of mercuric chloride by Koch; and (v) the dynamics of disinfection by Kronig, Paul and Ikeda.

In the early part of this century, other chlorine-releasing agents (CRAs) and some quaternary ammonium compounds (QACs) were introduced. By 1945, the biocides in common use were phenolics (phenol, cresols, chlorocresol), organomercurials (phenylmercuric nitrate, thiomersal), chlorine-releasing agents (hypochlorites, chloramines), iodine, alcohols, formaldehyde (methanal) in both liquid and gaseous forms, some QACs, organic acids and esters, the peroxygen hydrogen peroxide, silver compounds and dyes (triphenylmethanes and acridines).

The past 50 years

Compared with the large number of new antibiotics, few novel biocides have been introduced during the past half-century. However, considerable progress has been made in understanding how biocides act and how resistance can arise, in re-assessing their antimicrobial activity as hitherto unknown infectious agents arise, and in developing their applications in various fields. Space precludes a detailed consideration of all the currently available agents and thus a summary of their properties is presented in Table 1. Additional information can be found by consulting Gardner and Peel (1991), Block (1991*b*) and Russell, Hugo & Ayliffe (1992). In addition, combinations of some agents have been tested, and attempts have sometimes been made to improve activity, particularly against Gram-negative bacteria, by the use of so-called 'permeabilizers' which increase movement of biocides across the outer membrane.

Biocides are used widely as antiseptics and disinfectants in medical (including dental and veterinary) and domiciliary situations, as disinfectants in the food, water (including recreational) and cosmetic manufacturing industries, and as preservatives for a wide range of products such as food, pharmaceuticals, cosmetics, paper, fuels, lubricants, wood and textiles (Board, Allwood & Banks, 1987; Denyer & Hugo, 1990; Rutala, 1990; Favero, 1991; Russell & Gould, 1991).

In this paper, we discuss the types and antimicrobial properties of biocides that have been introduced within the last 50 years or so. In addition, more recent information about 'old' (that is, pre-1945) agents is considered where relevant.

SOME DEFINITIONS

Much argument has been generated over many years about the meanings of the terms 'disinfectant' and 'antiseptic'. In this paper, we define a disinfectant as a substance that is used in the process of disinfection for the purpose of removing micro-organisms, including potentially pathogenic ones, from the surfaces of inanimate objects. Bacterial spores are not necessarily destroyed. Disinfectants may achieve 'high level', 'intermediate level' or 'low level' disinfection, depending on the range of micro-organisms that are inactivated (Table 2). An antiseptic is an agent used in the process of antisepsis for the purpose of destroying or inhibiting micro-organisms on living tissues, thereby limiting or preventing the harmful results of infection. A preservative is a chemical agent included in a pharmaceutical, food or other product that prevents microbial spoilage.

Overall, the term biocide is usually preferable and will be adopted here, although it may be qualified where appropriate to consider one or more of the above types of activity, or where more specific usage is required as in

Table 1. *Some properties and uses of biocides introduced over the past 50 years*

Group	Biocidal agent	Antimicrobial activity	Applications and uses
Alkylating agents	Ethylene oxide	G+ G− M S V P?	Sterilization of some thermolabile pharmaceutical products; fumigation of egg shells
	Propylene oxide	G+ G− M S V	Sterilization of food products
Alcohols	Bronopol	G+ G− F	Preservative in various pharmaceutical and cosmetic products; leather preservative
Aldehydes	Glutaraldehyde	G+ G− M S F V P?	'Chemosterilization' or disinfection of medical equipment, especially endoscopes
	Succinaldehyde-based products	G+ G− M S F V	Disinfection of endoscopes
	Orthophthalaldehyde	G+ G− M S	Newly introduced; possible endoscope disinfection
Aldehydes (formaldehyde-releasers)[a]	Noxythiolin	G+ G−	Topically and as irrigation solution in treatment of peritonitis
	Taurolin	G+ G−	Treatment of peritonitis
	Imidazole derivatives	G+ G− F	Cosmetic preservatives
Amphoteric surfactants	Dodecyl-di(aminoethyl) glycine derivatives	G+ G−	Skin 'disinfection'; disinfection of surgical instruments; sanitizers and disinfectants in food industry
Antibiotic[b]	Nisin	S	Food preservative
Biguanides	Chlorhexidine	G+ G− F Some V	Antiseptic, disinfectant, preservative in some ophthalmic products; anti-plaque agent; veterinary teat dip
	Alexidine	G+ G−	Oral antiseptic; anti-plaque agent
	PHMB	G+ G−	Swimming pool disinfection; application to surfaces in food industry; preservation of leather

Continued

Table 1. *Continued*

Group	Biocidal agent	Antimicrobial activity	Applications and uses
Bisphenols	Hexachlorophane	G+ (G−)	Surgical scrubs; medicated soaps; limited use as preservative in cosmetics
	Triclosan	G+ (G−)	Surgical scrubs; soaps; deodorants; hand-cleansing gels
Halogen-releasing agents	Dichloro- and trichloro-isocyanuric acids	G+ G− M S F V	Disinfection of blood spillages containing HIV or HBV; industrial sanitizing compounds (food, dairy, restaurant, swimming pool); veterinary disinfection
	Iodophors[c] (including povidone-iodine)	G+ G− M S F V	Disinfection of hands pre-operatively; antiseptics; cleansing of dairy plant; veterinary teat dip
Isothiazolones	Mixture of chloromethyl and methyl derivatives	G+ G− F	Preservatives for cosmetics, toiletries and pharmaceuticals, fabrics
Peroxygens	Peracetic acid	G+ G− M S F V P	Disinfecting sewage sludge; widely employed in food processing and beverage industries
	Vapour phase hydrogen peroxide	G+ G− M S F V	Surface sterilizer
Phenolic antioxidants	Butylated hydroxyanisole	G+ (G−) F	Food preservative
	Butylated hydroxytoluene	G+ (G−) F	Food preservative
Quaternary ammonium compounds	Cetylpyridinium chloride	G+ (G−) F some V	Skin disinfection; antiseptic treatment of small wound surfaces; oral and pharyngeal antiseptic; preservative in emulsions; cosmetic preservative
Silver compounds	Silver sulphadiazine (AgSD)	G+ G−	Application to wounds

terms of an inhibition of multiplication, for example, bacteriostat, fungistat, or of an inactivating effect, such as bactericide, sporicide, viricide (virucide). A biocide is simply a chemical agent that inactivates micro-organisms. Chemosterilization implies that a chemical agent is used for sterilization purposes.

FACTORS AFFECTING ACTIVITY OF BIOCIDES

Factors which influence the activity of biocides include concentration, period of contact, temperature, pH, organic matter and the nature and condition of the organism being treated (Russell, 1992a; Russell & Day, 1993). Some of these parameters were examined and understood early this century and in various instances have been put on a sound mathematical basis. For example, the dilution coefficient n is defined as $C^n t = $ constant, where C is concentration and t is time. This is a useful parameter for determining how activity changes at two different concentrations, since $C_1{}^n t_1 = C_2{}^n t_2$ in which t_1 and t_2 represent the lethal exposure times of a biocide used at concentrations C_1 and C_2, respectively, at a specific temperature. A knowledge of n-values has been of considerable importance in evaluating biocidal activity (Russell, Ahonkhai & Rogers, 1979; Russell, 1981) and in undertaking sterility testing of sterilized pharmaceutical products containing preservatives (Russell et al., 1979). For instance, biocides such as phenolics with high n-values rapidly lose activity on dilution, whereas those like organomercurials which have low n-values retain much of their activity.

The activity of a biocide is usually enhanced at elevated temperatures, and thermal effects are usually represented by Q_{10} values, determined from inactivation rates (or the times to inactivate a fixed number of cells: Russell, 1992a). The Q_{10} values of biocides range from 1.5 for formaldehyde to 30–50 for aliphatic alcohols. This fact was taken into account when it was realized several years ago that the use of a biocide at high temperatures, 100°C, might be a suitable procedure for sterilizing some types of injectable products, and this method was incorporated into the British Pharmacopoeia of 1941, only finally being deleted in 1988.

The sporicidal activity obtained when formaldehyde is added to low-temperature steam at sub-atmospheric pressure was first described over 90

Footnotes to Table 1:

Abbreviations: G+, Gram-positive cocci; G−, Gram-negative bacteria; M, mycobacteria; S, bacterial spores; F, fungi; V, viruses in general; P, protozoa; (G−), Gram-negatives are less sensitive than Gram-positives.
[a]There is some doubt as to whether taurolin acts by virtue of formaldehyde release.
[b]Nisin is included here (a) because of its potential value as a food preservative and (b) because it is not a chemotherapeutic drug. It inhibits outgrowth of heat-damaged bacterial spores.
[c]Sporicidal activity of iodophors not shown at normal antiseptic concentrations.

Table 2. *Levels of disinfection*

Level of disinfection	Activity against
High	Spores[a], mycobacteria, non-sporulating bacteria, fungi[b] and viruses (lipid enveloped and non-lipid enveloped)
Intermediate	Mycobacteria, non-sporulating bacteria, fungi[b], non-lipid viruses[c] and lipid viruses
Low	Non-sporulating bacteria, fungi[b], non-lipid viruses[c] and lipid viruses

Based on data from Favero and Bond (1991, 1993).
[a]Prolonged periods of time may be necessary if large numbers of bacterial spores are present.
[b]Not necessarily chlamydospores or sexual spores.
[c]Viricidal activity may be limited.

years ago, but it was not until the 1960s that low temperature steam with formaldehyde was utilized as a sterilization process in this country. A temperature of 73°C is employed to minimize damage to heat-sensitive material whilst maintaining the sporicidal effect (Alder & Simpson, 1992).

Changes in pH can affect the efficacy of a biocide, either directly through changes in its ionization, or indirectly through changes in ionizable groups in the surface of a microbial cell, or both. For example, the active moiety of hypochlorites is undissociated hypochlorous acid (Trueman, 1971), although the sporicidal activity of sodium hypochlorite is enhanced in the presence of sodium hydroxide (Bloomfield & Arthur, 1994). Cationic biocides, such as chlorhexidine, QACs and diamidines, interact strongly with cell-surface groups and consequently their activity increases at alkaline pH. Glutaraldehyde interacts with amino groups in proteins to a much greater extent at alkaline pH than at acid pH (Eagar, Leder & Theis, 1986). Organic acids are widely employed as preservatives or acidulants in food-stuffs and some, such as benzoic and sorbic acids, are sometimes used as pharmaceutical preservatives. As pH increases, antimicrobial activity diminishes, although there has been much debate over the past 20 years or so as to the relative importance of undissociated and dissociated acid forms (Cherrington *et al.*, 1991). The parabens (methyl, ethyl, propyl and butyl esters of *para*[4]-hydroxybenzoic acid) have a pK_a of about 8, and consequently their activity is virtually stable over the pH range 4–8.

Organic matter in the form of blood, serum, pus, faeces, food residues or dirt, adversely affects the activity of many biocides, particularly those such as chlorine-releasing agents which are highly reactive chemically (Russell, 1992*a*). However, such biocides may still be employed, albeit at increased concentrations. For example, blood spillages containing human immuno-deficiency virus (HIV) or hepatitis B virus can be disinfected with sodium hypochlorite solutions containing 10 000 ppm available chlorine (Working Party, 1985). Alternatively, powders or granules of sodium dichloroisocya-nurate may be added to the spillage; this biocide may give a larger margin of safety because a higher concentration of available chlorine is achieved and is also less susceptible to inactivation by organic material (Coates, 1988; Coates & Wilson, 1989; Bloomfield, Smith-Burchnell & Dalgleish, 1990).

Other physical or chemical factors that influence biocidal activity include the presence of non-ionic surfactants, two-phase systems (oil:water) or three-phase systems (as in complex creams or emulsions), rubber closures into which preservatives may partition and polymeric materials such as polyvinyl chloride or nylon to which preservatives may bind (Kostenbauder, 1991; Russell, 1992a). In many of these examples, biocide activity is affected adversely, binding to polymers or partitioning into non-aqueous systems being responsible for reduced biocide availability for attacking micro-organisms.

The efficacy of a biocide may depend not only upon the type of micro-organism (vide infra) and the number of cells or infectious particles present but also upon the condition of the organism. There is currently much interest in the stress responses of micro-organisms and how these are modulated when cultures enter the stationary phase (Siegele & Kolter, 1992); such alterations have a bearing on the susceptibility to biocides. Changes in growth rate and nutrient limitation may alter the composition of both the outer and inner membranes of Gram-negative bacteria (Brown & Williams, 1985; Gilbert, 1988; Poxton, 1993), which in turn alters the ability of compounds to traverse the cell envelope, leading to changes in biocide susceptibility. Modulation of the cell wall of Gram-positive bacteria can be achieved under similar conditions by repeated subculture into glycerol-containing media, resulting in the production of so-called 'fattened' cells (Hugo, 1992b).

The association of microbes with solid surfaces leads to the generation of a biofilm, a consortium of bacteria organized within an extensive mucopoly-saccharide exopolymer known as the glycocalyx (Costerton et al., 1987, 1994). Biofilms are important in both medical (Gilbert, Evans & Brown, 1993) and industrial (Bryers, 1993) contexts. Bacteria existing within differ-ent parts of a biofilm experience different nutritional environments so that their physiology is affected (Geesey, Characklis & Costerton, 1992). Thus, sessile organisms found in biofilms may differ considerably phenotypically from planktonic cells and this, in turn, modifies their response to anti-bacterial agents. Changes in growth rate alter very markedly the activity of QACs, bisbiguanides and substituted phenols and this effect is clearly more evident with sessile than with planktonic organisms (Marrie and Costerton, 1981; Brown & Gilbert, 1993; Carpentier & Cerf, 1993; Poxton, 1993). The non-random distribution of microbes in biofilms has important applications in industry (biofouling, corrosion) and in medical practice (use of appliances in the body). Gram-negative pathogens growing as biofilms on catheter surfaces have a prolonged survival in chlorhexidine concentrations normally used for bladder instillation (Stickler et al., 1990, 1991). They have been

Table 3. *Relative sensitivity of micro-organisms and other infectious agents to biocides*

Infectious agent	Sensitivity to biocides	Comment
Bacteria	Non-sporing bacteria are most susceptible	*Ps. aeruginosa* may show high resistance
	Mycobacteria are less susceptible	Probable link with waxy cell envelope
	Spores are most resistant	Reduced biocide uptake?
Fungi	Fungal spores may be resistant	Reduced biocide uptake?
Viruses	Non-enveloped are more resistant than enveloped	Related to biocide uptake?
Prions	Usually highly resistant	Mechanism(s) unknown
Parasites	Coccidia may be highly resistant	Similar to bacterial spores

Based on data from Russell and Chopra (1990).

responsible for some surprising and spectacular sources of infection, such as the outbreak of *Serratia marcescens* infection derived from the inside surface of storage bottles of 2% chlorhexidine used for hand washing in a hospital (Marrie & Costerton, 1981).

RELATIVE SUSCEPTIBILITY OF DIFFERENT MICROBES TO BIOCIDES

A wide variety of techniques has been employed to quantitate microbial susceptibility to biocides, and the different values often obtained for a particular compound may relate to the numbers of cells, the inoculum history, the test strain, the biocide concentration, exposure time and recovery conditions (Russell, 1992a). It is, nevertheless, possible to reach valid conclusions about the relative sensitivity of microbes to biocides and then to consider reasons for the differing responses (Tables 3–5). A knowledge of such factors has profound importance in selecting anti-microbial agents for specific purposes, for example, in the development of hospital disinfection policies, which must be based on a knowledge of the spectrum of activity of individual agents and of the factors influencing their efficacy (Ayliffe, Coates & Hoffman, 1993; Coates & Hutchinson, 1994). In the food industry, where similar considerations apply, the three most common classes of chemical sanitizers are chlorine, iodine and QACs (Walker & LaGrange, 1991).

Non-sporulating, non-mycobacterial Gram-positive bacteria

Gram-positive cocci such as staphylococci and streptococci are generally more sensitive to biocides than are Gram-negative bacteria. Enterococci are

Table 4. *Comparative sensitivity of some Gram-positive and Gram-negative bacteria to some biocides*

Biocide	MIC (μg/ml) *versus*		
	S. aureus	*E. coli*	*Ps. aeruginosa*
Bronopol	62.5	31.25	31.25
Chlorhexidine	0.5–1	1	5–60
Cetylpyridinium chloride	0.5	100	>100
Hexachlorophane	0.05	12.5	250
Propamidine isethionate	2	64	256
Triclosan	0.1	5	>300

Based on Russell and Gould (1988).

Table 5. *Bactericidal, sporistatic and sporicidal chemical agents*

Bactericidal and sporistatic	Bactericidal and sporicidal	Comment
Phenols, organic acids and esters, QACs, bisbiguanides, alcohols, organomercurials		Low concentrations are sporistatic, but even high ones are not sporicidal
	Glutaraldehyde, formaldehyde, HRAs[a], peroxygens, ethylene oxide	Low concentrations are sporistatic; high concentrations are sporicidal (although long periods may be necessary)

[a]Halogen-releasing agents.

frequently antibiotic resistant, but are not more resistant to chlorhexidine than are streptococci (Baillie, Wade & Casewell, 1992). Methicillin-resistant *S. aureus* (MRSA) strains are rather more resistant than methicillin-sensitive (MSSA) strains to biocides (Day & Russell, 1992), and a clear soluble phenolic has been recommended for controlling outbreaks (Report, 1990).

Listeria monocytogenes is a widely distributed Gram-positive organism, of concern because of the serious nature of the food-borne disease that it causes. However, from a practical point of view, the organism is not particularly resistant to chlorine used as a sanitizer (Walker & LaGrange, 1991), although organic matter has a significant effect on activity of biocides against *Listeria* spp. (Best, Kennedy & Coates, 1990a).

Gram-negative bacteria

Generally, Gram-negative bacteria are less sensitive to many biocides than are Gram-positive cocci such as *S. aureus* (Table 4). This reduced effect is shown by organisms like *E. coli* and *S. marcescens* but is most marked with *Providencia stuartii*, *Proteus* spp., *Pseudomonas aeruginosa* and *Ps. cepacia* (but not by *Ps. stutzeri*, which is usually highly sensitive: Russell & Mills, 1974). *Ps. aeruginosa*, in particular, has long been considered a particularly troublesome organism, with above-average resistance to many biocides and antibiotics (Brown, 1975). A problem is sometimes found with contaminated biocide solutions: for instance, *Ps. cepecia* contamination of povidone-iodine solutions and *Ps. aeruginosa* in solutions of QACs such as benzalkonium chloride (Russell, Hammond & Morgan, 1986).

Hospital isolates of Gram-negative bacteria are often more resistant than culture collection strains to biocides (Russell *et al.*, 1986) and the resistance may be associated with antibiotic resistance (Stickler *et al.*, 1983).

A Gram-negative organism of potential significance is *Legionella pneumophila*, which first came to public prominence in 1977. In the laboratory, many biocides have been found to be active against legionellae, although *L. pneumophila* may show a response closer to *Ps. aeruginosa* than to *E. coli* (Elsmore, 1992). Furthermore, the presence of biofilms in practice may render hazardous the extrapolation of laboratory findings to in-use applications. *Acanthamoeba polyphaga* trophozoites support the intracellular replication of *L. pneumophila*, and *L. pneumophila* present within cysts of these free-living amoebae are protected from inactivation by chlorine (Kilvington & Price, 1990).

Mycobacteria

Early studies on the response of mycobacteria to biocides were reviewed by Croshaw (1971) who pointed out that mycobacteria were resistant to acids, alkalis, QACs, chlorhexidine, non-ionic and anionic surfactants, heavy metals and dyes, although many of these agents were mycobacteriostatic. The mycobactericidal agents listed were ampholytic (amphoteric) surfactants, ethylene oxide gas, iodine (considered to be more effective than hypochlorites), alcohols and especially phenolic compounds such as cresol-soap formulations. Croshaw (1971) concluded that comprehensive data on the effects of biocides on mycobacteria were lacking, that discrepancies existed, that many of the (then) newer biocides had not been examined and that most of the (then) published work referred only to the tubercle bacillus.

Notable commissions from the above list are formaldehyde and glutaraldehyde. Conflicting results have been obtained with the former, although alcoholic solutions are more potent than aqueous ones (Rubbo, Gardner & Webb, 1967). There has also been controversy as to whether or not glutaraldehyde is mycobactericidal. Bergan and Lystad (1972) claimed that the dialdehyde was surprisingly ineffective against tubercle bacilli. Recent studies, however, have demonstrated that 2% alkaline glutaraldehyde is effective against tubercle bacilli and atypical (non-tuberculous) mycobacteria (Collins, 1986a, b; van Klingeren & Pullen, 1987; Best et al., 1990b; Cole et al., 1990; Broadley et al., 1991). Significantly, in view of their association with respiratory complications in AIDS patients, M. avium-M. intracellulare (MAIC) demonstrate a higher resistance to glutaraldehyde (Collins, 1986b; Hanson, 1988; Russell, 1992b).

Bacterial spores

Bacterial spores are considerably less sensitive to biocides than are germinated spores or non-sporulating bacteria (Fig. 1(a), (b); Table 5). Phenolics, QACs and chlorhexidine are bactericidal for vegetative cells but only sporistatic (inhibiting spore germination and/or outgrowth). However, other chemical agents, such as glutaraldehyde, ethylene oxide, peracetic acid and CRAs are both bactericidal and sporicidal, although higher concentrations are needed to achieve the latter effect (Russell, 1990; Russell & Chopra, 1990; Bloomfield & Arthur, 1994). Comparatively few agents are, in fact, sporicidal (Table 5) but the activity of sporistatic chemicals against spores may be enhanced at elevated temperatures. Chemicals that are sporicidal also inhibit germination and/or outgrowth (Fig. 2), although the concentrations for these purposes are equivalent to those that are bacteriostatic and well below sporicidal levels.

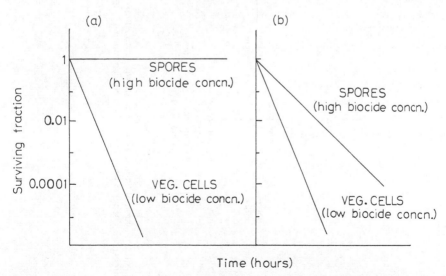

Fig. 1. Chemical agent showing (a) bactericidal and sporistatic activity, (b) bactericidal and sporicidal activity.

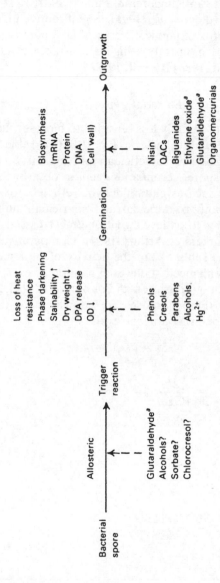

Fig. 2. Sites of action of antibacterial agents during germination and outgrowth. *Note:* some agents, e.g. glutaraldehyde, can act at more than one stage. [a]Denotes sporicidal agent when used at high concentration. ↑, increase; ↓ decrease; OD, optical density. (Reproduced with permission from Russell & Chopra, 1990.)

Spore destruction by chemical agents is important in chemosterilization by glutaraldehyde of instruments that cannot be sterilized by physical methods, for 'sterilization' by hydrogen peroxide of food contact surfaces, and for gaseous sterilization by ethylene oxide or low-temperature steam with formaldehyde (Russell, 1990). However, it must be added that these agents, although useful for treating thermolabile equipment, do not provide the same degree of safety as such thermal processes as autoclaving. In the food industry, moreover, heat (or possibly ionizing radiation) is employed to destroy spores, and preservatives are incorporated to prevent the germination and outgrowth (and associated toxin production) of spores of *Clostridium botulinum* and other clostridia and bacilli (Gould, 1983).

Yeasts and moulds

Compared with bacteria, data on the effects of biocides on yeasts and moulds is scarce (Wallhäuser, 1984; Russell & Gould, 1991; Russell, 1992c), but the following general conclusions may be drawn: (i) fungicidal concentrations are often much higher than concentrations needed to inhibit growth (Wallhäuser, 1984); (ii) moulds are sometimes, but not invariably, less sensitive than yeasts, for example, to chlorhexidine and organomercury compounds; (iii) compared with bacteria, fungi are more susceptible to inhibition by parabens, and these compounds show the classical response of increasing activity through the homologous series from methyl to butyl ester (Aalto, Firman & Rigler, 1993); because of their limited solubility, the methyl and propylparabens are often used in combinations and their effects are additive (Kabara & Eklund, 1991); and (iv) non-sporulating bacteria (except mycobacteria) are considerably more sensitive than moulds and yeasts to biocides (Karabit, Juneskans & Lundgren, 1989).

Chlorhexidine is less active against yeasts and moulds than towards cocci and Gram-negative bacteria. Glutaraldehyde is fungicidal or fungistatic, depending on its concentration, and peracetic acid is also an excellent fungicide although its activity decreases with increasing pH. Organic acids (benzoic, sorbic) and esters (parabens) are effective antifungal agents and are widely used as preservatives (Russell, 1992c).

Protozoa

Several distinctly different types of protozoa, including *Giardia*, *Cryptosporidium*, *Naegleria*, *Entamoeba* and *Acanthamoeba*, are potentially pathogenic and may be acquired from water. Furthermore, a resistant cyst stage is included in their life cycle (Jarroll, 1992). Agents, in ascending order of efficacy, that are cysticidal towards *G. muris* cysts are monochloramine, free chlorine, iodine, chlorine dioxide and ozone (Jarroll, 1988). Chlorine dioxide is more effective against *Cryptosporidium* oocysts than is either free chlorine or monochloramine; these cysts also are sensitive to ozone (Korich et al., 1990).

Acanthamoeba spp. are responsible for corneal keratitis, sometimes associated with the use of contaminated contact lenses or contact lens solutions. A special issue of *Reviews of Infectious Diseases* (1991, Vol. 13, Suppl. 5) deals with the cell cycle, encystment, ecology and biocide sensitivity of *Acanthamoeba* spp.. Whilst trophozoites are readily inactivated by contact-lens disinfecting solutions, cysts are more refractory and the presence of proteinaceous matter is an additional hazard (Seal & Hay, 1992).

Viruses

Early studies on the effects of biocides on viruses and bacteriophages are reviewed by Grossgebauer (1970). An important hypothesis was put forward in 1963 and modified in 1983 by Klein and Deforest (1983) who predicted that virus susceptibility to biocides could be based on whether viruses were 'lipophilic' because they possessed a lipid envelope (for example, herpes simplex) or 'hydrophilic' because they did not (for example, poliovirus). In this scheme, 'lipophilic biocides' such as 2-phenylphenol, and cationic surfactants, such as QACs, isopropanol, ether and chloroform were viricidal to the enveloped viruses. Curiously, the adenoviruses without a lipid envelope were lipophilic in nature and sensitive to these agents. A knowledge of the inactivation of viruses has, in the past, been important in the preparation of vaccines such as inactivated (Salk) poliomyelitis vaccine, in which formaldehyde was used as the viricidal agent. Nowadays, with the current importance of HIV and the hepatitis A (HAV) and B (HBV) viruses, plus the need to prevent the transmission of viral infection, it is essential to have a sound knowledge of viral inactivation. Important viricidal agents include CRAs, formaldehyde (in inactivated vaccine production), glutaraldehyde at alkaline pH and peracetic acid (Springthorpe & Sattar, 1990; Sattar et al., 1994). Bacteriophages are currently being considered as indicator 'organisms' for assessing the viricidal activity of biocides (Davies et al., 1993). It is likely that phages will play an increasingly important role in this context as well as providing additional insights into the mechanisms of action of viricides (Taylor, 1992).

Prions

Prions are responsible for the so-called 'slow virus diseases', a distinct group of transmissible degenerative encephalopathies (see chapter by Kerr and Lacey). Prions are remarkably resistant to inactivation by many chemical and physical agents, but as they have not been purified, it is difficult to state whether this is intrinsic or to what extent it is influenced by the protective effect of host tissue (Taylor, 1992). They are resistant to formaldehyde, glutaraldehyde, organic solvents, chlorine dioxide, hydrogen peroxide, iodine and strong acids but are inactivated by alkali.

MECHANISMS OF ANTIBACTERIAL ACTION

Much of the early work in the 1940s and 1950s by Gale, Hotchkiss, Hugo, McQuillen and Salton on phenols and QACs paved the way for later research. Most studies of the mode of action of biocides have examined non-sporulating bacteria, although recent work (for review see Bloomfield & Arthur, 1994) has also provided useful information about the ways in which spores are inactivated.

Non-sporulating bacteria

The initial interaction of a biocide with any cell clearly involves the cell surface. However, although useful, adsorption isotherms provide relatively little information about the site and action of an inhibitor (Hugo, 1992b). The target sites on or within the bacterial cell are shown in Fig. 3: these can be summarized by consideration of the effects of an antimicrobial agent on the outer cell layers, cytoplasmic membrane and cytoplasmic constituents.

Cell wall or outer membrane

Glutaraldehyde binds to several components in bacterial cell envelopes: proteins, peptide chains in peptidoglycan, teichoic acids in Gram-positive bacteria and lipoprotein in Gram-negatives (Russell & Chopra, 1990). The resulting high degree of cross-linking, particularly at alkaline pH (Eagar et al., 1986), means that the bacterial cell is unable to undertake most, if not all, of its essential functions.

Cationic bactericides, such as QACs and chlorhexidine, may damage the outer membrane of Gram-negative bacteria, thereby promoting their own entry into the cell (Hancock, 1984). Agents which increase the permeability of the outer membrane of Gram-negative bacteria include ethylenediamine and tetraacetic acid, sodium hexametaphosphate, organic acids (citric, malic, gluconic) at alkaline pH, polycations and iron-binding proteins (Nikaido & Vaara, 1985; Vaara & Jaakkola, 1989).

Cytoplasmic (inner) membrane

Damage to the bacterial membrane is often manifested as leakage of low-molecular-weight intracellular constituents, notably K^+ and amino acids, as well as macromolecules (Russell & Hugo, 1988; Hugo, 1992b). Leakage, as induced by phenolics, QACs, chlorhexidine and alcohols, is best considered as a measure of the generalized loss of function of the membrane as a permeability barrier and may be related to bacteriostasis but not necessarily

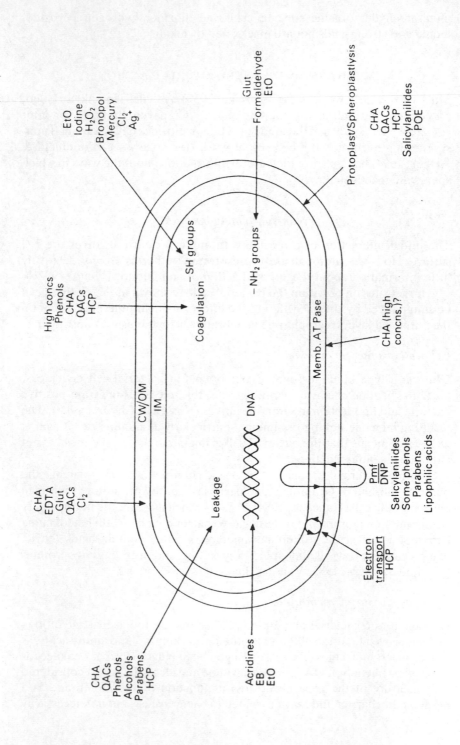

to bacterial death. Total membrane disruption of isolated spheroplasts and protoplasts is achieved by low concentrations of the cationic agents but at higher concentrations lysis is reduced because of intracellular coagulation (Hugo, 1992b). Although chlorhexidine collapses the membrane potential, it is membrane disruption rather than ATPase inactivation which is considered the lethal effect, despite earlier reports to the contrary (Kuyyakanond & Quesnel, 1992; Barrett-Bee, Newboult & Edwards, 1994). Alexidine and a polymeric biguanide, polyhexamethylene biguanide, but not chlorhexidine, promote the formation of domains of acidic phospholipids in the membrane. Permeability changes follow and it is likely that there is an altered function of some membrane-associated enzymes (Woodcock, 1988). Alcohols produce pleiotropic effects on a wide range of cellular functions, but the cytoplasmic membrane appears to be the main target site; some water is necessary (probably to aid membrane penetration) and 70% alcohol is commonly used as a biocide. Alcohols disrupt membrane structure by altering the packing and order of the fatty-acyl chains in lipids; they can inhibit membrane-bound enzymes, particularly ATPase and other transport proteins; and they can also influence membrane lipid acyl composition by modifying the properties of fatty acid synthetase (Seiler & Russell, 1991).

The action of organic acids, parabens, and dinitrophenol (DNP) on membranes appears to be specifically to dissipate the proton-motive force (pmf), although for individual compounds a range of activities has been ascribed: these include the inhibition of transport and enzymes in the cytoplasm as well as the membrane (Kabara & Eklund, 1991). Whether or not these other inhibitions are secondary to the primary effect on pmf is not at all clear. For instance, with organic acids it is uncertain whether acidification of the cytoplasm also has a primary role in growth inhibition.

Several biocides are known to act on membrane enzymes. Hexachlorophane has a specific effect on the electron transport chain, while bronopol, mercury compounds and isothiazolones interact with enzyme thiol groups. At growth-inhibitory concentrations, the isothiazolones have little effect on membrane integrity but a significant effect on active transport and the oxidation of glucose and a marked effect on thiol-containing enzymes (Collier et al., 1990a, b). Sulphite also acts on membranes, but this cannot be regarded as its primary target (except in the sense that it may interact first

---

Fig. 3. Target sites and effects of biocides (reprinted with permission from Russell & Chopra 1990). Many biocides have more than one target site and effects may be concentration dependent. Abbreviations: EDTA, ethylenediamine tetraacetic acid: Glut, glutaraldehyde; CHA, chlorhexidine diacetate (or gluconate); QACs, quaternary ammonium compounds; HCP, hexachlorophane; Cl_2, chlorine-releasing agents (e.g. hypochlorites); POE, phenoxyethanol; DNP, dinitrophenol; EB, ethidium bromide; EtO, ethylene oxide. CW, cell wall (Grampositive): OM, outer membrane (Gram-negative); IM, inner (or cytoplasmic) membrane; pmf, proton motive force.

with the exterior membrane of the cell) because of its action on such a wide range of cellular constituents (*vide infra*).

Cytoplasmic constituents

Several biocides interact with bacterial cytoplasmic constituents: agents such as ethylene oxide alkylate proteins and nucleic acids; glutaraldehyde and formaldehyde cross-link proteins, RNA and DNA; acridines intercalate into DNA; and nucleic acid-binding compounds, such as QACs and chlorhexidine, at high concentrations coagulate DNA, RNA and protein (Russell & Chopra, 1990; Hugo, 1992*b*).

Thiol groups form a prime target for biocidal action. Thus, hypochlorites progressively oxidize thiol groups to produce disulphides, sulphoxides and disulphoxides. The activity of hydrogen peroxide results from the formation of free hydroxyl radicals ($\cdot$OH) which oxidize thiol groups in enzymes and proteins. Peracetic acid is also believed to disrupt thiol groups. The mechanism of action of silver salts resides in the silver (Ag^+) ion, which interacts strongly with $-SH$ groups in proteins, to produce mercaptides, and also with DNA (Hamilton-Miller, Shah & Smith, 1993; Russell & Hugo, 1994).

Sulphite attacks a plethora of targets within the cell, including the membrane as well as proteins, lipids and cofactors within the cytoplasm, through its chemical reaction with carbonyl, thiol and disulphide groups and the inactivation of such cofactors as pyridoxal phosphate, thiamine pyrophosphate, folate, haem and flavin (Gould & Russell, 1991). This results in the inhibition of intermediary metabolism, energy production, and protein and nucleic acid biosynthesis.

Bacterial spores

As noted above, some biocides are bactericidal but sporistatic, whereas others are bactericidal and sporicidal (Table 5). Virtually all the important findings on the mechanisms of sporicidal and sporistatic action have been made within the past 25–30 years or so and have been reviewed by Russell (1990) and Bloomfield & Arthur (1994).

Sporicidal action

Glutaraldehyde at both acid and alkaline pH interacts with the outer layers of spores; it is considered that the acid form resides at the cell surface, whereas an alkalinating agent such as sodium bicarbonate assists in the increased penetration of the alkaline form into the spore where it interacts strongly with the cortex and core (Power & Russell, 1990). Formaldehyde, which is much less effective than glutaraldehyde, probably penetrates into the spore where it interacts with RNA, DNA and the amino groups in protein. A new aldehyde, orthophthalaldehyde, is claimed to be sporicidal (Alfa & Sitter, 1994), but its mechanism of action is unknown.

CRAs induce the release of dipicolinic acid from spores, which suggests an increase in spore permeability and a site of action in the cortex. In addition, they release coat protein and increase spore sensitivity to lysozyme; the extent of coat protein release correlates with sporicidal activity, sodium hypochlorite being more effective than sodium dichloroisocyanurate on both counts. Hypochlorite also induces the substantial release of cortex hexosamine, sodium dichloroisocyanurate again being less effective. Thus, CRAs cause coat and cortex degradation, thereby enabling chlorine to reach the core and probably produce inactivation by oxidation of essential proteins (Bloomfield & Arthur, 1994).

Hydrogen peroxide removes coat protein from *Clostridium bifermentans*. Activation to hydroxyl radicals ($\cdot$OH) is necessary, and peroxide in the presence of Cu^{2+} ions produces lysis of cortical fragments and also of spore protoplasts (Russell, 1990; Bloomfield & Arthur, 1994).

Sub-lethal spore injury (as opposed to inactivation) may be manifested as an increased susceptibility to 'stressing agents', which are not themselves sporicidal (Williams & Russell, 1992, 1993). It has also been shown that it is possible to revive some spores of *Bacillus* spp. that have been exposed to high concentrations of some sporicides even for prolonged periods of time (Power *et al.*, 1989; Williams & Russell, 1992, 1993). The exact mechanisms of repair are unknown.

Sporistatic action

Most, if not all, biocides at low concentrations inhibit spore germination and/or outgrowth (Fig. 2). Phenolics, parabens, alcohols, glutaraldehyde and formaldehyde all inhibit germination, with glutaraldehyde probably acting at a very early stage (Power & Russell, 1990). The inhibitory effects of the first three agents and of low concentrations of formaldehyde, but not of glutaraldehyde, are reversible, probably indicative of a fairly loose binding to a site(s) on the spore surface. For unknown reasons, QACs and chlorhexidine have no effect at this stage.

Spore outgrowth is inhibited by glutaraldehyde, organomercurials, QACs, chlorhexidine, CRAs and ethylene oxide, but not by phenolics and the parabens (except at high concentrations) (Fig. 2).

Nitrite is particularly useful as a preservative of cured meats because it not only is effective against spores of potential pathogens but also contributes to the flavour of the product (Roberts *et al.*, 1991). Unfortunately, the progress in understanding the chemistry of such compounds as iron–nitrosyl complexes, which nitrite might form in food through its interaction with iron–sulphur proteins, has not revealed the primary action of this biocide. Its action is increased at pH values below 6 and is synergistic with ascorbate which converts nitrite to nitric oxide; but again the role of the latter compound in growth inhibition is unclear (Roberts *et al.*, 1991).

MECHANISMS OF ACTION AGAINST OTHER MICROBES

Antifungal action

Possible mechanisms of the antifungal action of biocides have been reviewed by Lyr (1987) and Russell (1992c). It is often assumed that the same effects as those in non-sporulating bacteria are responsible for fungal inactivation but in view of the considerable structural and biochemical differences between these microbes this is undoubtedly an over-simplification.

An initial interaction at the cell surface is followed by passage of a biocide across the fungal cell wall to reach its target site(s), but little information is available about the ways whereby uptake into fungal cells is achieved despite long-standing studies of biocide adsorption to yeasts and moulds (Gadd & White, 1989).

Cell wall as target

Although the fungal cell wall may be a prime target for developing new antifungal antibiotics (Hector, 1993), few biocides are likely to have the wall as a sole target. Chitin has been suggested as being a potentially reactive site for glutaraldehyde action in yeasts (Gorman & Scott, 1977). Glutaraldehyde is also known to cause agglutination of yeast cells (Navarro & Monsen, 1976).

Plasma membrane as target

Chlorhexidine induces K^+ release from baker's yeast (Elferink & Booij, 1974) and affects the ultrastructure of budding *C. albicans* with the loss of cytoplasmic constituents (Bobichon & Bouchet, 1987). QACs also affect membrane integrity. Heavy metals probably bind to key functional groups of enzymes (Lyr, 1987).

Other target sites

Very few relevant studies have been made, but DNA and RNA would be expected to be targets for a number of biocides, such as cationic agents that interact strongly with nuclcic acids (Hugo, 1992b). A particularly sensitive target for the action of sulphite in the yeast, *Saccharomyces cerevisiae*, appears to be the enzyme glyceraldehyde-3-phosphate dehydrogenase, but it is unclear whether this results from inactivation of the enzyme *per se*, its substrate or its cofactor, all of which contain sensitive groups (Gould & Russell, 1991).

Antiviral action

Information concerning the mechanisms of viricidal activity is still sketchy. Grossgebauer (1970) considered the possible effects of biocides on viruses and proposed that water-saturated phenol caused separation of protein from infectious RNA in poliovirus, that low formaldehyde concentrations produced an antigenic but non-infectious particle (for use in a vaccine) whereas higher aldehyde concentrations gave a non-infectious destroyed particle. Thurman & Gerba (1988, 1989) considered the possible interactions between viruses and biocides as being: (i) adsorption to capsid receptors, (ii) conformational changes in virus, (iii) destruction of capsid leading to release of infectious nucleic acid, (iv) as for (iii) but nucleic acid rendered non-infectious, (v) capsid remaining intact, but nucleic acid rendered non-infectious and they pointed out that CRAs could inactivate viruses by attacking either capsid proteins or nucleic acid. Although evidence has been put forward in support of both theories, the precise mechanism of inactivation is unresolved.

Glutaraldehyde reduces the activity of hepatitis B surface antigen and especially core antigen in hepatitis B virus (Adler-Storthz et al., 1983) and interacts with lysine residues on the surface of hepatitis A virus (Passagot et al., 1987). Low concentrations (<0.1%) of alkaline glutaraldehyde act against purified poliovirus whereas poliovirus RNA is highly resistant to higher concentrations (Bailly et al., 1991). From this, it may be inferred that changes to the capsid are responsible for loss of infectivity. Support for this contention has obtained by demonstrating that the capsid proteins of poliovirus and echovirus react with low concentrations (0.05 and 0.005%, respectively) of the dialdehyde, the ten-fold difference in aldehyde concentration probably reflecting major structural alterations in the two viruses (Chambon, Bailly & Peigue-Lafeuille, 1992). Some biocides (hypochlorite, 70% ethanol and cetrimide) induce a rapid loss of the outer capsid layer, whereas chlorhexidine and phenol affect morphology only after extended periods of exposure (Rodgers et al., 1985).

Little is known about the manner in which biocides penetrate viruses and it could well be that bacteriophages, currently used in laboratories in this country and elsewhere, could provide some valuable information. Certainly phages have been shown to be of value as models of human enteric viruses (Havelaar, 1993).

Antiprotozoal action

Very little is known about the ways in which biocides, as opposed to chemotherapeutic agents, act against trophozoites and cysts. Leland (1991) describes antiprotozoal compounds, including drugs, but does not consider their mechanism of action. Likewise, Jarroll (1988, 1992) discusses the sensitivity of protozoa, including *Giardia* cysts, to biocides but again does not deal with the ways in which protozoa are inactivated.

It is, difficult, therefore, to reach any conclusions about mechanisms of anti-protozoal activity of biocides.

MECHANISM OF BACTERIAL RESISTANCE

In contrast to antibiotics, the mechanisms of resistance to biocides are comparatively poorly understood, although progress in the last 10–20 years has been encouraging. Resistance mechansims can be considered as being either intrinsic (natural, innate) or acquired.

Intrinsic resistance

This form of resistance is usually considered to involve exclusion of a biocide as a consequence of impermeability and examples are provided by Gram-negative bacteria, mycobacteria and bacterial spores. Additionally, physiological (phenotypic) adaptations of an organism in response to changes in growth environment can modulate sensitivity to biocides.

Gram-negative bacteria

These organisms are usually more resistant than Gram-positive cocci to many biocides (see Table 4 above). The outer membrane (OM) of Gram-negative bacteria plays an important role in conferring resistance to many biocides and antibiotics (Brown, 1975; Brown & Williams, 1985; Gilbert, 1988; Heinzel, 1988). Phenols, parabens, QACs and diamidines are all more effective against deep-rough mutants than parent strains of *E. coli* and *Salmonella typhimurium* (Russell & Gould, 1988); in the case of chlorhexidine salts, the OM of wild-type *S. typhimurium*, but not of *E. coli*, confers intrinsic resistance. It must be added, however, that phenotypic adaptation does produce changes in cell envelope composition that modulates sensitivity to chlorhexidine also in *E. coli*. The Gram-negative cell envelope acts as a barrier to the entry of high molecular weight polyhexamethylene biguanides (Gilbert, Pemberton & Wilkinson, 1990).

Recent observations on phenotypic acid tolerance in enterobacteria have a bearing on possible resistance to those biocides which have their effect directly or indirectly through altering intracellular pH. Exposure of *E. coli* and *S. typhimurium* to the moderately low pH values of 5.5–6.0 protects them from subsequent challenge of a low pH of 3.5–4.0, giving them up to 100-fold increase in resistance (Foster & Hall, 1990, 1991; Raja *et al.*, 1991). This response is called habituation in *E. coli* and acid tolerance response in *S. typhimurium*, and in each micro-organism it requires protein synthesis. During the exposure to moderately low pH, gene expression changes and on subsequent exposure to low pH the synthesis of specific proteins is either increased or decreased greatly. Most of these proteins have not been identified, but some are membrane proteins and others are examples of heat-shock proteins, indicating that their likely functions are in the immediate sensing of pH changes (at the membrane surface) and in dealing with acid-damage to intracellular proteins (Olson, 1993). Recent evidence suggests that the rate of growth has a marked effect on habituation in *E. coli* (Small *et al.*, 1994) and the acid tolerance response in *S. typhimurium* (Lee, Slonczewski & Foster, 1994). Future work should include study of how these effects influence biocide sensitivity.

Notoriously resistant organisms are *Pseudomonas aeruginosa* to, for example, QACs (Brown, 1975; Sakagami *et al.*, 1989) and *Ps. cepacia*, *Proteus* spp. and *Providencia stuartii* to chlorhexidine (Russell & Gould, 1988). In *Ps. aeruginosa*, a factor of considerable importance is the high cation content of the OM which helps to produce strong inter-LPS (lipolysaccharide) linkages, thereby hindering biocide access but conversely being responsible for the high sensitivity of the organism to chelating agents. In *Proteus* spp. and *Ps. cepacia*, the resistance to cationic biocides may be due to changes in LPS composition that decrease the affinity of these inhibitors (Cox & Wilkinson, 1991; Vaara, 1992). In general terms, hydrophobic biocides are largely excluded (Nikaido & Vaara, 1985), whereas cationic biocides might be able to promote their own entry into cells by damaging the outer membrane (Hancock, 1984).

There are several possible reasons for the reduced sensitivity of sessile bacteria in biofilms (Brown & Gilbert, 1993): biocide access is prevented or otherwise modified (this depends on the nature of the biocide, its binding capacity to the glycocalyx and the rate of growth of the biofilm microcolony); modulation of the micro-environment may occur, involving nutrient limitation and growth rate changes; and there might be increased production of degradative enzymes by attached cells.

Gram-positive cocci

The plasticity of the cell envelope of bacteria is well known (Poxton, 1993). The physiological state of the cells, which is influenced by growth rate and by any growth-limiting nutrient, affects the thickness and degree of cross-linking of peptidoglycan and hence the sensitivity to antibacterial agents such as chlorhexidine and phenoxyethanol. Since lysozyme-induced protoplasts remain sensitive to these 'membrane-active' agents, the cell wall must be responsible for the modified response in whole cells (Gilbert, 1988). Likewise, 'fattened' cells of *S. aureus* are less sensitive to higher phenols than normal, unfattened cells (Hugo, 1992*b*).

Mycobacteria

Mycobacteria are much more resistant to a variety of biocides than are other non-sporulating bacteria, although differences in sensitivity exist between different mycobacterial species (Russell, 1992*b*). It was proposed as early as the 1930s that this resistance was related to the content of waxy material in the cell wall, and today we still seem little nearer to understanding the mechanisms of biocide uptake into mycobacteria. Certainly the lipid content of the wall acts as a barrier and the activity of chlorhexidine against some mycobacteria can be improved by the use of agents that inhibit the synthesis of specific wall components (Russell *et al.*, 1994). Further investigations of the wall as a permeability barrier are required.

Bacterial spores

It has been known for more than a century that bacterial spores are more resistant than vegetative forms, and studies in the early and mid-parts of the present century demonstrated that bactericides such as phenolics were sporistatic and not sporicidal (see also Table 6). Mechanisms whereby biocides are taken up by spores are poorly understood, but the coat(s) and cortex appear to prevent many compounds from reaching the spore core.

Several techniques have been employed for studying spore resistance to biocides, three of which have been found to be of particular use. The first involves the removal of outer spore layers; the second the use of mutants of *Bacillus subtilis* 168 which cannot genetically proceed beyond a certain point in spore development; and the third a 'step-down' procedure in which vegetative cells growing in a rich medium are suddenly transferred into a nutritionally poor medium, when during subsequent incubation there is about 90% synchronous sporulation (Power *et al.*, 1988; Russell, 1990; Bloomfield & Arthur, 1994). Coat-removal methods which utilize a combination of urea plus dithiothreitol and sodium dodecyl sulphate at alkaline pH are now known to remove a substantial amount of cortex also (Bloomfield & Arthur, 1994) and the subsequent addition of lysozyme further breaks down this layer.

There are seven well-defined stages in the development of a bacterial spore from a vegetative cell (Russell, 1990). Of these, the laying-down of the cortex and spore coats in the later stages (IV–VI) are associated with reduced sensitivity to biocides. On the basis of the methods outlined above, it has been demonstrated that: (i) resistance during sporulation to chlorhexidine and QACs develops at Stage IV and is completed at Stage V; (ii) resistance during sporulation to glutaraldehyde develops much later during sporulation, commencing at late Stage V and being fully developed by the end of Stage VI; and (iii) the spore coat(s) and cortex are involved in resistance to CRAs and iodine and probably to other biocides as well. Small, acid-soluble proteins (SASPs) might play a role in peroxide resistance (Setlow, 1994).

Acquired resistance

Acquired resistance to biocides arises either as a consequence of chromosomal gene mutation or by the acquisition of plasmids and transposons.
Detailed knowledge of mutation and of these genetic elements has clearly
been derived only during the last 30 years or so and consequently progress in
understanding acquired resistance to biocides is within this time-span.

Mercury resistance is plasmid- or transposon-borne, is inducible and may
be transferred by conjugation or transduction. Inorganic (Hg^{2+}) mercury
resistance is a common property of clinical isolates of *S. aureus* containing
penicillinase plasmids. Plasmids are either (a) narrow spectrum, specifying
resistance to Hg^{2+} and some organomercurials, or (b) broad spectrum,
encoding additional resistance to other organomercurials. The enzymes
involved are mercuric reductase (Hg^{2+}) and lyase (hydrolase) and reductase
(organo-compounds) (Foster, 1983). Resistance to other metals has been
described (Foster, 1983; Silver *et al.*, 1989) and often involves efflux
mechanisms. Silver (Ag^{+}) reduction may not form the basis of silver or
silver sulphadiazine (AgSD) resistance (Belly & Kydd, 1982; Foster, 1983;
Silver *et al.*, 1989) which may be associated with accumulation (Trevor,
1987; Russell & Hugo, 1994). Resistance to cationic biocides could arise as a
result of the presence of an efflux mechanism (Midgley, 1987). Mutants of
Enterococcus hirae (formerly *Streptococcus faecalis*) have been isolated in
which increased resistance to ethidium bromide is associated with constitut-
ive expression of an efflux system which is inducible in the sensitive parent
strain (Midgley, 1994).

Some methicillin-resistant strains of *S. aureus* (MRSA) containing plasmids encoding gentamicin resistance (MGRSA) also have increased MIC
values towards such biocides as QACs, chlorhexidine, ethidium bromide
(EB), acridines and propamidine isethionate (Emslie, Townsend & Grubb,
1985a; Emslie *et al.*, 1985b; Lyon & Skurray, 1987; Cookson & Phillips,
1990). At least three determinants are responsible for biocide resistance in
clinical isolates of *S. aureus*: these are *qacA*, which specifies resistance to
QACs, acridines, EB, propamidine isethionate and low-level chlorhexidine
resistance; *qacB*, which is similar; and the genetically unrelated *qacC*
encoding QAC and low-level EB resistance.

These biocides bind strongly to DNA and have been termed nucleic acid-
binding (NAB) compounds. Those MGRSA strains without NAB plasmids
are more sensitive to chlorhexidine than are MSSA strains, whereas MRSA
with such plasmids (termed GNAB) encoding resistance to gentamicin and
NAB compounds are more resistant to chlorhexidine (Cookson, Bolton &
Platt, 1991). Curing of methicillin resistance does not produce changes in
chlorhexidine sensitivity in MRSA strains with or without NAB plasmids,
whereas curing of GNAB plasmids produces a reduced MIC of chlorhexi-

Table 6. *Possible mechanisms of plasmid-mediated biocide resistance*

Mechanism	Example(s)	Comment	References
Inactivation	Mercurials	Hydrolases and reductases	Foster (1983), Silver *et al*. (1989)
	Chlorhexidine?	Possible chromosomally-mediated mechanism: not shown to be plasmid-encoded	Ogase *et al*. (1992)
	Formaldehyde	Formaldehyde dehydrogenase responsible	Heinzel (1988)
Decreased uptake	Silver nitrate	Inactivation unlikely	Russell & Hugo (1994)
	Chlorhexidine?	Unproven	Russell & Day (1993)
	QACs?	Unproven	Russell & Day (1993)
Cell surface alterations	Formaldehyde	Outer membrane proteins appear to be involved	Kaulfers, Karch & Laufs (1987)
Efflux	Acridines, crystal violet, diamidines, ethidium bromide, QACs	MRSA strains	Emslie *et al*. (1985a, b) Midgley (1986, 1987) Lyon & Skurray (1987)
	Chlorhexidine?	MRSA strains	Russell & Day (1993)

dine (Cookson *et al.*, 1991). From this, it is clear that GNAB is related to chlorhexidine resistance. There is evidence that there is an efflux of cationic agents (including chlorhexidine?) from MRSA strains (Midgley, 1986, 1987).

Recombinant *S. aureus* plasmids transferred into *E. coli* are responsible for conferring resistance in the Gram-negative cells (Yamamoto, Tamura & Yokota, 1988). A plasmid-borne, EB-resistance determinant from *S. aureus* cloned in *E. coli* encoded resistance to EB and to QACs which were expelled from the cells (Midgley, 1986, 1987).

In Gram-negative bacteria, the introduction of the RP1 plasmid into *Ps. aeruginosa* does not alter the response to biocides except, apparently, hexachlorophane (Russell, 1985). In some instances, the introduction of a plasmid may even increase sensitivity. Plasmid R124 alters the OmpF OM protein in *E. coli* and cells containing this plasmid are more resistant to cetrimide and other agents (Roussow & Rowbury, 1984). Formaldehyde resistance has been claimed to be plasmid encoded in some cases (Kaulfers, Karch & Laufs, 1987; Heinzel, 1988). Generally, however, there is little experimental evidence that plasmids are responsible for the development or spread of biocide resistance (Russell, 1985). Possible mechanisms of plasmid-mediated resistance are summarized in Table 6.

MECHANISMS OF RESISTANCE OF OTHER MICROBES

Fungi

Two basic mechanisms of fungal resistance to biocides can be envisaged: intrinsic and acquired (Dekker, 1987; Russell, 1992c). The fungal cell wall contains various types of polymers, including chitin and chitosan (Zygomycetes), chitin and glucan (mycelial forms of Ascomycetes and Deuteromycetes), glucan and mannan (yeast forms of Ascomocytes and Deuteromycetes). Thus, there is ample opportunity for a cell to exclude biocide molecules. Studies on the sensitivity of C. albicans to the polyene antibiotic, amphotericin B, suggests that glucan may have a role to play in limiting drug uptake (Gale, 1986). No significant studies have been made to date with biocides. Clearly, further studies are needed to substantiate this contention and to examine the mechanisms involved in the uptake of biocides into yeasts and moulds.

There is no evidence linking the presence of plasmids in fungal cells and the ability of the organisms to acquire resistance to fungistatic or fungicidal agents, although acquired resistance of yeasts to organic acids is known (Warth, 1977).

Viruses

Thurman and Gerber (1988) pointed out that conflicting results were often reported about the action of biocides on different virus types. They suggested that the structural integrity of a virus was altered by an agent that reacted with viral capsids to increase viral permeability, so that a 'two-stage disinfection' could offer an efficient means of viral inactivation whilst overcoming the possibility of multiplicity reactivation first put forward by Luria in 1947 to explain an initial reduction and then an increase in titre of biocide-treated bacteriophage.

Klein & Deforest (1983) have attempted to explain the relative sensitivities and resistance to biocides on the basis of whether they do or do not contain lipid, and classified viruses into three groups (A, lipid-containing; B, non-lipid picornaviruses; C, other non-lipid viruses, larger than those in group B) and biocides into two classes (broad-spectrum biocides that inactivate all viruses and lipophilic biocides that fail to inactivate picornaviruses and parvoviruses).

The penetration of biocides into viruses and phages of different types has not been examined in depth, nor has interaction with viral protein and nucleic acid. It is, therefore, difficult at present to provide adequate reasons to explain the relative response, or resistance, of different virus types to biocides.

Protozoa

As pointed out above, the cyst form represents the stage in the life cycle which is resistant to biocides. No significant studies appear to have been undertaken about biocide sensitivity or resistance during encystment and excystment or about the relative uptake of biocides into cysts and trophozoites. It must, at present, be concluded tentatively that the outer regions of cysts limit entry of biocides, thus providing one mechanism of intrinsic resistance.

ANTIBIOTIC AND BIOCIDE RESISTANCE IN BACTERIA; POSSIBLE LINKED MECHANISMS?

In view of the fact that bacterial resistance can develop to both antibiotics and biocides, it is necessary to consider whether there is a possible relationship in the susceptibility to the two types of agents. Where antibiotic resistance is due to enzymic inactivation, it seems unlikely that biocide resistance would follow, and co-resistance seems even more unlikely when antibiotic resistance is mediated via specific alterations in a target protein (enzyme). In comparison, resistance to both types of agent is more likely where less specific mechanisms of action are involved. For example, in Gram-negative bacteria, it would be expected that the outer membrane acts as a non-specific exclusion blanket to prevent the entry of several chemically unrelated molecules, be they antibiotics or biocides. Stickler *et al.* (1983) observed antibiotic and biocide (chlorhexidine, QAC) resistance in Gram-negative bacteria causing urinary tract infections but could obtain no evidence of a plasmid-linked resistance association. They proposed that the extensive use of these cationic biocides was responsible for the selection of these resistant strains. Likewise, Dance *et al.* (1987) isolated a strain of *P. mirabilis* responsible for a hospital outbreak that was resistant to antibiotics and chlorhexidine. There was no evidence of a genetic linkage between these resistances, however, which were considered to be intrinsic rather than plasmid-mediated.

Conversely to the general statement noted above, resistance to propamidine isethionate, QACs, ethidium bromide and aminoglycoside antibiotics always transferred together in MRSA (Cookson & Phillips, 1990). There is some evidence of a linkage here and the possibility always exists that extensive biocide usage, particularly of the cationic type, could lead to the selection of strains showing resistance to antibiotics and biocides. It must, however, be added that *Enterococcus faecium* strains showing resistance to vancomycin (MIC >64 μg/ml) or to gentamicin (MIC >1000 μg/ml) or to both antibiotics are not chlorhexidine resistant, which implies that the use of this antiseptic does not select for the antibiotic-resistant strains (Baillie *et al.*, 1992). Furthermore, ciprofloxacin-resistant mutants of *Ps. aeruginosa*

Table 7. *Biocidal activity:significant progress over the past 50 years*

Item	Progress
Specific activity *versus* emerging and other pathogens	Biocidal activity known *versus* MRSA, HIV, hepatitis B, non-enveloped viruses, fungi, protozoa
Applications	Greater knowledge obtained about use as antiseptics, disinfectants and preservatives in health (e.g. disinfection policies) and industry (e.g. preservatives)
Mechanisms of action	Significant progress in understanding mechanisms of antibacterial action. Less satisfactory is a knowledge of antifungal, antiviral and antiprotozoal action
Mechanisms of resistance	Satisfactory knowledge about Gram-negative bacteria and bacterial spores. Progress with viruses, fungi and protozoa is spasmodic and patchy
Enhancement of activity	Combinations of biocides or biocide plus permeabilizer

strain PAO have been found to be supersensitive to chlorhexidine (Baillie, Power & Phillips, 1993).

CURRENT KNOWLEDGE SUMMARIZED

Before 1945, the spectrum of activity and many of the factors influencing efficacy of the comparatively few biocides then in force had been well established. On the other hand, very few investigations had been undertaken of the ways in which micro-organisms were inactivated or inhibited, why different degrees of response were exhibited by diverse organisms, reasons for resistance to particular biocides and the possible means of enhancing activity. Of course, as knowledge of morphology, chemical constitution, biosynthetic processes and microbial genetics has grown so it has become possible to obtain more relevant and advanced data about mechanisms of action of, and resistance to, biocides.

Thus, present knowledge of mechanisms of bactericidal (except mycobactericidal) and sporicidal action and of bacterial and spore resistance are generally proceeding satisfactorily, if slowly, because biocides do not usually generate the same degree of interest as antibiotics. In contrast, information about antiviral, antifungal and antiprotozoal action is often sparse (Table 7).

FUTURE PERSPECTIVES

Very few new biocide-type antimicrobial agents are being produced, or are likely to be produced in the foreseeable future because of the high financial costs involved in their development. New formulations of existing biocides

Table 8. *Biocides:future perspectives*

Perspective	Comment
New biocidal molecules	Possible, but doubtful (high cost of development)
Specific design of new biocides	Better understanding needed of (a) mechanisms of action (b) uptake into micro-organisms (c) mechanisms of resistance
Combinations of biocides	Possible, but limit already probably reached
Biocides plus permeabilizers	Improved activity against Gram-negative bacteria and especially against mycobacteria and bacterial spores
Bacterial spore germination and outgrowth inhibitors	More detailed information necessary
Prions	Biocides with improved activity needed
Repair of injury in sublethally damaged micro-organisms	Little currently known: of potential significance in medical and food microbiology
Biofilm control	Better understanding needed of heterogeneity within biofilms
Endoscope disinfection	Better control of mycobacterial inactivation and prevention of microbial transmission
Emerging technologies	(1) Ozone: potential sterilization of medical devices (2) Vapour phase hydrogen peroxide: rapid, low-temperature technique for sterilization of surfaces

do appear from time to time, but it may be difficult to know whether these represent any significant improvement. One new biocide of potential value, however, is orthophthalaldehyde (Alfa & Sitter, 1994) which is claimed to be bactericidal, mycobactericidal, sporicidal and viricidal towards non-enveloped as well as to enveloped viruses. This agent clearly merits further investigation. Combinations of biocides to produce enhanced antimicrobial activity are always a possibility, but it is doubtful whether much additional benefit is likely to accrue by such means.

A number of perspectives are summarized in Table 8. A key element remains the goal of a better understanding of the action of biocides and their uptake into micro-organisms and of the ways in which resistance is expressed. Special mention should also be made of biofilms, which are likely to assume even greater importance than at present in both the medical and industrial contexts and their control will depend upon further efforts to understand the heterogeneity of micro-organisms within them. New medical and veterinary problems will continue to challenge: for instance, the current problem of BSE (see chapter by Lacey) highlights the need for disinfection processes directed against such novel disease agents as prions. Finally, what are termed 'emerging technologies' (Table 8) may play a role in future disinfection and/or sterilization processes.

REFERENCES

Aalto, T. R., Firman, M. C. & Rigler, N. E. (1993). p-Hydroxybenzoic acid esters as preservatives. I. Uses, antibacterial and antifungal studies, properties and determination. *Journal of the American Pharmaceutical Association Scientific Edition*, 42, 449–57.

Adler-Storthz, K., Schultster, L. M., Dreesman, G. R., Hollinger, F. B. & Melnick, J. L. (1983). Effect of alkaline glutaraldehyde on hepatitis B virus antigens. *European Journal of Clinical Microbiology*, 2, 316–20.

Alder, V. G. & Simpson, R. A. (1992). Sterilization and disinfection by heat methods. In (Russell, A. D., Hugo, W. B. & Ayliffe, G. A. J., eds) *Principles and Practice of Disinfection, Preservation and Sterilization*. 2nd edn, pp. 483–498. Oxford: Blackwell Scientific Publications.

Alfa, M. J. & Sitter, D. L. (1994). In-hospital evaluation of orthophthalaldehyde as a high level disinfectant for flexible endoscopes. *Journal of Hospital Infection*, 26, 15–26.

Ayliffe, G. A. J., Coates, D. & Hoffman, P. N. (1993). *Chemical Disinfection in Hospitals*. London: Public Health Laboratory Service.

Baillie, L. W. J., Power, E. G. M. & Phillips, I. (1993). Chlorhexidine hypersensitivity of ciprofloxacin-resistant variants of *Pseudomonas aeruginosa* PAO. *Journal of Antimicrobial Chemotherapy*, 31, 219–25.

Baillie, L. W. J., Wade, J. J. & Casewell, M. W. (1992). Chlorhexidine sensitivity of *Enterococcus faecium* resistant to vancomycin, high levels of gentamicin, or both. *Journal of Hospital Infection*, 20, 127–8.

Bailly, J.-L., Chambon, M., Peigue-Lafeuille, H., Laveran, H., De Champs, C. & Beytout, D. (1991). Activity of glutaraldehyde at low concentrations (<2%) against poliovirus and its relevance to gastrointestinal endoscope disinfection procedures. *Applied and Environmental Microbiology*, 57, 1156–60.

Barrett-Bee, K., Newboult, L. & Edwards, S. (1994). The membrane destabilising action of the antibacterial agent chlorhexidine. *FEMS Microbiology Letters*, 119, 249–54.

Belly, R. T. & Kydd, G. C. (1982). Silver resistance in microorganisms. *Developments in Industrial Microbiology*, 23, 567–77.

Bergan, T. & Lystad, A. (1972). Antitubercular action of disinfectants. *Journal of Applied Bacteriology*, 34, 751–6.

Best, M., Kennedy, M. E. & Coates, F. (1990a). Efficacy of a variety of disinfectants against *Listeria* spp. *Applied and Environmental Microbiology*, 56, 377–80.

Best, M., Sattar, S. A., Springthorpe, V. S. & Kennedy, M. E. (1990b). Efficacies of selected disinfectants against *Mycobacterium tuberculosis*. *Journal of Clinical Microbiology*, 28, 2234–9.

Block, S. S. (1991a). Historical review. In (Block, S. S., ed.) *Disinfection, Sterilization and Preservation*. 4th edn, pp. 3–17. Philadelphia: Lea and Febiger.

Block, S. S., ed. (1991b). *Disinfection, Sterilization and Preservation*. Philadelphia: Lea and Febiger.

Bloomfield, S. F. & Arthur, M. (1994). Mechanisms of inactivation and resistance of spores to chemical biocides. *Journal of Applied Bacteriology, Symposium Supplement*, 76, 91S–104S.

Bloomfield, S. F., Smith-Burchnell, C. A. & Dalgleish, A. G. (1990). Evaluation of hypochlorite-releasing agents against the human immunodeficiency virus (HIV). *Journal of Hospital Infection*, 15, 273–8.

Board, R. G., Allwood, M. C. & Banks, J. G., eds (1987). *Preservatives in the Food*,

Pharmaceutical and Environmental Industries. Society of Applied Bacteriology Technical Series No. 22. Oxford: Blackwell Scientific Publications.

Bobichon, H. & Bouchet, P. (1987). Action of chlorhexidine on budding *Candida albicans*: scanning and transmission electron microscopic study. *Mycopathologia*, **100**, 27–35.

Broadley, S. J., Jenkins, P. A., Furr, J. R. & Russell, A. D. (1991). Antimycobacterial activity of biocides. *Letters in Applied Microbiology*, **13**, 118–22.

Brown, M. R. W. (1975). The role of the cell envelope in resistance. In (Brown, M. R. W., ed.) *Resistance of* Pseudomonas aeruginosa. pp. 71–107. London: Wiley.

Brown, M. R. W. & Gilbert, P. (1993). Sensitivity of biofilms to antimicrobial agents. *Journal of Applied Bacteriology, Symposium Supplement*, **74**, 87S–97S.

Brown, M. R. W. & Williams, P. (1985). The influence of the environment on envelope properties affecting survival of bacteria in infections. *Annual Review of Microbiology*, **39**, 527–56.

Bryers, J. D. (1993). Bacterial biofilms. *Current Opinion in Biotechnology*, **4**, 197–204.

Carpentier, B. & Cerf, O. (1993). Biofilms and their consequences, with particular reference to hygiene in the food industry. *Journal of Applied Bacteriology*, **75**, 499–511.

Chambon, M., Bailly, J.-L. & Peigue-Lafeuille, H. (1992). Activity of glutaraldehyde at low concentrations against capsid proteins of poliovirus type 1 and echovirus type 25. *Applied and Environmental Microbiology*, **58**, 3517–21.

Cherrington, C. A., Hinton, M., Mead, G. C. & Chopra, I. (1991). Organic acids: chemistry, antibacterial activity and practical applications. *Advances in Microbial Physiology*, **32**, 87–108.

Coates, D. (1988). Comparison of sodium hypochlorite and sodium dichloroisocyanurate disinfectants: neutralization by serum. *Journal of Hospital Infection*, **11**, 60–7.

Coates, D. & Hutchinson, D. N. (1994). How to produce a hospital disinfection policy. *Journal of Hospital Infection*, **26**, 57–68.

Coates, D. & Wilson, M. (1989). Use of sodium dichloroisocyanurate granules for spills of body fluids. *Journal of Hospital Infection*, **13**, 241–51.

Cole, E. C., Rutala, W. A., Nessen, L., Wannamaker, N. S. & Weber, D. J. (1990). Effect of methodology, dilution and exposure time on the tuberculocidal activity of glutaraldehyde-based disinfectants. *Applied and Environmental Microbiology*, **56**, 1813–17.

Collier, P. J., Ramsey, A. J., Austin, P. & Gilbert, P. (1990*a*). Growth inhibitory and biocidal activity of some isothiazoline biocides. *Journal of Applied Bacteriology*, **69**, 569–77.

Collier, P. J., Ramsey, A., Waight, R. D., Douglas, K. T, Austin, P. & Gilbert, P. (1990*b*). Chemical reactivity of some isothiazolone biocides. *Journal of Applied Bacteriology*, **69**, 578–84.

Collins, F. M. (1986*a*). Kinetics of the tuberculocidal response by alkaline glutaraldehyde in solution and on an inert surface. *Journal of Applied Bacteriology*, **61**, 87–93.

Collins, F. M. (1986*b*). Bactericidal activity of alkaline glutaraldehyde solution against a number of atypical mycobacteria. *Journal of Applied Bacteriology*, **61**, 247–51.

Cookson, B. D. & Phillips, I. (1990). Methicillin-resistant staphylococci. *Journal of Applied Bacteriology, Symposium Supplement*, **69**, 55S–70S.

Cookson, B. D., Bolton, M. C. & Platt, J. H. (1991). Chlorhexidine resistance in methicillin-resistant *Staphylococcus aureus* or just an elevated MIC? An *in vitro* and *in vivo* assessment. *Antimicrobial Agents and Chemotherapy*, **35**, 1997–2002.

Costerton, J. W., Cheng, K.-J., Geesey, G. G., Ladd, T. I., Nickel, J. C., Dasgupta, M. & Marrie, T. J. (1987). Bacterial biofilms in nature and disease. *Annual Review of Microbiology*, **41**, 435–64.

Costerton, J. D., Lewandowski, Z., DeBeer, D., Caldwell, D., Korber, D. & James, G. (1994). Biofilms, the customized niche. *Journal of Bacteriology*, **176**, 2137–42.

Cox, A. D. & Wilkinson, S. G. (1991). Ionising groups of lipopolysaccharides of *Pseudomonas cepacia* in relation to antibiotic resistance. *Molecular Microbiology*, **5**, 641–646.

Croshaw, B. (1971). The destruction of mycobacteria. In (Hugo, W. B., ed.) *Inhibition and Destruction of the Microbial Cell*. pp. 429–449. London: Academic Press.

Dance, D. A. B., Pearson, A. D., Seal, D. V. & Lowes, J. A. (1987). A hospital outbreak caused by a chlorhexidine and antibiotic resistant *Proteus mirabilis*. *Journal of Hospital Infection*, **10**, 10–16.

Davies, J. G., Babb, J. R., Bradley, C. R. & Ayliffe, G. A. J. (1993). Preliminary study of test methods to assess the virucidal activity of skin disinfectants using poliovirus and bacteriophages. *Journal of Hospital Infection*, **25**, 125–31.

Day, M. J. & Russell, A. D. (1992). Bacterial sensitivity and resistance. F. Methicillin-resistant staphylococci. In (Russell, A. D., Hugo, W. B. & Ayliffe, G. A. J., eds) *Principles and Practice of Disinfection, Preservation and Sterilization*. 2nd edn, pp. 264–273. Oxford: Blackwell Scientific Publications.

Dekker, J. (1987). Development of resistance to modern fungicides and strategies for its avoidance. In (Lyr, H., ed.) *Modern Selective Fungicides*. pp. 39–52. Harlow: Longman.

Denyer, S. P. & Hugo, W. B., eds (1990). *Mechanisms of Action of Chemical Biocides, Their Study and Exploitation*. Society for Applied Bacteriology Technical Series No. 27. Oxford: Blackwell Scientific Publications.

Eagar, R. G., Leder, J. & Theis, A. B. (1986). Glutaraldehyde: factors important for microbiocidal efficacy. *Proceedings of the 3rd Conference on Progress in Chemical Disinfection*, pp. 32–49. Binghamton: State University of New York.

Elferink, J. G. R. & Booij, H. L. (1974). Interaction of chlorhexidine with yeast cells. *Biochemical Pharmacology*, **23**, 1413–19.

Elsmore, R. (1992). Bacterial sensitivity and resistance. E. Legionella. In (Russell, A. D., Hugo, W. B. & Ayliffe, G. A. J., eds) *Principles and Practice of Disinfection, Preservation and Sterilization*. 2nd edn, pp. 254–263. Oxford: Blackwell Scientific Publications.

Emslie, K. R., Townsend, D. E. & Grubb, W. B. (1985*a*). A resistance determinant to nucleic acid-binding compounds in methicillin-resistant *Staphylococcus aureus*. *Journal of Medical Microbiology*, **20**, 139–145.

Emslie, K. R., Townsend, D. E., Bolton, S. & Grubb, W. B. (1985*b*). Two distinct resistance determinants to nucleic acid-binding compounds in *Staphylococcus aureus*. *FEMS Microbiology Letters*, **27**, 61–4.

Favero, M. S. (1991). Practical applications of liquid sterilants in health care facilities. In (Morrissey, R. F. & Prokopenko, Y. I., eds) *Sterilization of Medical Products*. Vol. V, pp. 397–405. Morin Heights, Canada: Polyscience Publications.

Favero, M. S. & Bond, W. W. (1991). Sterilization, disinfection and antisepsis in the hospital. In (Balows, A., Hausler, W. J. Jr., Hermann, K. L., Isenberg, H. D. & Shadomy, H. J., eds) *Manual of Clinical Microbiology*. 5th edn, pp. 183–200. Washington, DC: American Society for Microbiology.

Favero, M. S. & Bond, W. W. (1993). The use of liquid chemical germicides. In (Morrissey, R. F. & Phillips, G. B., eds) *Sterilization Technology:A Practical Guide for Manufacturers and Users of Health Care Products*. pp. 309–334. New York: Van Nostrand Reinhold.

Foster, J. W. & Hall, H. K. (1990). Adaptive acidification tolerance response of *Salmonella typhimurium*. *Journal of Bacteriology*, **172**, 771–8.

Foster, J. W. & Hall, H. K. (1991). Inducible pH homeostasis and the acid tolerance response of *Salmonella typhimurium*. *Journal of Bacteriology*, **173**, 5129–35.

Foster, T. J. (1983). Plasmid-determined resistance to antimicrobial drugs and toxic metal ions in bacteria. *Microbiological Reviews*, **47**, 361–409.

Gadd, G. M. & White, C. (1989). Heavy metal and radionuclide accumulation and toxicity in fungi and yeasts. In (Poole, R. K. & Gadd, G. M., eds) *Metal–Microbe Interactions*. Special Publications of the Society for General Microbiology, No. 26, pp. 19–38. Oxford: Oxford University Press.

Gale, E. F. (1986). Nature and development of phenotypic resistance to amphotericin B in *Candida albicans*. *Advances in Microbial Physiology*, **27**, 277–320.

Gardner, J. F. & Peel, M. M. (1991). *Introduction to Sterilization, Disinfection and Infection Control*. Edinburgh: Churchill Livingstone.

Geesey, G. G., Characklis, W. G. & Costerton, J. W. (1992). Centers, new technologies focus on biofilm heterogeneity. *ASM News*, **58**, 546–7.

Gilbert, P. (1988). Microbial resistance to preservative systems. In (Bloomfield, S. F., Baird, R., Leak, R. E. & Leech, R., eds) *Microbial Quality Assurance in Pharmaceuticals, Cosmetics and Toiletries*. pp. 171–194. Chichester: Ellis Horwood.

Gilbert, P., Evans, D. J. & Brown, M. R. W. (1993). Formation and dispersal of bacterial biofilms *in vivo* and *in situ*. *Journal of Applied Bacteriology, Symposium Supplement*, **74**, 67S–78S.

Gilbert, P., Pemberton, D. & Wilkinson, D. E. (1990). Barrier properties of the Gram-negative cell envelope towards high molecular weight polyhexamethylene biguanides. *Journal of Applied Bacteriology*, **69**, 585–92.

Gorman, S. P. & Scott, E. M. (1977). A quantitative evaluation of the antifungal properties of glutaraldehyde. *Journal of Applied Bacteriology*, **43**, 83–9.

Gould, G. W. (1983). Mechanisms of resistance and dormancy. In (Hurst, A. & Gould, G. W., eds) *The Bacterial Spore*. Vol. 2, pp. 173–209. London: Academic Press.

Gould, G. W. & Russell, N. J. (1991). Sulphite. In (Russell, N. J. & Gould, G. W., eds) *Food Preservatives*. pp. 72–88. Glasgow and London: Blackie.

Grossgebauer, K. (1970). Virus disinfection. In (Benarde, M. A., ed.) *Disinfection*. pp. 103–148. New York: Marcel Dekker.

Hamilton-Miller, J. M. T., Shah, S. & Smith, C. (1993). Silver sulphadiazine: a comprehensive *in vitro* reassessment. *Chemotherapy*, **39**, 405–9.

Hancock, R. E. W. (1984). Alterations in membrane permeability. *Annual Review of Microbiology*, **38**, 237–64.

Hanson, P. J. V. (1988). Mycobacteria and AIDS. *British Journal of Hospital Medicine*, **40**, 149.

Havelaar, A. H. (1993). Bacteriophages as models of human enteric viruses in the environment. *ASM News*, **59**, 614–18.

Hector, R. F. (1993). Compounds active against cell walls of medically important fungi. *Clinical Microbiology Reviews*, **6**, 1–21.

Heinzel, M. (1988). The phenomena of resistance to disinfectants. In (Payne, K. R., ed.) *Industrial Biocides*. pp. 52–67. Chichester: Wiley.

Hugo, W. B. (1991). A brief history of heat and chemical preservation and disinfection. *Journal of Applied Bacteriology*, **71**, 9–18.

Hugo, W. B. (1992*a*). Historical introduction. In (Russell, A. D., Hugo, W. B. & Ayliffe, G. A. J., eds) *Principles and Practice of Disinfection, Preservation and Sterilization*. 2nd edn, pp. 3–6. Oxford: Blackwell Scientific Publications.

Hugo, W. B. (1992*b*). Disinfection mechanisms. In (Russell, A. D., Hugo, W. B. & Ayliffe, G. A. J., eds) *Principles and Practice of Disinfection, Preservation and Sterilization*. 2nd edn, pp. 187–210. Oxford: Blackwell Scientific Publications.

Jarroll, E. (1988). Effect of disinfectants on *Giardia* cysts. *CRC Reviews in Environmental Control*, **18**, 1–28.

Jarroll, E. L. (1992). Sensitivity of protozoa to disinfectants. In (Russell, A. D., Hugo, W. B. & Ayliffe, G. A. J., eds) *Principles and Practice of Disinfection, Preservation and Sterilization*. 2nd edn, pp. 180–186. Oxford: Blackwell Scientific Publications.

Kabara, J. J. & Eklund, T. (1991). Organic acids and esters. In (Russell, N. J. & Gould, G. W., eds) *Food Preservatives*. pp. 44–71. Glasgow and London: Blackie.

Karabit, M. S., Juneskans, O. T. & Lundgren, P. (1989). Factorial designs in the evaluation of preservative efficiency. *International Journal of Pharmaceutics*, **56**, 169–74.

Kaulfers, P.-M., Karch, H. & Laufs, R. (1987). Plasmid-mediated formaldehyde resistance in *Serratia marcescens* and *Escherichia coli*: alterations in the cell surface. *Zentralblatt für Bakteriologie und Hygiene*, **A226**, 239–48.

Kaye, S. & Phillips, C. R. (1949). The sterilizing action of gaseous ethylene oxide. IV. The effect of moisture. *American Journal of Hygiene*, **50**, 296–306.

Kilvington, S. & Price, J. (1990). Survival of *Legionella pneumophila* within cysts of *Acanthamoeba polyphaga* following chlorine exposure. *Journal of Applied Bacteriology*, **68**, 519–25.

Klein, M. & Deforest, A. (1983). Principles of viral inactivation. In (Block, S. S., ed.) *Disinfection, Sterilization and Preservation*. 3rd edn, pp. 422–434. Philadelphia: Lea and Febiger.

Korich, D. G., Mead, J. R., Madore, M. S., Sinclair, N. A. & Sterling, C. R. (1990). Effects of ozone, chlorine dioxide, chlorine and monochloramine on *Cryptosporidium parvum* oocyst viability. *Applied and Environmental Microbiology*, **56**, 1423–8.

Kostenbauder, H. B. (1991). Physical factors influencing the activity of antimicrobial agents. In (Block, S. S., ed.) *Disinfection, Sterilization and Preservation*. 4th edn, pp. 59–71. Philadelphia: Lea and Febiger.

Kuyyakanond, T. & Quesnel, L. B. (1992). The mechanism of action of chlorhexidine. *FEMS Microbiology Letters*, **100**, 211–16.

Lee, S., Slonczewski, J. L. & Foster, J. W. (1994). A low-pH-inducible, stationary-phase acid tolerance response in *Salmonella typhimurium*. *Journal of Bacteriology*, **176**, 1422–6.

Leland, S. E., Jr (1991). Antiprotozoan, anthelmintic and other pest management compounds. In (Block, S. S., ed.) *Disinfection, Sterilization and Preservation*. 4th edn, pp. 482–491. Philadelphia: Lea and Febiger.

Lyon, B. R. & Skurray, R. A. (1987). Antimicrobial resistance of *Staphylococcus aureus*: genetic basis. *Microbiological Reviews*, **51**, 88–134.

Lyr, H. (1987). Selectivity in modern fungicides and its basis. In (Lyr, H., ed.) *Modern Selective Fungicides*. pp. 31–38. Harlow: Longman.

Marrie, T. J. & Costerton, J. W. (1981). Prolonged survival of *Serratia marcescens* in chlorhexidine. *Applied and Environmental Microbiology*, **42**, 1093–102.

Midgley, M. (1986). The phosphonium ion efflux system of *Escherichia coli*:

relationship to the ethidium efflux system and energetic studies. *Journal of General Microbiology*, **132**, 1387–93.

Midgley, M. (1987). An efflux system for cationic dyes and related compounds in *Escherichia coli*. *Microbiological Sciences*, **14**, 125–7.

Midgely, M. (1994). Characteristics of an ethidium efflux system in *Enterococcus hirae*. *FEMS Microbiology Letters*, **120**, 119–24.

Navarro, J. M. & Monsen, P. (1976). Étude du mechanisme d'interaction du glutaraldehyde avec les micro-organismes. *Annales Microbiologie (Paris)*, **127B**, 295–307.

Nikaido, H. & Vaara, M. (1985). Molecular basis of bacterial outer membrane permeability. *Microbiological Reviews*, **49**, 1–32.

Ogase, H., Nagai, I., Kameda, K., Kume, S. & Ono, S. (1992). Identification and quantitative analysis of degradation products of chlorhexidine with chlorhexidine-resistant bacteria with three-dimensional high performance liquid chromatography. *Journal of Applied Bacteriology*, **73**, 71–8.

Olson, E. R. (1993). Influence of pH on bacterial gene expression. *Molecular Microbiology*, **8**, 5–14.

Passagot, J., Crance, J. M., Biziagos, E., Laveran, H., Agbalika, F. & Deloince, R. (1987). Effect of glutaraldehyde on the antigenicity and infectivity of hepatitis A virus. *Journal of Virological Methods*, **16**, 21–8.

Power, E. G. M. & Russell, A. D. (1990). Uptake of L-^{14}C-alanine to glutaraldehyde-treated and untreated spores of *Bacillus subtilis*. *FEMS Microbiology Letters*, **66**, 271–6.

Power, E. G. M., Dancer, B. N. & Russell, A. D. (1988). Emergence of resistance to glutaraldehyde in spores of *Bacillus subtilis* 168. *FEMS Microbiology Letters*, **50**, 223–6.

Power, E. G. M., Dancer, B. N. & Russell, A. D. (1989). Possible mechanisms for the revival of glutaraldehyde-treated spores of *Bacillus subtilis* NCTC 8236. *Journal of Applied Bacteriology*, **67**, 91–8.

Poxton, I. R. (1993). Prokaryote envelope diversity. *Journal of Applied Bacteriology, Symposium Supplement*, **74**, 1S–11S.

Raja, N., Goodson, M., Chui, W. C. M., Smith, D. G. & Rowbury, R. J. (1991). Habituation to acid in *Escherichia coli*: conditions for habituation and its effects on plasmid transfer. *Journal of Applied Bacteriology*, **70**, 59–65.

Report (1990). Revised guidelines for the control of epidemic methicillin-resistant *Staphylococcus aureus*. *Journal of Hospital Infection*, **16**, 351–7.

Roberts, T. A., Woods, L. F. J., Payne, M. J. & Cammack, R. (1991). Nitrite. In (Russell, N. J. & Gould, G. W., eds) *Food Preservatives*. pp. 89–110. Glasgow and London: Blackie.

Rodgers, F. G., Hufton, P., Kurzawska, E., Molloy, C. & Morgan, S. (1985). Morphological response of human rotavirus to ultraviolet radiation, heat and disinfectants. *Journal of Medical Microbiology*, **20**, 123–30.

Roussow, F. T. & Rowbury, R. J. (1984). Effects of the resistance plasmid R124 on the level of OmpF outer membrane protein and on the response of *Escherichia coli* to environmental agents. *Journal of Applied Bacteriology*, **56**, 63–79.

Rubbo, S. D., Gardner, J. F. & Webb, R. L. (1967). Biocidal activities of glutaraldehyde and related compounds. *Journal of Applied Bacteriology*, **30**, 78–87.

Russell, A. D. (1981). Neutralization procedures in the evaluation of bactericidal activity. In (Collins, C. H., Allwood, M. C., Bloomfield, S. F. & Fox, A., eds) *Disinfectants, Their Use and Evaluation of Effectiveness*. Society for Applied Bacteriology Technical Series No. 16, pp. 45–59. London: Academic Press.

Russell, A. D. (1985). The role of plasmids in bacterial resistance to antiseptics, disinfectants and preservatives. *Journal of Hospital Infection*, **6**, 9–19.

Russell, A. D. (1990). The bacterial spore and chemical sporicidal agents. *Clinical Microbiology Reviews*, **3**, 99–119.

Russell, A. D. (1992*a*). Factors influencing the efficacy of antimicrobial agents. In (Russell, A. D., Hugo, W. B. & Ayliffe, G. A. J., eds) *Principles and Practice of Disinfection, Preservation and Sterilization*. 2nd edn, pp. 89–113. Oxford: Blackwell Scientific Publications.

Russell, A. D. (1992*b*). Mycobactericidal agents. In (Russell, A. D., Hugo, W. B. & Ayliffe, G. A. J., eds) *Principles and Practice of Disinfection, Preservation and Sterilization*. 2nd edn, pp. 246–253. Oxford: Blackwell Scientific Publications.

Russell, A. D. (1992*c*). Antifungal activity of biocides. In (Russell, A. D., Hugo, W. B. & Aylifffe, G. A. J., eds) *Principles and Practice of Disinfection*, Preservation and Sterilization. 2nd edn, pp. 134–149. Oxford: Blackwell Scientific Publications.

Russell, A. D. (1993). Microbial cell walls and resistance of bacteria to antibiotics and biocides. *Journal of Infectious Diseases*, **168**, 1339–40.

Russell, A. D. & Chopra, I. (1990). Antiseptics, disinfectants and preservatives: their properties, mechanism of action and uptake into bacteria. In *Understanding Antibacterial Action and Resistance*, pp. 95–131. Chichester: Ellis Horwood.

Russell, A. D. & Day, M. J. (1993). Antibacterial activity of chlorhexidine. *Journal of Hospital Infection*, **25**, 229–38.

Russell, A. D. & Gould, G. W. (1988). Resistance of Enterobacteriaceae to preservatives and disinfectants. *Journal of Applied Bacteriology, Symposium Supplement*, **65**, 167S–95S.

Russell, A. D. & Hugo, W. B. (1988). Perturbation of homeostatic mechanisms in bacteria by pharmaceuticals. In (Whittenbury, R., Gould, G. W. & Board, R. G., eds) *Homeostatic Mechanisms in Micro-organisms*. FEMS Symposium No. 44, pp. 206–219. Bath: Bath University Press.

Russell, A. D. & Hugo, W. B. (1994). Antimicrobial activity and action of silver. In (Ellis, G. P. & Luscombe, D. K., eds) *Progress in Medicinal Chemistry*. Vol. 31, pp. 351–371. Amsterdam: Elsevier.

Russell, A. D. & Mills, A. P. (1974). Comparative sensitivity and resistance of some strains of *Pseudomonas aeruginosa* and *Pseudomonas stutzeri* to antibacterial agents. *Journal of Clinical Pathology*, **27**, 463–6.

Russell, A. D., Ahonkhai, I. & Rogers, D. T. (1979). Microbiological applications of the inactivation of antibiotics and other antimicrobial agents. *Journal of Applied Bacteriology*, **46**, 207–45.

Russell, A. D., Broadley, S. J., Furr, J. R. & Jenkins, P. A. (1994). Potentiation of the antimycobacterial activity of biocides. *Journal of Infection*, **28**, 108–9.

Russell, A. D., Hammond, S. A. & Morgan, J. R. (1986). Bacterial resistance to antiseptics and disinfectants. *Journal of Hospital Infection*, **7**, 213–25.

Russell, A. D., Hugo, W. B. & Ayliffe, G. A. J., eds (1992). *Principles and Practice of Disinfection, Preservation and Sterilization*, 2nd edn. Oxford: Blackwell Scientific Publications.

Russell, N. J. & Gould, G. W., eds (1991). *Food Preservatives*. Glasgow and London: Blackie.

Rutala, W. A. (1990). APIC guidelines for selection and use of disinfectants. *American Journal of Infection Control*, **18**, 99–117.

Sakagami, Y., Yokagama, H., Nishimura, H., Osey, Y. & Tashima, T. (1989). Mechanism of resistance to benzalkonium chloride by *Pseudomonas aeruginosa*. *Applied and Environmental Microbiology*, **55**, 2036–40.

Sattar, S. A., Springthorpe, V. S., Conway, B. & Xu, Y. (1994). Inactivation of the human immunodeficiency virus: an update. *Reviews in Medical Microbiology*, **5**, 139–50.

Seal, D. V. & Hay, J. (1992). Contact lens disinfection and *Acanthamoeba*: problems and practicalities. *Pharmaceutical Journal*, **248**, 717–19.

Seiler, D. A. L. & Russell, N. J. (1991). Ethanol as a food preservative. In (Russell, N. J. & Gould, G. W., eds) *Food Preservatives*. pp. 153–171. Glasgow and London: Blackie.

Setlow, P. (1994). Mechanisms which contribute to the long-term survival of spores of *Bacillus* species. *Journal of Applied Bacteriology, Symposium Supplement*, **76**, 49S–60S.

Siegele, D. A. & Kolter, R. (1992). Life after log. *Journal of Bacteriology*, **174**, 345–8.

Silver, S., Nucifora, G., Chu, L. & Misra, T. K. (1989). Bacterial ATPases: primary pumps for exporting toxic cations and anions. *Trends in Biochemical Sciences*, **14**, 76–80.

Small, P., Blankenhorn, D., Welty, D., Zinser, E. & Slonczewski, J. L. (1994). Acid and base resistance in *Escherichia coli* and *Shigella flexneri*: role of *rpoS* and growth pH. *Journal of Bacteriology*, **176**, 1729–37.

Springthorpe, V. S. & Sattar, S. A. (1990). Chemical disinfection of virus-contaminated surfaces. *Critical Reviews in Environmental Control*, **20**, 169–229.

Stickler, D. J., Dolman, J., Rolfe, S. & Chawla, J. (1990). Activity of antiseptics against *Escherichia coli* growing as biofilms on silicone surfaces. *European Journal of Clinical Microbiology*, **8**, 974–8.

Stickler, D. J., Dolman, J., Rolfe, S. & Chawla, J. (1991). Activity of some antiseptics against urinary tract pathogens growing on biofilms on silicone surfaces. *European Journal of Clinical Microbiology and Infectious Diseases*, **10**, 410–15.

Stickler, D. J., Thomas, B., Clayton, J. C. & Chawla, J. A. (1983). Studies on the genetic basis of chlorhexidine resistance. *British Journal of Clinical Practice, Symposium Supplement*, **25**, 23–8.

Taylor, D. M. (1992). Inactivation of unconventional agents of the transmissible degenerative encephalopathies. In (Russell, A. D., Hugo, W. B. & Ayliffe, G. A. J., eds) *Principles and Practice of Disinfection, Preservation and Sterilization*. 2nd edn, pp. 171–179. Oxford: Blackwell Scientific Publications.

Thurman, R. B. & Gerba, C. P. (1988). Molecular mechanisms of viral inactivation by water disinfectants. *Advances in Applied Microbiology*, **33**, 75–105.

Thurman, R. B. & Gerba, C. P. (1989). The molecular mechanisms of copper and silver ion disinfection of bacteria and viruses. *CRC Critical Reviews in Environmental Control*, **18**, 295–315.

Trevor, J. T. (1987). Silver resistance and accumulation in bacteria. *Enzyme and Microbial Technology*, **9**, 331–3.

Trueman, J. R. (1971). The halogens. In (Hugo, W. B., ed.) *Inhibition and Destruction of the Microbial Cell*. pp.135–183. London: Academic Press.

Vaara, M. (1992). Agents that increase the permeability of the outer membrane. *Microbiological Reviews*, **56**, 395–411.

Vaara, M. & Jaakkola, J. (1989). Sodium hexametaphosphate sensitizes *Pseudomonas aeruginosa*, several other species of *Pseudomonas* and *Escherichia coli* to hydrophobic drugs. *Antimicrobial Agents and Chemotherapy*, **33**, 1741–7.

van Klingeren, B. & Pullen, W. (1987). Comparative testing of disinfectants against *Mycobacterium tuberculosis* and *Mycobacterium terrae* in a quantitative suspension test. *Journal of Hospital Infection*, **10**, 292–8.

Walker, H. W. & LaGrange, W. S. (1991). Sanitation in food manufacturing operations. In (Block, S. S., ed.) *Disinfection, Sterilization and Preservation*. 4th edn, pp. 791–801. Philadelphia: Lea and Febiger.

Wallhäuser, K. H. (1984). Antimicrobial preservatives used by the cosmetic industry. In (Kabara, J. J., ed.) *Cosmetic and Drug Preservation. Principles and Practice*. pp. 605–745. New York: Marcel Dekker.

Warth, A. D. (1977). Mechanism of resistance of *Saccharomyces bailii* to benzoic, sorbic and other weak acids used as food preservatives. *Journal of Applied Bacteriology*, **43**, 215–30.

Williams, N. D. & Russell, A. D. (1992). The nature and site of biocide-induced sublethal injury in *Bacillus subtilis* spores. *FEMS Microbiology Letters*, **99**, 277–80.

Williams, N. D. & Russell, A. D. (1993). Injury and repair in biocide-treated spores of *Bacillus subtilis*. *FEMS Microbiology Letters*, **106**, 183–6.

Woodcock, P. M. (1988). Biguanides as industrial biocides. In (Payne, K. R., ed.) *Industrial Biocides*. pp. 19–36. Chichester: Wiley.

Working Party (1985). Acquired immune deficiency syndrome: recommendations of a Working Party of the Hospital Infection Society. *Journal of Hospital Infection*, (*Supplement C*), **6**, 67–80.

Yamamoto, T., Tamura, Y. & Yokota, T. (1988). Antiseptic and antibiotic resistance plasmid in *Staphylococcus aureus* that possesses ability to confer chlorhexidine and acrinol resistance. *Antimicrobial Agents and Chemotherapy*, **32**, 932–5.

INDEX